Nanomaterial-based Sensors for Healthcare Application

This book primarily focuses on nanomaterial synthesis, properties, sensor fabrication, and their uses for disease diagnosis and therapy. Major topics covered include synthesis of different dimensional nanomaterials, characterizations and analysis of nanomaterials, procedure for non-invasive disease diagnosis, and measurement principles of medical devices or sensor devices. It also illustrates state-of-the-art nanomaterials for drug delivery supported by practical examples and case studies.

Features:

Explores different dimensional (0D, 1D, 2D, 3D) nanomaterial synthesis techniques.

Introduces various nanomaterials for fabrication sensors.

Discusses the applications of sensors in disease diagnostics and therapy.

Covers nanomaterial characterizations techniques such as XRD, FESEM, AFM, and TEM.

Describes state of the art nanomaterials for drug delivery.

This book is aimed at graduate students and researchers in materials science, nanomaterials, sensors, and biosensors.

Emerging Materials and Technologies

Series Editor: Boris I. Kharissov

The *Emerging Materials and Technologies* series is devoted to highlighting publications centered on emerging advanced materials and novel technologies. Attention is paid to those newly discovered or applied materials with potential to solve pressing societal problems and improve quality of life, corresponding to environmental protection, medicine, communications, energy, transportation, advanced manufacturing, and related areas.

The series takes into account that, under present strong demands for energy, material, and cost savings, as well as heavy contamination problems and worldwide pandemic conditions, the area of emerging materials and related scalable technologies is a highly interdisciplinary field, with the need for researchers, professionals, and academics across the spectrum of engineering and technological disciplines. The main objective of this book series is to attract more attention to these materials and technologies and invite conversation among the international R&D community.

Nanomaterials for Electrochemical Sensing
Edited by Syed Shahabuddin, Rama Gaur, and Nandini Mukherjee

Organic-Inorganic Hybrid Nanomaterials: Energy Harvesting, Storage, and Advanced Applications
Edited by Srikanta Moharana, Bibhuti B. Sahu, and Suresh Sagadevan

Nanocarbon to Polymer-Nanocarbon Composites: From Synthesis to Breakthrough Performance
Edited by Srikanta Moharana, Bibhuti B. Sahu, and Suresh Sagadevan

Materials for Dye Degradation
Edited by Tarun Parangi and Soyebkhan Pathan

Circularity of Cementitious Materials: A Practical Approach
Christian Paglia

Machine Learning for Semiconductor Materials
Edited by Neeraj Gupta, Rashmi Gupta, Rekha Yadav, Sandeep Dhariwal, and RajKumar Sarma

Nanomaterial-based Sensors for Healthcare Application
Basanta Bhowmik

For more information about this series, please visit: www.routledge.com/Emerging-Materials-and-Technologies/book-series/CRCEMT

Nanomaterial-based Sensors for Healthcare Application

Edited by
Basanta Bhowmik

CRC Press
Taylor & Francis Group
Boca Raton London New York

CRC Press is an imprint of the
Taylor & Francis Group, an **informa** business

Designed cover image: Basanta Bhowmik

First edition published 2026
by CRC Press
2385 NW Executive Center Drive, Suite 320, Boca Raton FL 33431

and by CRC Press
4 Park Square, Milton Park, Abingdon, Oxon, OX14 4RN

CRC Press is an imprint of Taylor & Francis Group, LLC

ISBN: 9781032581910 (hbk)
ISBN: 9781032592220 (pbk)
ISBN: 9781003453635 (ebk)

DOI: 10.1201/9781003453635

Typeset in Times
by Newgen Publishing UK

Dedicated

to

My Elder Brother Sri Krishna Pada Bhowmik

Contents

Preface

In the first chapter, the nano-materialistic aspect in the field of biosensors is discussed with an intervention toward the applicability and choice of vivid dimensionality of the sensing material. Transduction phenomena are highly dependent on the material, morphology, and orientation of the nanostructured sensing layer in addition to the advents of nano-sized materials (free energies, short or confined carrier transportation, catalytic effects, etc.). These points are discussed with a focus on making the choice of nanostructure for targeted bio-species. Immobilization protocols are excluded here as the choice of nanostructure is more important for response time. Moreover, the variety of targeted stimulus is another important feature that should be kept in consideration prior to the choice of sensing nanostructured material.

The second chapter deals with a brief history of nanomaterials and their uses to trigger advances in nanotechnology development. In particular, nanomaterials can be produced with outstanding magnetic, electrical, optical, mechanical, and catalytic properties that are substantially different from their bulk counterparts. Nanomaterial properties can be tuned as desired via precisely controlling the size, shape, synthesis conditions, and appropriate functionalization.

For diagnosing a disease and its therapy, a number of nanotechnology tools are available, which is discussed in the third chapter. This chapter provides an overview of nanoparticles and their types. The use of nanoparticles in nanotechnology to obtain different tools in the medical field is discussed. A number of research studies have demonstrated the potential growth of nanotheranostics, which has attracted considerable attention in clinics due to its diagnostic potential in multimodal imaging and therapy platforms. The integration of nanotechnology into medical practice holds immense promise for improving patient outcomes, introducing a new era of precision medicine and transformative healthcare solutions.

The fourth chapter is mainly focused on the oxide-based sensor used for medical and health diagnosis employing binary/ternary or more than ternary composite materials. Then after different types of synthesis techniques like the hydrothermal technique, the sol-gel method, and electrochemical anodization of different materials are examined, characterization techniques like FESEM, XRT, XPS, Raman spectra, and so forth are discussed. The physics behind the absorption and desorption of electrons when they come into contact with the target material is also covered under the sensing mechanism topic. Finally, the chapter discusses different sensors that are used for diagnosing diseases like diabetes, kidney disease, blood cholesterol, halitosis, lung cancer, and so on.

Graphene as 2D material in the field of sensor research is discussed in the fifth chapter as it shows excellent performance in terms of sensitivity, fast response time, high conductivity, good longevity, and recoverability. Ultrathin, flexible, and lightweight electronic transducers can act as invasive and nonintrusive human health monitoring, which is investigated. It is able to provide real-time measurements of body temperature, heart rate, blood pressure, blood glucose level, pulse oxygenation, respiration rate, and so forth. Synthesis and characterization of graphene-based materials in in-vitro and in-vivo model systems can be used to understand the biological effects of these materials on humans and are explained in the chapter. Not only that but also how graphene-based materials can be used for applications for therapeutics and biodegradation is discussed.

In chapter six, the role of nanoparticles to overcome the limitations of free therapeutics and navigate biological barriers that are heterogeneous across patients and diseases is broadly discussed. It is observed how lipid-based, polymeric and inorganic nanoparticles can be optimized for drug delivery in a more personalized manner for precision medicine. Cancer nano-therapeutics is discussed as it can solve several limitations of conventional drug delivery systems such as nonspecific biodistribution and targeting. In order to improve the biodistribution of cancer drugs, nanoparticles are designed for optimal size and surface characteristics to increase their circulation time in the bloodstream. They are also able to carry loaded active drugs to cancer cells by selectively using the unique pathophysiology of tumors.

In the seventh chapter, various 2D material-based sensors for healthcare applications are examined. Materials like flexible polymer, paper, sponge, and fabric substrates are mainly used to deposit the 2D material as an active layer for flexible sensors. Since flexible substrates cannot withstand high-temperature fabrication processes, most researchers use low-temperature deposition techniques to transfer active material on their flexible substrates, like drop casting, screen printing, RF sputtering, and so on. Continuous efforts are being made to improve or tune the performance of sensors in terms of selectivity, sensitivity, response time, and recovery time by using different device structures and functionalizing pristine 2D material with other organic or inorganic molecules. The chapter reviews sensors for detecting glucose, lactose, and PH levels and identifying specific biomarkers like isoprene, acetaldehyde, toluene, NO_2, and so forth. since most VOCs in breath require ppb-level detection. Sometimes, that kind of high sensitivity is not achievable at room temperature.

In chapter eight, the basic working operation and performance analysis of field effects transistor (FET) employing nanomaterial-based biosensors are discussed. The quantitative and qualitative performance metrics challenges in designing FET devices such as short-channel effects (SCEs), immobilization technique, signal-to-noise ratio (SNR), and so on are examined. The role of different nanostructure based nanomaterial FET biosensor and their impact on performance is presented. The application of a titanium dioxide-based FET biosensor for detecting different biomolecules is studied. This chapter is dedicated to the state-of-the-art analysis in the fields of biosensors.

In the ninth chapter, the design and analysis of MEMS-based 2D finite element method (FEM) simulation, analysis, and optimization of surface acoustic wave (SAW)-based gas sensor is presented. The simulation of SAW gas sensors with and without the presence of nanorods is shown. Gas adsorption on a surface causes a change in the mass, modulus, and conductivity of the sensing layer, which can be accurately detected using SAW-based gas sensors. Simulations in COMSOL multiphysics are performed using eigen frequency, time-dependent, and frequency domain analysis. The materials of the intermediate layer, as well as integrated transducers (IDTs) and the height of the nanorods, are varied for the optimization of the device. The frequency shift and total displacement were significantly improved for the nanorod-based SAW gas sensor device.

In the tenth chapter, a complete history of wearable sensors in the biomedical field is highlighted with in-depth analysis of the use of wearable sensors in healthcare applications. Special emphasis is placed on what will be the general sensor architecture and their design methodologies for detecting different human activity. Various types of wearable devices used in health monitoring, their applications, challenges, and the prospects they hold in transforming the healthcare landscape are discussed. The challenges related to the design trade-offs, improved sensors, power, size, computation algorithms, and security are explored. In reading this chapter, readers will get a glance at the development stages: mathematical modeling, optimization research, and cost-effective timely production of present and future wearable devices. Further, the impact of these devices on the human body due to continuous all-time usage is also explored and certain guidelines for minimizing their ill-effects have also been listed.

Electrochemical and fluorescent sensors are widely employed in the diverse domains of healthcare, biosensing, drug administration, and so forth due to their special features like high sensitivity, stability, and selectivity. In the eleventh chapter, the uses of electrochemical sensors embedded in contact lenses, which provide health-related information through electrochemical detection in tears based on non-invasive real-time monitoring, are discussed first. The electrochemical-based cardiac biosensor, which is widely used for the monitoring and prognosis of cardiovascular illness, has also been highlighted in this chapter. The application of a fluorescent sensor for glutathione detection is taken into consideration. A 5G-enabled fluorescence sensor for the quantitative detection of SARS CoV-2 spike and nucleocapsid proteins is covered in detail. The technological challenges for the development of electrochemical and fluorescent sensors for fast diagnosis, early cancer detection, and detection of infectious disease have been discussed in the penultimate section.

In the twenty-first century, human civilization witnessed the most dangerous fatal pandemic Covid-19. Covid-19, a novel coronavirus also known as SARS-CoV-2, was found to be similar (50%–80%) in genome structure to the earlier SARS and MERS diseases. Alongside of the Covid-19 discussion regarding symptoms, transmission process, and genome sequencing, the recent progress made by the nanotechnology community in the detection and treatment of Covid-19 patients is discussed in detail. Point-of-care testing methodology like biosensors, electrochemical sensors, quantum dots, and so forth are better in terms of cost, turnaround time and efficiencies compared to the conventional test methods like real time-polymerase chain reaction (RT-PCR), computed tomography (CT), and so forth. Further, the role of nanomedicine along with various other proposed possible Covid-19 treatments, as well as necessary preventive safeguards for society against Covid-19 are considered carefully.

About the Editor

Basanta Bhowmik is currently Assistant Professor in the Department of Electronics and Communication Engineering, NIT Jamshedpur. He was a PhD fellow at the Indian Institute of Engineering Science and Technology (IIEST), Shibpur, India. His research interests includes semiconducting materials and devices for sensing and disease diagnosis. He is the author and coauthor of 52 research articles in journals and conferences of international repute. He received a Council of Scientific and Industrial Research (CSIR) Senior Research Fellowship award to pursue his PhD. He was given the best poster award in the event Research Scholar Colloquium organized by IIEST Shibpur in 2014. He was awarded second-best PhD thesis by the Institute of Smart Structure and System, IISc Bangalore. He is a member of the Institute of Electrical and Electronics Engineers (IEEE), IEEE Electron Device Society, IEEE Sensor Council, IEEE Nanotechnology Council, and a life member of the Institute of Smart Structure and System, India. He presently serves as an active reviewer for many journals.

Contributors

Rejaul Ali, Department of Electrical and Electronics Engineering, Birla Institute of Technology and Science (BITS)-Pilani, Vidya Vihar, Rajasthan 333031, India

Indranil Banerjee, Department of Pharmaceutical Technology, JIS University, Kolkata-700109, India

Nabaneeta Banerjee, Electronics and Communication Engineering Department, Guru Nanak Institute of Technology, Kolkata, India

Sabyasachi Banerjee, Department of Pharmaceutical Chemistry, Gupta College of Technological Sciences, Asansol-713301, West Bengal, India

Basudeba Behera, Department of Electronics and Communication Engineering, National Institute of Technology Jamshedpur, India

Basanta Bhowmik, Thin Film Devices Laboratory, Department of Electronics and Communication Engineering, National Institute of Technology, Jamshedpur

Subhrajit Chowdhury, Department of Electronics and Communication Engineering, Guru Nanak Institute of Technology, Kolkata-700114, India

Priyanka Daw, Department of Electronics and Communication Engineering, Guru Nanak Institute of Technology, Kolkata-700114, India

Arnab De, Department of Pharmaceutics, Bharat Technology, Uluberia, Howrah–711316, West Bengal, India

Sourav De, Department of Pharmaceutical Chemistry, Eminent College of Pharmaceutical Technology, Barasat-700126, West Bengal, India

Biplab Debnath, Department of Pharmaceutical Chemistry, Bharat Technology, Uluberia, Howrah–711316, West Bengal, India

Koushik Dutta, Department of ECE IIITDM, Jabalpur, India

Santanu Ghosh, Department of Pharmaceutical Sciences & Technology, Birla Institute of Technology, Mesra, Ranchi-835215, Jharkhand, India

Souvik Ghosh, Thin Film Devices Laboratory, Department of Electronics and Communication Engineering, National Institute of Technology, Jamshedpur

Arnab Hazra, Department of Electrical and Electronics Engineering, Birla Institute of Technology and Science (BITS)-Pilani, Vidya Vihar, Rajasthan 333031, India

Aditya Kumar Singh, Department of Computer Science and Engineering, Parul Institute of Technology, Waghodia, Vadodara, 391760, India

Dilip Kumar, Thin Film Device Laboratory, Department of Electronics and Communication Engineering, National Institute of Technology, Jamshedpur, 831014, India

Pankaj Kumar, Department of Electronics and Communication Engineering, National Institute of Technology Jamshedpur, India

Prashant Kumar, Department of Electronics and Communication Engineering, National Institute of Technology Jamshedpur, India

Suman Kumar, Thin Film Devices Laboratory, Department of Electronics and Communication Engineering, National Institute of Technology, Jamshedpur

Vikash Kumar, Thin Film Device Laboratory, Department of Electronics and Communication Engineering, National Institute of Technology, Jamshedpur, 831014, India

Amit Kundu, GITAM School of Pharmacy, Gandhi Institute of Technology and Management, Visakhapatnam – 530045, Andhra Pradesh, India

Anurup Nag, Department of Electronics and Communication Engineering, Guru Nanak Institute of Technology, Kolkata-700114, India

Aneek Nath, Department of Electronics and Communication Engineering, Guru Nanak Institute of Technology, Kolkata-700114, India

Ashish Pathak, Department of Electronics and Communication Engineering, National Institute of Technology, Jamshedpur-831014, Jharkhand, India

Susobhan Paul, Department of Physics Scottish Church College, Kolkata, India

Soumik Podder, Department of Basic Sciences and Humanities, Institute of Engineering and Management (IEM), University of Engineering and Management, Kolkata

Aman Raj, Department of Electronics and Communication Engineering, National Institute of Technology Jamshedpur, India

Polychetty Veenasheela Rao, Department of Electronics and Communication Engineering, National Institute of Technology Jamshedpur, India

Mrutyunjay Rout, Department of Electronics and Communication Engineering, National Institute of Technology Jamshedpur, Jharkhand-831014, India

Sunipa Roy, Electronics and Communication Engineering Department, Guru Nanak Institute of Technology, Kolkata, India

Bishal Sarkar, Department of Electronics and Communication Engineering, Guru Nanak Institute of Technology, Kolkata-700114, India

Baruna Kumar Turuk, Department of Electronics and Communication Engineering, National Institute of Technology Jamshedpur, India

Vikash Verma, Department of Electronics and Communication Engineering, National Institute of Technology Jamshedpur, India

1 Nanomaterials

What Determines Our Choice for Sensing?

Koushik Dutta

1.1 INTRODUCTION

The scientific records of nanotechnology were identified by the color-changing ability of the gold colloidal suspensions through an experiment by Michael Faraday in 1850.[1] The sensor device is one of the application-oriented technologies using nanomaterials, and a vivid variety of stimuli are handled via different physical and/or chemical mechanisms.[2] A representative historical aspect of glucose sensors is summarized in Table 1.1.[3]

However, the present discussion is restricted only to bio-molecule sensing, with a special tendency for application in health care, as per the main objective of the volume proposal. The application of glucose sensors, via electrochemical mode, was the first instance of the introduction of sensors into health care, and it has been considered a diagnostic sensor device since its commercialization in 1975.[4,5] Such sensor devices provide a ready-to-understand reading disengaging the expert requirement as well as time constraints. In this verge, over-the-counter testing kits to diagnose different diseases and medical situations (i.e., pregnancy tests, allergy detection, etc.) became feasible with appreciable accuracy.[4] In addition, the physics behind the sensing relies on different techniques, like optical, amperometric, electrochemical, MEMS-based, and so forth.[6,7] Further, the sensing is not limited only to these stimuli or the kit-based measurement. For this sensing, nanomaterials are one of the popular choices for cost-effective solutions.[4–7] Whatever the case, the bio-sensors have three major parts, namely, the Sensing element (recognizes the presence of biomarkers), the Transduction element (provides measurable electrical signal), and the Signal Processing system (process to decision-making and display).[3,8]

Under such consideration, nanomaterials have already proven their vitality for sensing via the chemo-physical framework of interaction with the stimulus.[7,9] Nanomaterials are unique in their attributes of free energies, which is the pristine source for enthalpy justification for a particular sensing mechanism.[7,9–11] Nanomaterials show a wide variety of applications according to their efficacy of sensing different stimuli (which component has to be sensed), mainly mapped through the response magnitude, which is directly related to the potential of nanomaterial.[7,9–11] In other words, the chemo-physical ability of the nanomaterial toward the stimuli is the prime point of interest behind the choice of material and nanostructures. In such cases, sensor

DOI: 10.1201/9781003453635-1

TABLE 1.1
Historical Perspective of Glucose Sensors as Instigating Biosensors[3]

Year	Event
1962	First description of a biosensor by Clark and Lyons
1967	First practical enzyme electrode by Updike and Hicks
1973	Glucose enzyme electrode based on detection of hydrogen peroxide [36]
1975	Relaunch of first commercial biosensor, i.e., YSI analyzer
1976	First bedside artificial pancreas (Miles)
1982	First needle-type enzyme electrode for subcutaneous implantation by Shichiri
1984	First ferrocene mediated amperometric glucose biosensor by Cass
1987	Launch of the MediSense ExacTech blood glucose biosensor
1999	Launch of a commercial in vivo glucose sensor (MiniMed)
2000	Introduction of wearable noninvasive glucose monitor (GlucoWatch)

technology evolves around some variational points, that is, target species (stimulus), sensing material, morphology, and orientation of the nanomaterials (array or haphazard), and so forth. A short description of these points may aid in understanding the scenario behind the choice of sensing.

1.2 STIMULUS VARIETY

In health-care applications, the target species can be categorized into two broad areas: (i) biologically derived substances (like, sugar, acetone, etc.) and (ii) biological substances (like, enzymes, lipids, bacteria, etc.).[6,7,9] However, vivid testing, correlation, and instrumentation are involved in identifying the stimulus and its concentration. A descriptive sensing behavior of different stimuli is presented here in two phases.

(i) Biologically Derived Substances: These substances are generally found in body fluids (blood, urine, tears, breath, etc.) and found in very trace quantities. A prominent example of such stimuli is glucose in the blood. The sensor devices, generally, work via the electrochemical principle where the chemical reaction (Reduction-Oxidation reaction) is translated to electron transfer (for electrical signal generation) for further instrumentation.[3,6,7,9,12] Briefly, the measurement of the glucose is presented for ready reference. For selective sensing of glucose, the sensing element (popularly, glucose oxidase enzyme with flavin adenine dinucleotide redox-cofactor) is immobilized on the transduction element.[3] The selective oxidation, of a glucose molecule, was catalyzed by the enzyme resulting in the generation of gluconic acid and hydrogen peroxide. Further, the electron transfer by the H_2O_2 at the transducing element (Pt electrode) provides a readable electrical signal (as the number of glucose molecules is proportional to the effective number of electrons) for signal processing.[3] In this context, it is clear that the choice of the transducing element is governed by the catalytic effectiveness required for electron transfer from the by-product of the first-hand chemical reaction.

In this row, the composition of particles and water in tears, human chorionic gonadotropin in urine, cytokines and chemokines in tears, the composition of urine, and so on are the different targeted stimuli to diagnose different abnormalities or medical conditions.[13–16] Sensing of such biomarkers is dominated by electrochemical or photochemical route, targeting fast sensing, low cost, and appreciable accuracy.[16]

(ii) *Biological Substances*: Sensing of such substances includes diagnosis, prevention, and bio-engineering applications.[17] Substances including bacteria, viruses, nucleic acids, and so forth come under this category.[4,18–21] The presence of the substance is generally detected by electrochemical or piezoelectric principle adding up with different immobilization protocols for enhanced selectivity.[9,22] Electrochemical means transduce the chemical changes, whereas mass or elastic properties are translated into electrical signals for the piezoelectric detection route.[4] In some cases, rapid detection is required of such substances where the presence of the target species is enough to identify the disorder (deviation from the normal biological concern).[13,19,23,24] Otherwise, laboratory-level detection may be time-consuming to infer the quantitative presence and its consequences.[25]

Common for these categories, biosensor technology always focuses on automated display so that chances of human error may be curtailed.[13,19,23–25] In this connection, the immobilization protocols play a pivotal role in selective detection even with a trace amount.[9,22,26,27] The immobilization protocols, with the advent of enzymes, DNA protein, antibodies, and so on to a particular target species, are convenient through linkers on the transducing layer.[26,28,29] The transduction layer should be efficient enough to provide high homage for the immobilized bio-molecules (functionalized group) equipped with efficient and rapid electron transportation.[4,18,22] Under such consideration, the effective surface of the transducing layer is a vital figure of merit.[4,18,22,26,27] At this point, the surface states (morphology) of the transducing layer became important where the nanostructured surface morphology is popularly undertaken for providing higher effective surface area and other nano-materialistic properties (e.g., free energy, confined electronic transportation, catalytic effect, etc.).[10,11,18,21,30,31] In the next subsection, the varieties of surface morphology will be discussed by categorizing them in terms of dimensionality with a target for practical applicability.

1.3 MORPHOLOGICAL VARIETY

The heart of bio-sensing relies on the coverage and adaptability of the species on the immobilized functionalization groups coupled with the carrier transportation path and catalytic properties.[10,11,18,21,30,31] The sufficed criteria are attributed to the effective surface area of the transduction layer, and it can be availed by the nanomat erials.[10,11,18,21,30,31] Nanomaterials have unique properties via their higher surface area, higher free energies, as well as cost-effectiveness.[32] In such a context, the morphology of the nanomaterial plays a vital role in providing an appreciable effective reactive surface.[10,11,18,21,30,32] In this connection, according to the dimensionality, the nanomaterial can be categorized into (i) 0D (zero-dimensional), (ii) 1D (one-dimensional), (iii) 2D (two-dimensional), and (iv) 3D (three-dimensional). A representative schematic is illustrated in Figure 1.1 for a quick ideation. The effectiveness of each category is described here with a focus on applications in health care.

(i) *0D Nanomaterials*: Nanoparticles are generally considered under this category.[22,33] 0D nanomaterials provide higher interaction sites due to the higher exposed surface area.[8,25] In this category, the noble material nanoparticles have a dynamic role through their catalytic properties.[21,30,31,34,35] A vigorous effort has to be employed to coat the nanoparticles with bio-conjunctive functional groups.[7,34,35] In this connection, the requirement of the substrate became mandatory for most of the cases.[7,21,30,31,34,35] However, this fact opens up the chance for flexible sensors arrowing toward smart applications.[6] Further, a nanoparticles-assisted drug-delivery system is another utilization of 0D nanomaterial in health care that is practiced popularly.[13,21,35]

(ii) *1D Nanomaterials*: Nanorods and nanotubes are generally referred to as 1D nanostructures, which cut down the number of grain boundaries in the carrier transportation path.[36–39] As a result, a faster transduction can be expected having a consequence of rapid sensing of the bio-species.[36–39] Under this category carbon nanotubes are an emerging material due to their high biocompatibility and ease of functionalization.[17]

(iii) *2D Nanomaterials*: With a focus on efficient transduction and/or catalytic effect, 2D nanomaterials are preferred as the presence of edges is more prone to provide free energy.[10,23] Interlayer stacking, of 2D nanomaterials, by van der Waals force enables higher effective surface area with numerous sites for easy functionalization combined with efficient carrier transportation ability and with acceptable stable chemical condition.[23,26,40]

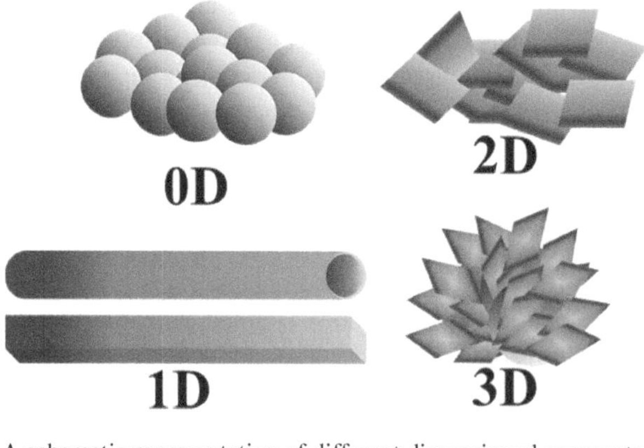

FIGURE 1.1 A schematic representation of different dimensioned nanomaterials.

(iv) *3D Nanomaterials*: Combining the advent of a higher amount of porosity, intrinsic conductivity, and favorable sites for functionalization, 3D nanostructures (e.g., nanospheres, nanoflowers, etc.) became a popular choice for highly responsive and faster biosensor material.[17,18] Enhanced diffusivity is the dominant advantage over the 2D structure along with a hosting capability for other dimensional nanomaterials to combine corresponding prospects.[20]

The choice of dimensionality of the nanomaterials depends predominantly on the target species (with its limit of detection), required response magnitude and response time, and the ease of fabrication (related to cost).[17,18,23,26,36–40] Another vital point of nanostructure orientation is presented next before the fabrication strategies and their corresponding prospects and constraints are discussed. This orientation variety provides insight into the sensing mechanism as well as the measurement techniques.

1.4 ORIENTATION VARIETY

The distribution of nanomaterial for possible biosensing should include good conductivity and acceptable coverage of immobilized protocols.[4,32] This distribution of nanomaterials depends mainly on the synthesis process adopted along with the choice of substrate.[13,21,35] The orientation of nanomaterials is broadly divided into two segments: (i) Unordered Orientation and (ii) Well-ordered Orientation.

(i) *Unordered Orientation*: The orientation of nanomaterials is not ordered and sometimes variable sizes can be observed.[7,10,21,23,26,30,31,34,35,40] By fabrication, the oriented nanostructure can be obtained through a one-step method. Additionally, the use of flexible substrates is feasible for creating biosens ors.[7,21,30–32,34,35] This orientation includes each of the dimensions affecting the intra-nanostructure connection.[10,23,26,32,40] In these cases, the modeling (estimation) of electron transportation has to be done crucially to explain the mechanism. Dominantly, electron transport occurs via grain-boundary control under such orientation, which may increase the sensing time.[41–44] Further, the distribution/coverage of the immobilized functional groups results in an inadvertent manner, leading to less efficacy of the particular sensor.[32,35]
(ii) *Well-ordered Orientation*: Vertical alignment of the nanomaterial or nanomaterial array is recognized as a well-ordered orientation.[29,36,39] The confinement of carrier transportation is the main feature of these nanomaterials, ensuring a faster response.[7,39] Also, the recoverability of the sensor's probe can possibly be ascertained in special cases.[22,29,36] Moreover, the aligned nature guarantees the uniformity of the immobilization protocols, in the case of biosensors.[7,9,21,35]

To understand the carrier transportation for each of the dimensionalities of nanostructures, Figure 1.2 may be referred to. The grain of the nanostructures is fully or partially depleted by the ligand (functional groups) attachment and by the ambient oxygen through the Debye length.[4,41,42,45] On the interaction of the bio-species with the ligands, carrier exchange (increase of decrease of depletion width or Debye length)

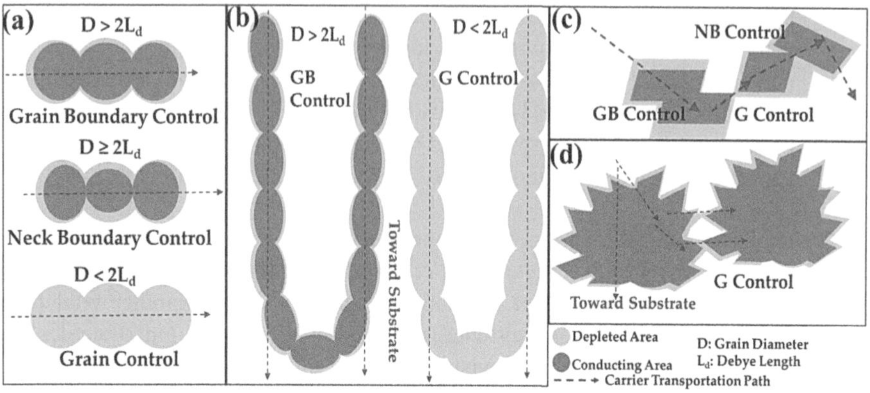

FIGURE 1.2 Schematic representation of carrier transportation through the (a) 0D, (b) 1D, (c) 2D, and (d) 3D nanomaterial. The prevalent transportation mechanism for different conditions of the grain size and corresponding Debye length is highlighted [39–42,45].

changes the chemical and/or physical properties of the nanostructures.[9,22] Information on the presence and quantity of targeted species is inferred through the excess carrier transportation. In the case of 0D nanomaterials, this transportation is modeled through Grain Boundary (GB) control (for grain size greater than twice Debye length), Neck Boundary (NB) Control (for grain size comparable to twice Debye length), or Grain (G) Control (for grain size lesser than twice Debye length), as depicted in Figure 1.2 (a).[41,42] Further, for 1D carrier transportation, considering GB control and G control, two types of carrier transportation may be observable.[39] The confined path of electron transportation is evident for such nanostructures. Depending on the ligand attachment of the available surfaces of each 2D nanosheet and the interconnectivity between the sheets (as depicted in Figure 1.2 (c)), the carrier transportation of each type (GB, NB, G control) is feasible.[26,45,46] As the 3D nanostructures are formed by the assembly of either the nanoparticles (resulting in nanospheres) or nanosheets (resulting in nanoflower), the carrier transportation is dominated by the G control mechanism at the interconnects.[17,18,20,45] Additionally, the transportation in the core part may be quick as the GB control is prevalent for that case (Figure 1.2 (d)).

Apart from the orientation, the choice of nanomaterials also depends on the chemical and physical properties of the sensing materials themselves. The experimental synthesis procedures also vary concerning the material choice. However, the following point is focused on the variety of sensing materials as suitable biosensor candidates.

1.5 MATERIAL VARIETY

As the biosensor material, various elements, compounds, and composites are being used in the commercial field as well as in the emerging research arena. Biosensing materials have two major segments, namely, (a) bio-functional groups (to enhance the selectivity), and (b) organic/inorganic transducing layer.[4,17,32] The former is well defined and has less room for engineering as these functional groups have special biological signatures. However, as a sensor technologist, the scope for engineering is only left to the transducing material to which the excavation of the properties of nanostructures can be applied. Biosensing nanomaterials can crudely be divided into four categories: (i) Pure Semiconductor, (ii) Noble metal, (iii) Metal chalcogenides, and (iv) Carbon-based.[4,7,9,17] A short overview is presented point-wise to identify the corresponding prospects and constraints.

(i) Pure Semiconductor Nanomaterials: Mainly silicon (Si) is identified as the pure semiconductor sensing layer and is compatible with the standard semiconductor fabrication strategies.[7,9,47,48] The control over the surface condition by the etching process(es) established the applicability of this material. The porous nature of the surface of the silicon enriches the effective surface area for favorable interaction alongside the resonant frequency (crucial for electrochemical measurement) by modulating the electric field lines passing through pores and interfaces with the analytes.[7] Reproducibility is enhanced for the measurement through resonant frequency capping with higher longevity.[7] Germanium (Ge) is also a good choice to support biosensor electronics due to its biodegradable nature.[47,48] Further, the well-characterized spectroscopic analysis of pure semiconductors is helpful in identifying the interference of bio-species, which laid the base behind the spectroscopic measurement of biosensor response.[48]

(ii) Noble Metal Nanomaterials: Utilizing the catalytic effect of nanoforms of noble metals (e.g., Au, Ag, Pd, Pt, etc.), the transduction of the biosensors is effectively enhanced resulting in a higher response rate.[1,8,21,23,25,29] Unique chemical properties added up with the free energy of nanostructures, the noble metal nanomaterials provide a suitable platform for ligand attachment and transduction due to extensive surface area and good conductivity, respectively.[8,21] However, in most cases, 0D nanoparticles are evidenced as a sensing layer.[1,18,23,25]

(iii) Metal chalcogenides: The compound of metal with the molecules of the oxygen family in periodic tables are generalized as metal chalcogenides.[26,40] Under this category, metal oxides, metal (di)sulfides, and so forth are prominent examples of nanomaterials.[26,40] The freedom available for cost-effective chemical routes for nanostructure fabrication and engineering have made metal chalcogenides a well-explored material in nanotechnology.[26,40] A wide variety of dimensionality is expected with such materials and is useful to deploy various materialistic properties for better biosensors.[36,37,40,49]

From the viewpoint of various nanomaterials, the biosensors rely not only on the morphological structures and carrier transportation efficacy but also on the chemical and catalytic perspective of the particular material.[8,23,50] However, the chemical and catalytic properties are tunable through the nanostructured theme (0D, 1D, 2D, or 3D), and the same decides the carrier transportation in an essential attire.[6,8,10,17,18,23,50] Thus, the synthesis process has to be explored to identify the repeatability, tunability, and cost-effectiveness with the ease of experimental steps. The following subsection concentrates on these features, with the aim of indicating the irresistible settings of experiments for efficient nanomaterials for efficient biosensors.

1.6 EXPERIMENTAL

In this subsection, different experimental setups will be discussed in a representative manner for 0D, 1D, 2D, and 3D nanomaterials, respectively, with the intention of determining tunability in the cost-effective process flow. The process flow is categorized into three classes: (i) Standard Semiconducting Fabrication Process, (ii) Open Air Synthesis Process, and (iii) Vacuum Assisted Process.

(i) *Standard Semiconducting Fabrication Process*: The process used in the present semiconductor technology includes single crystal growth, epitaxy, oxidation, lithography, and so on.[7,9,47,48] Generally, pure semiconductor material is used with this fabrication technique, but the method is not only limited to the pure semiconductors.[7,9,47,48] Further, the device structure (especially FET-based devices) highly depends on such technology.[7,9,22,47,48] Through this process, the porous silicon structure can generally be fabricated, and in a few instances, the array of pores (to support aligned nanostructures) is synthesized via engineering the process.[22,47–49] However, the field effect transistor (FET) type device structure and its applications in bio-sensing are favorable with this fabrication process.[22,49]

(ii) *Open Air Synthesis Process*: Those chemical synthesis processes that can be done under normal atmospheric pressure, that is, there is no need for low pressure or vacuum, are considered in this category. The main prospects of such processes are their cost-effectiveness and preferably one-step and/or one-pot synthesis mechanism.[10,37,51,52] The variation in temperature is done while synthesis to obtain the desired nanostructure.[4,53] Also, heat treatment (annealing), for material sustainability, became essential.[10,16,37,51,52] Sol-gel preparation, dip coating, spin coating, chemical bath deposition, and their combinations are possibly considered here.[10,37,51–54] In these cases, variations in temperature, chemical constituents, and/or the nature of substrates are the prominent determining factors behind the dimensionality of nanomaterials. In short, it can be emphasized that 0D and 1D (nonaligned) nanostructures can easily be fabricated through the mentioned route. Moreover, electrochemical anodization is the most commonly and reliably used method to establish the aligned nature in nanostructures. This method is also performed under normal atmospheric pressure with preferable tunability and reliability.[4,51–53] Aligned 1D nanostructures are formed in the direction of electric field lines and thus, uniformity is expected with this technique. In a few instances, the template-assisted alignment of nanorods or nanotubes may be possible, but is restricted due to usage of extra materials (adding cost element).[50] Further, the cost-effectiveness is an attribute to such conditions at the cost of repeatability issues.

(iii) *Vacuum Assisted Fabrication Process*: Fabrication processes like Chemical Vapor Deposition (CVD), Atomic Layer Deposition (ALD), and so forth come under this condition.[4,48,55] Here, the controllability can be attained to the highest degree via tuning the synthesis parameters.[4,48,55] Thus, repeatable nanostructures can be evidenced. The cost element is the main concern for these processes as the requirement of the evacuated chamber demands a high technological concern.[4,26,40,48,55] Generally, in the field of sensors, these fabrication processes are neglected for cost-cutting purposes.

Different nanostructures can be formed by adopting various synthesis routes, and by tuning the synthesis parameters engineering on the nanostructures became possible. Engineering on the morphology of the nanomaterial is beneficial as discussed above and for self-catalytic sites (edges corners).[10] Introducing self-catalytic sites may be favorable for better immobilization protocols and hence the response.[10,18] The details of different synthesis procedures can be found in the published chapter.[56] Further, the sensing mechanism is vital for different nanostructures along with their corresponding carrier transportation phenomena. A brief overview of modeling and its explanation is presented in the next subsection.

1.7 MODELING AND EXPLANATION

The sensing mechanism is vital to infer the physical framework behind the device performance. To probe this, different types of scientific explanations and modeling are used by researchers. In general, the performance of biosensors depends on the attachment probability of functionalization protocols and corresponding electron transportation phenomena.[7,17,35,57] Also, the type of device structure (planar, vertical, FET) plays a pivotal role in the sensitivity of the biosensors.[24,49,58,59] In this verge, the repeatable usability of the biomarkers is another important factor.[24,49] In this connection, the surface-to-volume ratio (SVR) is the leading figure of merit to decide the number of functionalization groups, as these are attached to the surface of nanom aterials.[7,17,24,35,49,57] However, in the connection of the nanomaterials, carrier transportation is mainly focused, as per the chapter objective. Here, carrier transportation in 0D nanostructure is identified through the grain boundary model.[41,42,58] In this case, the fully depleted case is the preferable condition for better sensitivity. But not in all cases can fully depleted conditions be achieved. From the perspective of the device structure, FET-based modeling follows the standard FET modeling combined with the grain boundary model.[58] A representative simulation result for FET biosensors is presented in Figure 1.3. Here, the symbols are of standard meaning for transistor modeling.[58]

On the other hand, 1D carrier transportation is achievable in the confined path through the semiconducting path.[39] Hence, the vertical device structure is popular with such nanostructure. Further, 2D nanostructured material transports the carrier through the quantum phenomena and is generally simulated with density functional theory (DFT).[60,61]

In a similar manner, the carrier transportation in 3D nanostructure is modeled through the quantum phenomena, and the collection is dominantly the 1D counterpart.[17] It should be mentioned that the modeling is purely a device-oriented (constituent-specific) course to explain the mechanism. However, under this scope,

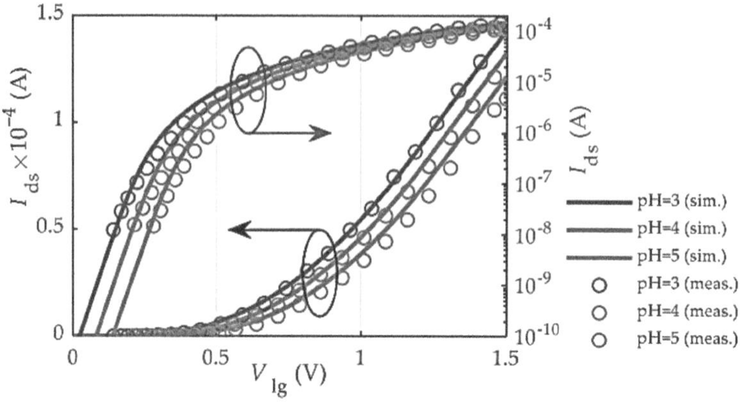

FIGURE 1.3 Simulated FET characteristics of biosensor where the pH changes due to presence of bio-molecules.[58]

the modeling of the biosensors is not elaborated to excise the published reviews of the corresponding nanostructures.

1.8 SUMMARY

The present chapter is intended to present the choice of different dimensional (0D, 1D, 2D, and 3D) nanostructures for better biosensors converging the target material variation and orientation. Further, the device structure is another aspect of the biosensors where the specific sensing strategies (electrochemical, optical, piezoelectric, etc.) define the reusability and quality of the biosensors. On the other hand, from a research point of view, an arrow is indicated for modeling and explanation. This chapter may also be considered as an instigating conceptual part for a wise choice of nanostructure for better biosensors.

Further, the carrier transportation of the transducing part, in which the involvement and determining factors of nanomaterial are most important, should be considered in a significant manner. The different strategies of controlled carrier transportation mechanisms paved the path for device-level modeling and explanation. The application of the nanostructure opened up a new arena for smart biosensors, which is essential for the continuous diagnosis for the human being.

REFERENCES

1. Kheirandish A, Sepehri Javan N, Mohammadzadeh H. (2020) Modified Drude model for small gold nanoparticles surface plasmon resonance based on the role of classical confinement. *Sci Rep.*; 10(1):1–10. doi:10.1038/s41598-020-63066-9
2. Ida N. (2014) *Sensors, Actuators, and Their Interfaces: A Multidisciplinary Introduction.* SciTech Publishing- Institution of Engineering and Technology; doi:10.1049/PBCE127E

3. Yoo E, Lee S. (2010) Glucose biosensors: An overview of use in clinical practice. *Sensors*; 10(5):4558–4576. doi:10.3390/s100504558
4. McGrath MJ, Scanaill CN. (2013) *Sensor Technologies*. Vol 6. Apress. doi:10.1007/978-1-4302-6014-1
5. Harper A, Anderson MR. (2010) Electrochemical glucose sensors-developments using electrostatic assembly and carbon nanotubes for biosensor construction. *Sensors*; 10(9):8248–8274. doi:10.3390/s100908248
6. Knopf GK, Bassi AS. (2018) *Smart Biosensor Technology*. CRC Press. doi:10.1201/9780429429934
7. Roychaudhuri C. (2015) A review on porous silicon based electrochemical biosensors: Beyond surface area enhancement factor. *Sensors Actuators, B Chem.*; 210:310–323. doi:10.1016/j.snb.2014.12.089
8. Zhu Z. (2016) Gold nanoparticle based biosensors. *New Dev Gold Nanomater Res.*; 43(1):95–116.
9. Ghosh H, Roychaudhuri C. (2016) Pore Geometry Optimization of Nanocrystalline Silicon Oxide Impedance Biosensor. *IEEE Sens J.*; 16(22):7843–7852. doi:10.1109/JSEN.2016.2604854
10. Dutta K, Chattopadhyay PP, Bhattacharyya P. (2016) Voltage controlled rupturing of TiO_2 nanotubes for gas sensor device applications: correlation with surface and edge energy. *IEEE Trans Electron Devices*; 63(12):4933–4938. doi:10.1109/TED.2016.2620560
11. Roduner E. (2006) Size matters: Why nanomaterials are different. *Chem Soc Rev.*; 35(7):583. doi:10.1039/b502142c
12. Fleming L, Gibson D, Song S, Li C, Reid S. (2018) Reducing N_2O induced cross-talk in a NDIR CO_2 gas sensor for breath analysis using multilayer thin film optical interference coatings. *Surf Coatings Technol.*; 336(June 2017):9–16. doi:10.1016/j.surfcoat.2017.09.033
13. Vaidya VS, Ford GM, Waikar SS, et al. (2009) A rapid urine test for early detection of kidney injury. *Kidney Int.*; 76(1):108–114. doi:10.1038/ki.2009.96
14. Bajkowska D, Szelachowska M, Buczyńska A, Krętowski AJ, Siewko K. (2022) Tears as a source of biomarkers in the diagnosis of Graves' Orbitopathy. *Biomolecules*; 12(11):1620. doi:10.3390/biom12111620
15. Gnoth C, Johnson S. (2014) Strips of hope: Accuracy of home pregnancy tests and new developments. *Geburtshilfe Frauenheilkd*; 74(07):661–669. doi:10.1055/s-0034-1368589
16. Fan J, Wang M, Wang C, Cao Y. (2017) Advances in human chorionic gonadotropin detection technologies: A review. *Bioanalysis*; 9(19):1509–1529. doi:10.4155/bio-2017-0072
17. Claussen JC, Hengenius JB, Wickner MM, Fisher TS, Umulis DM, Porterfield DM. (2011) Effects of carbon nanotube-tethered nanosphere density on amperometric biosensing: Simulation and experiment. *J Phys Chem C.*; 115(43):20896–20904. doi:10.1021/jp205569z
18. Huang H, Liu M, Wang X, et al. (2016) Label-Free 3D Ag nanoflower-based electrochemical immunosensor for the detection of Escherichia coli O157:H7 Pathogens. *Nanoscale Res Lett.*; 11(1):0–7. doi:10.1186/s11671-016-1711-3
19. Li Y, Xie G, Qiu J, et al. (2018) A new biosensor based on the recognition of phages and the signal amplification of organic-inorganic hybrid nanoflowers for discriminating and quantitating live pathogenic bacteria in urine. *Sensors Actuators, B Chem.*; 258:803–812. doi:10.1016/j.snb.2017.11.155

20. Muttaqien SEI, Khoris IM, Pambudi S, Park EY. (2023) Nanosphere structures using various materials: A strategy for signal amplification for virus sensing. *Sensors*; 23(1). doi:10.3390/s23010160
21. Hua Z, Yu T, Liu D, Xianyu Y. (2021) Recent advances in gold nanoparticles-based biosensors for food safety detection. *Biosens Bioelectron.*; 179(January):113076. doi:10.1016/j.bios.2021.113076
22. Basu J, Roychaudhuri C. (2018) Graphene nanoporous FET biosensor: Influence of pore dimension on sensing performance in complex analyte. *IEEE Sens J.*; 18(14):5627–5634. doi:10.1109/JSEN.2018.2841060
23. Chia HL, Mayorga-Martinez CC, Pumera M. (2021) Doping and decorating 2D materials for biosensing: Benefits and drawbacks. *Adv Funct Mater.*; 31(46):1–31. doi:10.1002/adfm.202102555
24. Ertürk G, Lood R. (2017) Bacteriophages as biorecognition elements in capacitive biosensors: Phage and host bacteria detection. *Sensors Actuators, B Chem.*; 258:535–543. doi:10.1016/j.snb.2017.11.117
25. Sanvicens N, Pastells C, Pascual N, Marco MP. (2009) Nanoparticle-based biosensors for detection of pathogenic bacteria. *TrAC – Trends Anal Chem.*; 28(11):1243–1252. doi:10.1016/j.trac.2009.08.002
26. Zhang W, Yang J, Wu D. (2020) Surface-functionalized MoS_2 nanosheets sensor for direct electrochemical dtection of PIK3CA gene related to lung cancer. *J Electrochem Soc.*; 167(2):027501. doi:10.1149/1945-7111/ab61ec
27. Kim HS, Kim JH, Park SY, et al. (2019) Carbon nanotubes immobilized on gold electrode as an electrochemical humidity sensor. *Sensors Actuators, B Chem.*; 300(April):127049. doi:10.1016/j.snb.2019.127049
28. Gasparyan L, Mazo I, Simonyan V, Gasparyan F. (2020) Noises and Signal-to-Noise ratio of nanosize EIS and ISFET biosensors. *Open J Biophys.*; 10(01):1–12. doi:10.4236/ojbiphy.2020.101001
29. Ramulu S, Reddy V, Soo S, Kim C. (2016). Biosensors and bioelectronics electrochemical biosensor for Mycobacterium tuberculosis DNA detection based on gold nanotubes array electrode platform. *Biosens. Bioelectron.*; 78:483–488. doi:10.1016/j.bios.2015.11.098
30. Liu F, Li K, Zhang Y, et al. (2020) An electrochemical DNA biosensor based on nitrogen-doped graphene nanosheets decorated with gold nanoparticles for genetically modified maize detection. *Microchim Acta.*; 187(10):574 (1–8). doi:10.1007/s00604-020-04511-4
31. Farooq S, Wali F, Zezell DM, de Araujo RE, Rativa D. (2022) Optimizing and quantifying gold nanospheres based on LSPR label-free biosensor for dengue diagnosis. *Polymers (Basel)*; 14(8):1–12. doi:10.3390/polym14081592
32. Yao S, Swetha P, Zhu Y. (2018) Nanomaterial-enabled wearable sensors for healthcare. *Adv Healthc Mater.*; 7(1). doi:10.1002/adhm.201700889
33. Su X, Wu Q, Li J, et al. (2014) Silicon-Based nanomaterials for lithium-ion batteries: A review. *Adv Energy Mater.*; 4(1). doi:10.1002/aenm.201300882
34. Alba-Patiño A, Russell SM, Borges M, Pazos-Pérez N, Álvarez-Puebla RA, De La Rica R. (2020) Nanoparticle-based mobile biosensors for the rapid detection of sepsis biomarkers in whole blood. *Nanoscale Adv.*; 2(3):1253–1260. doi:10.1039/d0na00026d
35. Singh SB, Prabhakar NR, Pentyala SN. (2014) *Translational Research in Environmental and Occupational Stress*. Springer India. doi:10.1007/978-81-322-1928-6
36. Yang K, She G, Wang H, et al. (2009) ZnO nanotube arrays as biosensors for glucose. *J Phys Chem C.*; 113(47):20169–20172. doi:10.1021/jp901894j

37. Cipriano AF, Miller C, Liu H. (2014) Anodic growth and biomedical applications of TiO_2 nanotubes. *J Biomed Nanotechnol.*; 10(10):2977–3003. doi:10.1166/jbn.2014.1927
38. Mitra S, Basak M. (2022) Diverse bio-sensing and therapeutic applications of plasmon enhanced nanostructures. *Mater Today.*; 57(August):225–261. doi:10.1016/j.mattod.2022.05.023
39. Hazra A, Bhattacharyya P. (2014) Tailoring of the gas sensing performance of TiO_2 nanotubes by 1-D vertical electron transport technique. *IEEE Trans Electron Devices*; 61(10):3483–3489. doi:10.1109/TED.2014.2342277
40. Lu J, Chen M, Dong L, et al.(2020) Molybdenum disulfide nanosheets: From exfoliation preparation to biosensing and cancer therapy applications. *Colloids Surfaces B Biointerfaces*; 194(May). doi:10.1016/j.colsurfb.2020.111162
41. Sahay PP, Nath RK. (2008) Al-doped ZnO thin films as methanol sensors. *Sensors Actuators, B Chem.*; 134(2):654–659. doi:10.1016/j.snb.2008.06.006
42. Barsan N, Koziej D, Weimar U. (2007) Metal oxide-based gas sensor research: How to? *Sensors Actuators, B Chem.*; 121(1):18–35. doi:10.1016/j.snb.2006.09.047
43. Nilsson D, Kugler T, Svensson PO, Berggren M. (2002) An all-organic sensor-transistor based on a novel electrochemical transducer concept printed electrochemical sensors on paper. *Sensors Actuators, B Chem.*; 86(2–3):193–197. doi:10.1016/S0925-4005(02)00170-3
44. Judy JW. (2001) Microelectromechanical systems (MEMS): Fabrication, design and applications. *Smart Mater Struct.*; 10(6):1115–1134. doi:10.1088/0964-1726/10/6/301
45. Ghosal S, Dutta K, Bhattacharyya P. (2020) Understanding the improved vapor sensor device performance of dual surface engineered WO_3 nanospheres using semi-quantitative energy band model. *IEEE Electron Device Lett.*; 41(6):912–915. doi:10.1109/LED.2020.2991085
46. Wang P, Wang A, Hassan MM, Ouyang Q, Li H, Chen Q. (2020) A highly sensitive upconversion nanoparticles-WS2 nanosheet sensing platform for Escherichia coli detection. *Sensors Actuators, B Chem.*; 320(June):128434. doi:10.1016/j.snb.2020.128434
47. Zhao H, Xue Z, Wu X, Wei Z, Guo Q, Xu M. (2022) Biodegradable germanium electronics for integrated biosensing of physiological signals. *NPJ Flex Electron.*; 6(1):63. doi:10.1038/s41528-022-00196-2
48. Blond P, Bevernaegie R, Troian-gautier L, et al. (2020) Ready-to-Use Germanium surfaces for the development of FTIR-based biosensors for Proteins. *Langmuir.*; 36(40):12068–12076. doi:10.1021/acs.langmuir.0c02681
49. Chakraborty B, Chakrabarty S, Ghosh S, Chaudhuri CR. (2018) High Performance Biosensor Based on RGO/ZnO Thin Film Transistor. In: *2018 IEEE SENSORS*. Vol 2018-Octob. IEEE:1–4. doi:10.1109/ICSENS.2018.8589592
50. Das S, Jayaraman V. (2014) SnO_2: A comprehensive review on structures and gas sensors. *Prog Mater Sci.*; 66:112–255. doi:10.1016/j.pmatsci.2014.06.003
51. Bhowmik B, Dutta K, Bhattacharyya P. (2015) n-TiO_2/p-Si heterojunction devices as a potential ethanol sensor. In: *2015 6th International Conference on Computers and Devices for Communication (CODEC)*. IEEE:1–4. doi:10.1109/CODEC.2015.7893190
52. Banerjee N, Roy S, Sarkar CK, Bhattacharyya P. (2013) High dynamic range methanol sensor based on aligned ZnO nanorods. *IEEE Sens J.*; 13(5):1669–1676. doi:10.1109/JSEN.2013.2237822
53. Dutta K, Hazra A, Bhattacharyya P. (2016) Ti/TiO_2 Nanotube Array/Ti capacitive device for non-polar Aromatic Hydrocarbon Detection. *IEEE Trans Device Mater Reliab.*; 16(2):235–242. doi:10.1109/TDMR.2016.2564447

54. Dutta K, Bhowmik B, Hazra A, Chattopadhyay PP, Bhattacharyya P. (2015) An efficient BTX sensor based on p-type nanoporous titania thin films. *Microelectron Reliab.*; 55(3–4):558–564. doi:10.1016/j.microrel.2014.12.010
55. Martinson ABF, Elam JW, Hupp JT, Pellin MJ. (2007) ZnO nanotube based dye-sensitized solar cells. *Nano Lett.*; 7(8):2183–2187. doi:10.1021/nl070160+
56. Prasad S, Kumar V, Kirubanandam S, Barhoum A. (2018) *Engineered Nanomaterials: Nanofabrication and Surface Functionalization.* Elsevier Inc. doi:10.1016/B978-0-323-51254-1.00011-7
57. Martín-Gracia B , Martín-Barreiro A, Cuestas-Ayllón C, et al. (2020) Nanoparticle-based biosensors for detection of extracellular vesicles in liquid biopsies. *J Mater Chem B.*; 8(31):6710–6738. doi:10.1039/d0tb00861c
58. Pasadas F, Grour T El, Marin EG, et al. (2023) Compact modeling of Two-Dimensional Field-Effect biosensors. *Sensors*; 23(4):1840 (1–16). https://doi.org/10.3390/s23041840
59. Gudiksen MS, Lauhon LJ, Wang J, Smith DC, Lieber CM. (2002) Growth of nanowire superlattice structures for nanoscale photonics and electronics. *Nature.*; 415(6872):617–620. doi:10.1038/415617a
60. Vaidyanathan A, Mathew M, Radhakrishnan S, Rout CS, Chakraborty B. (2020) Theoretical insight on the biosensing applications of 2D materials. *J Phys Chem B.*; 124(19):11098–11122. doi:10.1021/acs.jpcb.0c08539
61. Chen F, Tang Q, Ma T, et al. (2022) Structures, properties, and challenges of emerging 2D materials in bioelectronics and biosensors. *InfoMat.*; 4(5):1–37. doi:10.1002/inf2.12299

2 Properties of Materials for Micro/Nanoscale Devices

Sunipa Roy, Nabaneeta Banerjee, and Susobhan Paul

2.1 INTRODUCTION

The properties of materials for micro/nanoscale devices play a fundamental role in the design, functionality, and performance of these miniature technologies. Any material is composed of atoms, and the smallest atom is typically of the dimension 0.1nm. This is the fundamental limit of making any device arbitrarily small. We cannot fabricate transistors smaller than the size of an atom. The material properties change drastically as we decrease the size of any object due to the onset of quantum effects. In bulk matter, the quantum mechanical effects of the constituent atoms may not be readily apparent due to the statistical averaging of a large number of particles, typically about $\sim 10^{23}$. However, when we decrease the size of the object, the surface area increases compared to its volume. If we consider a spherical nanoparticle of radius r, then its surface-to-volume ratio (s/v) would be $3/r$. Therefore, decreasing the size (r) of the sphere results in a significant increase in its surface area. Due to the large surface area of any nanostructured material $(L \sim 1-100nm)$, a significant number of atoms will be on the surface, leading to enhanced surface energy, and significant changes will appear in the physicochemical properties. Apart from that, due to reduced dimensionality, the behavior of the electron is longer like a classical particle but rather its nature is more quanta mechanical. Due to the confinement of quantum mechanical electrons in small dimensions, the electronic band structure drastically changes from the bulk counterpart. This is called the quantum confinement effect. The changed electronic band structure affects the mechanical, electrical, optical, and thermal properties of the nanomaterials.

Nanoparticles can be employed for testing biomolecules as biomarkers and tumor labels. Nano medicinal products range from chemotherapy to biological agents to immunotherapy and more. Nanotech can also be utilized by molecules identifying illness to improve selective diagnosis.

A nanomaterial has at least one dimension within 1–100 nm. Nanomaterials are fascinating because of their unique properties that arise from their small size. Nanostructured materials can include things like nanoparticles, nanotubes, and nanowires, where the structure itself is within the nanoscale range. Nanostructured elements, on the other hand, can refer to individual atoms, molecules, or clusters that

DOI: 10.1201/9781003453635-2

exhibit nanoscale properties. This distinction helps in understanding and categorizing the diverse range of materials that fall under the umbrella of nanotechnology.

2.2 ELECTRONIC PROPERTIES

The energy spectrum of an isolated atom is essentially discrete in nature, as dictated by the Schrödinger equation. When two atoms form a molecule, the energy spectrum becomes discrete doublets, where each line in the doublet comes from individual atoms. Now, in a situation in which atoms coalesce to form solid material, individual sublevels of energy (mostly the outer shell electrons) broaden to form an electronic band structure,[1–3] as illustrated in Figure 2.1.

The band energy (E) is a function of the wave vector ($\vec{k}$). The highest occupied energy level in the band structure is known as the valence band.[2] However, the range of energy levels above the valence band, where electrons are not bound to the atom, is called the conduction band. The energy difference between the highest occupied energy state of the valence band and the lowest unoccupied state of the conduction band is known as the bandgap (E_g). Now we need to define density of states (DOS). DOS, $g(E)$ for electrons in a band is the number of electron states per unit volume in the energy range. If we know the energy band spectrum or dispersion relation $E(\vec{k})$, DOS may be calculated as follows:[4]

$$g(E) = \frac{1}{v} \int \delta\left(E - E\left(\vec{k}\right)\right) d^3k, \quad (2.1)$$

where v is the volume of the material, δ is the Dirac-delta function, and $\vec{k}$, space integration, is performed for all possible values of k in the Brillouin zone.

Nanomaterials can be fabricated in 0D, 1D, 2D, and 3D.[5] DOS will dramatically change for different dimensions as shown in Figure 2.2. For 3D conventional bulk material, DOS is generally proportional to $\sqrt{E}$. Here we have assumed the parabolic energy dispersion relation, i.e., $E(k) \propto k^2$.

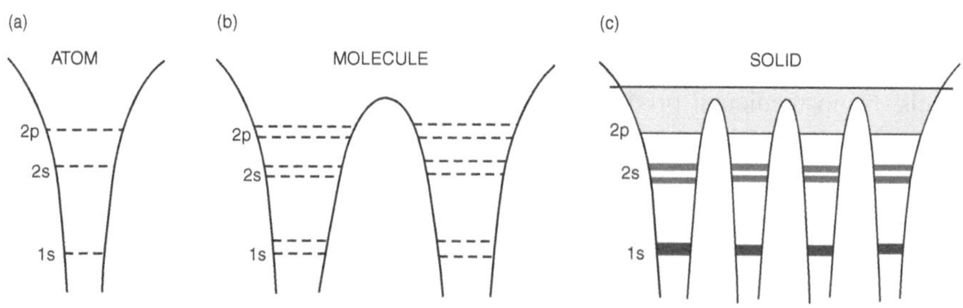

FIGURE 2.1 Formation of energy bands in solids shows (a) the line of spectrum of atoms, (b) energy spectrum of molecule, and (c) energy band in solids.

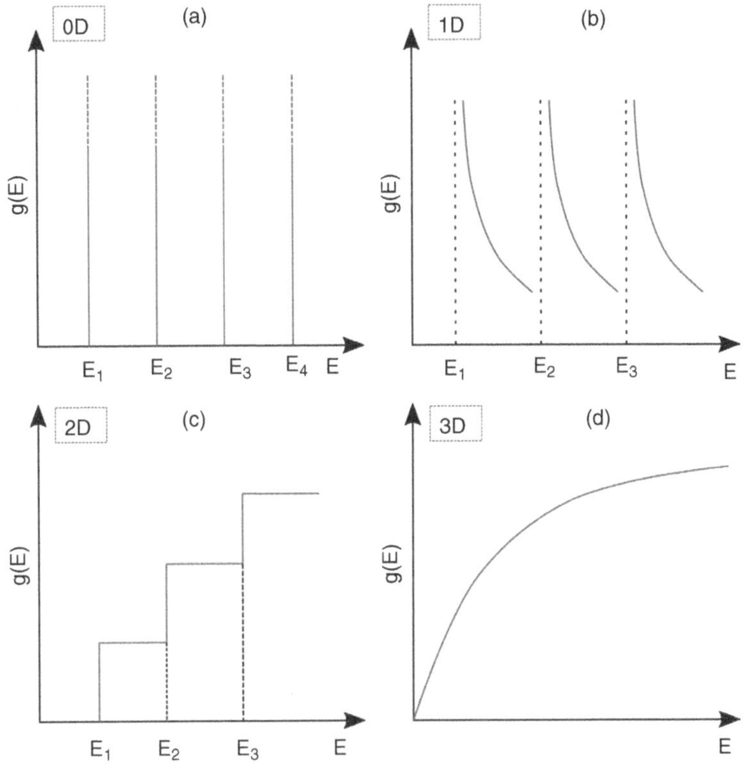

FIGURE 2.2 DOS for (a) 0D (b) 1D (c) 2D (d) 3D nanostructures.

i. **Zero-Dimensional Nanostructures:** Quantum dots, or 'artificial atoms', are a few nanometer-size semiconductor-based nanostructures, which belong to the zero-dimensional nanostructure.[2,3] Quantum dots have found numerous applications in optoelectronics, quantum computing, and many more. In this zero-dimensional (0D) structure, $\vec{k}$ is quantized in all three directions. Its DOS is like a discrete line, like atomic spectra, and it is given by[4,5]

$$g(E) = 2\sum_{n_x, n_y, n_z} \delta\left(E - E_{n_x, n_y, n_z}\right) \tag{2.2}$$

where E_{n_x, n_y, n_z}, the confinement energies, are associated with the x, y, and z directions, and $\delta\left(E - E_{n_x, n_y, n_z}\right)$ is the Dirac-delta function. In this case, all three directionsare comparable to the de Broglie wavelength, and therefore the spectrum is purely discrete.

ii. **One-Dimensional Nanostructures:**

In one-dimensional (1D) nanostructures, like quantum wire, the motion of electron motion is continuous along the longitudinal direction (z) only. The other two transverse direction (x-y) motions of the electron are quantized. The expression for DOS for a 1Dnanostructure will be[4,5]

$$g(E)=\frac{1}{\pi}\sqrt{\frac{2m^*}{\hbar^2}}\sum_{n_x,n_y}\frac{1}{\sqrt{E-E_{n_x,n_y}}}\Theta\left(E-E_{n_x,n_y}\right) \quad (2.3)$$

where E_{n_x,n_y}, the confinement energies, are associated with the x and y directions, and

$\Theta\left(E-E_{n_x,n_y}\right)$ is the Heaviside unit step function. The inverse square root of energy is the hallmark of a 1D nanowire structure.

iii. **Two-Dimensional Nanostructures:**

For two-dimensional (2D) nanostructures, like thin film, the only confinement direction is the z axis. The electron can have large degrees of freedom along the x-y plane of the film. In this situation, DOS can be written as[4,5]

$$g(E)=\frac{m^*}{\pi\hbar^2}\sum_{n_z}\Theta\left(E-E_{n_z}\right) \quad (2.4)$$

where E_{n_z} represents the confined energy along the z direction.

iv. **Three-Dimensional Nanostructure:**

Three-dimensional (3D) nanomaterials are not restricted to the nanoscale (<100 nm) in any dimension. Nevertheless, 3D nanomaterials have a nanocrystalline structure or involve the presence of features at the nanoscale. In terms of the nanocrystalline structure, bulk nanomaterials can be composed of multiple arrangements of nano-size crystals in different orientations. With respect to the presence of features at the nanoscale, 3D nanomaterials can contain dispersions of nanoparticles, bundles of nanowires, and nanotubes as well as multi-nanolayers. The DOS of any nano-composite material cannot be modeled mathematically in generic form because it depends on the compositions. However, due to the quantum-confinement effect, DOS of the electrons will change from the bulk 3D material (i.e., $g(E)\propto\sqrt{E}$) significantly.

Effective Mass: The mass of the electron is different from the rest mass (m_0) within the solid material due to the presence of different interactions. It is related to the curvature of the band structure diagram (or E-k diagram). In an isotropic solid it is defined as[1,2]

$$m^* = \hbar^2 \Big/ \frac{d^2E}{dk^2} \tag{2.5}$$

It is evident that, where the curvature is large –i.e., $\frac{d^2E}{dk^2}$ is large, m^* is small – a small curvature implies large mass. Depending on the curvature of the E-k diagram, m^* can be negative also. This means that in this region of k-space, the lattice exerts a large retarding force on the charge carrier.

In general, any solid material is anisotropic and a more general definition of m^* is

$$m^* = \left(\frac{\partial^2 E}{\hbar^2 \partial k_i \partial k_j}\right)^{-1}, \quad i, j = x, y, z \tag{2.6}$$

Therefore, the effective mass is a tensor quantity (second order) that contains nine components.

The mobility of charge carriers (electron or hole) is intimately dependent on the effective mass and thereby influences the conductivity of the material.

2.3 ELECTRICAL PROPERTIES

In this section, we will briefly discuss the conductivity, or dielectric property, of a material. It is intimately connected to the design of the nano-/micro-level device fabrication.

i. **Electrical Conductivity:** If we apply some voltage across a sample, the response of the sample would be to produce current. Depending on the amount of current produced, material can be macroscopically classified as a conductor, insulator, or semiconductor. Explanation of this phenomenon can be done by using a band structure. However, macroscopically, this current-voltage response in a conductor can be written in a simple form, i.e., $V = iR$ – Ohm's law. Here, R represents the resistance offered by the conductor. Most metals are good conductors of electricity. For a conductor in the form of a cylinder, R can be written as $R = \frac{\rho l}{A}$, where ρ is the resistivity of the conductor, l represents the length of the cylinder, and A is the cross-sectional area of the cylinder. Now ρ can be written as $\sigma = 1/\rho$, where σ is the conductivity of the material. Ohm's law can be cast as $\vec{j} = \sigma\vec{E}$, where $\vec{j}$ is the current density vector and $\vec{E}$ be is the electric field impressed upon the material. The microscopic nature of conductivity for solid material depends on many factors –like carrier density, mobility of charge carriers, and so forth. These factors also depend on the band structure, DOS, the dimensionality of the material, and so on. The modulation of conductivity is the main task of the material scientist. The whole semiconductor industry is based on the engineering of conductivity of semiconducting materials. Nanoscience and nanotechnology open up a new paradigm for conductivity engineering.[6] In the light of free electron theory, conductivity can be expressed as $\sigma = \frac{ne^2\tau}{m^*}$, where n represents carrier concentration, e is the electronic charge, τ is the collision time for the charge carriers at the Fermi surface, and m^* be is the effective mass of the carrier. The most important parameter is the collision time, τ, which is very hard to model mathematically as well as difficult to measure experimentally. The scattering of charge carriers by phonons, impurities, and other lattice defects is an extremely complicated phenomenon in general. In the framework of the band theory of solids, conductivity can also be expressed as[1,2]

$$\sigma = \frac{1}{3}e^2 v_F \tau g\left(E_F\right) \tag{2.7}$$

Here, v_F is the Fermi velocity and $g\left(E_F\right)$ is the DOS at the Fermi surface. In this expression, it is clear that DOS predominantly controls the conductivity of the material.

ii. **Dielectric Property:** The dielectric constant (k) is an important parameter for designing microelectronic devices. High-k materials are used in the fabrication of MOS transistors, memory cells, and capacitors, whereas low-k materials are utilized in high-speed communication devices and insulation where the effect of capacitance needs to be minimized. A dielectric is an electrical insulator. However, when subjected to an external electric field, the positively charged nucleus of each atom displaces in one direction, while the electron cloud shifts in the opposite direction. This displacement results in dielectric polarization within the material. When a medium is polarized, the internal electric field changes, and it is expressed as[1,2,7]

$$\vec{D} = \varepsilon_0 \overrightarrow{E_i} + \vec{P}, \tag{2.8}$$

where $\vec{D}$ is the electric displacement vector, ε_0 is the permittivity of free space, $\overrightarrow{E_i}$ is the electric field inside the material, and $\vec{P}$ is the electric polarization vector. Now the internal field can be written as

$$\vec{E_i} = \vec{E_e} - \frac{1}{\varepsilon_0}\vec{P}, \tag{2.9}$$

since $\vec{D} = \varepsilon_0 \vec{E_e}$, $\vec{E_e}$ being is the external field. It is evident that the effect of the external field diminishes the internal field. In the literature, $\vec{D}$ is usually written as

$$\vec{D} = \epsilon \vec{E_i} = \varepsilon_0 \varepsilon_r \vec{E_i}, \tag{2.10}$$

where, ε_r is the relative permittivity or dielectric constant of the medium and ϵ is the absolute permittivity of the material.[8] In practice, we always use relative permittivity (ε_r or k) as the dielectric constant.

In the presence of a time-dependent electric field, a dielectric medium would not be polarized instantly and incur energy loss. The frequency dependent complex permittivity, $\varepsilon^*(\omega)$, of a material in the presence of an alternating field, is given by

$$\varepsilon^*(\omega) = \epsilon_\infty + \left(\frac{\epsilon_s - \epsilon_\infty}{1 + j\omega\tau}\right), \tag{2.11}$$

where ϵ_∞ is the dielectric constant at high frequency, ϵ_s is the dielectric constant at low frequency, $\omega = 2\pi f$, τ is the relaxation time, and $\varepsilon^*(\omega) = \varepsilon' - i\varepsilon''$. Here ε' = real part of permittivity and ε'' = imaginary part of permittivity. The dielectric loss, a measure of dielectric relaxation, can be defined as

$$\tan\delta = \frac{\varepsilon''}{\varepsilon'}. \tag{2.12}$$

A frequency-dependent dielectric constant is often determined via dielectric spectroscopy, which examines the sample's interaction with a time-dependent electric field.

The dielectric constant of the material is frequency dependent because polarization of the medium comes from the various grounds. They are as follows:

(a) **Dipolar polarization** (α_d)**:** When the field is applied to the material, the molecular dipole tends to align along the field lines and results in non-vanishing polarization in the material. It operates at the frequency range: $10^8 \sim 10^{10}$ Hz.
(b) **Ionic polarization** (α_i)**:** If the molecule contains ionic bonds, then the field tends to stretch the length of the bond, which in effect produces a net dipole moment in the unit cell of the lattice. Ionic compounds react to the external fields up to frequencies $\sim 10^{12}$ Hz.

(c) **Electronic polarization (α_e):** This arises when individual atoms or ions in a molecule are individually polarized by the field. Electronic polarization is the most significant source of polarization in the material. It acts up to frequency 10^{15} Hz.

The polarization effects are additive in nature, $\alpha = \alpha_d + \alpha_i + \alpha_e$.

Factors influencing a material's dielectric constant (k):

(a) **Crystallite size**: The large crystalline size contains a greater number of dipoles, leading to a high dielectric constant.
(b) **Porosity**: Highly porous material contains voids, which have a low dielectric constant because vacuums have a low k-value, resulting in an overall decrease in the dielectric constant.
(c) **Lattice Defects:** The presence of various sorts of defects or vacancies creates charge trapping inside the material, which enhances the overall polarization.

Nanotechnology opens up the possibility of manipulating the different factors affecting the dielectric constant as per the requirements.

2.4 OPTICAL PROPERTIES

Nanomaterials exhibit a diverse array of unconventional and captivating optical characteristics, including reflection, transmission, absorption, and emission, which significantly deviate from those of the corresponding bulk material. By meticulously controlling the dimensions, morphology, and surface attributes of nanoparticles, a broad spectrum of optical effects can be harnessed, offering numerous practical applications in photonics, sensors, displays, and imaging systems.[9] As optical properties are intricately linked with the electronic structure of the material, investigation of these gives a comprehensive understanding of low-dimensional nano structures.

Absorption of Light: The light absorption is one of the most fundamental aspects of a semiconductor or insulator. The important absorption process involves the transition of electrons from the valence band (VB) to the conduction band (CB), and this is referred to as fundamental absorption. The incoming photon energy ($h\nu$) must be equal to or larger than the energy bandgap (E_g). In this process, the total energy and momentum of the electron-photon system must be conserved. This process occurs in the case of direct band-gap material (DBM), where the top of the valence band and the bottom of the conduction band occur at the same k-value. We can derive the following form of the absorption coefficient (α_d) by using the quantum mechanical perturbation theory[10–12]

$$\alpha_d = A\left(h\nu - E_g\right)^{\frac{1}{2}}, \tag{2.13}$$

where A is the constant determined by the index of refraction, and electron and hole effective masses. Common examples of DBMs are GaAs, GaN, CdS, and so forth, which are utilized for LEDs, laser diodes, photovoltaics, and so on. There are some materials where the top of the VB and the bottom of the CB do not lie in the same k-value, which is termed indirect band-gap material (IBM). In this scenario, photon absorption is a two-step process: the phonon assistance, or emission, supplies the necessary k-value to the photon momentum, thereby conserving the momentum conservation law. Now the absorption coefficient (α_i) is given by[12]

$$\alpha_i = A\frac{\left(h\nu - E_g + E_p\right)^2}{e^{\frac{E_p}{KT}} - 1} + A\frac{\left(h\nu - E_g - E_p\right)^2}{1 - e^{-\frac{E_p}{KT}}}, \tag{2.14}$$

where Ep is the phonon energy. Here, the first term indicates the phonon assistance, while the second term stands for the photon emission. The above equations are commonly used for determining the nature of bandgaps in different materials.

Another kind of absorption involves the exciton-bound state formation of electron and hole. The photon energy involved in this process is slightly less than the band-gap energy. In the nanostructure of transition metal dichalcogenide, excitonic transition was observed recently.

A typical nanoparticle has a high (s/v) ratio, which causes many surface atoms to have distinct energy states, resulting in an energy state in between band-gap energy. This is known as the trap state, and it is essential to understanding the optical properties of the low-dimensional nanostructure.

In some cases, electrons and holes may absorb radiation without becoming excited in the other band, but rather transit into the different energy states of the same band. Such a process is called intra band transition. In quantum dots, intra band transitions play a crucial role in determining their optical behavior.

2.5 MECHANICAL PROPERTIES

Mechanical properties[13,14] of materials are of great importance when designing micro- or nanoscale devices, as these properties directly influence the device's structural integrity, reliability, and functionality. Here are some key mechanical properties of materials for designing micro- and nanoscale device fabrication.

Elastic Moduli: Elasticity describes a material's capacity to withstand elastic deformation. It indicates how atomic forces bind the atoms together. The relationships between stress and strain are represented by three moduli, viz. Young's Modulus (E), Shear Modulus (G), and Bulk Modulus (K). In addition, Poisson's ratio (ν) is also an important parameter. As the size of the material decreases to the nanoscale, the elastic constants do not remain constant like in bulk material. Depending on the nanocrystalline material, with the reduction in size, some materials become harder while others are soft.

(i) **Young's Modulus (E):**This is the ratio of tensile or compressive stress (σ) and axial strain (ε) within the elastic limit. Mathematically, it can be expressed as:

$$E = \frac{\sigma}{\varepsilon} \tag{2.15}$$

It is found that when the size of the nanoparticle reduces, so does Young's modulus.[15] It is known that the Young's modulus and the phonon vibration frequency (v) are both related with the force constant, k, in the following manner:

$$E = \frac{k}{r_0}\ ;\ v = Ck^{\frac{1}{2}}, \tag{2.16}$$

where, r_0 is the atomic distance in equilibrium, or the bond length, and C is a constant. The force constant can be determined from the interatomic potential, $u(r)$, of the material:

$$k = \left.\frac{d^2u(r)}{dr^2}\right|_{r=r_0}. \tag{2.17}$$

Thus, the atomic interaction energy and the atomic distance are intrinsically linked to both the Young's modulus and the vibration frequency. The size-dependent value of Young's modulus is observed in various nanostructured systems because of quantum size effects.[16]

(ii) **Shear Modulus (G):** Shear modulus, also known as the modulus of rigidity, is the measure of the rigidity of the body, given by the ratio of shear stress to shear strain. It is often denoted by G, and sometimes by S or μ.

(iii) **Bulk Modulus (K):** This is a measure of a resistance to the compressible stress. It is define as

$$K = -V\frac{dP}{dV}, \tag{2.18}$$

where $\frac{dP}{dV}$ is the derivative of pressure with respect to volume, and V is the initial volume. The inverse of bulk modulus is compressibility.

(iv) **Poisson's ratio** (v): This the negative ratio of transverse strain to the axial strain.

The relationships between the three moduli and Poisson's ratio are

$$E = 2(1+v)G \tag{2.19}$$

and

$$E = 3(1 - 2\nu)K. \tag{2.20}$$

A stress-strain curve (Figure 2.3) indicates how a material behaves under stress, which provides information about the material's strength, stiffness, ductility, and failure limits. From Figure 2.3,it is clear that stress and strain are proportional up to certain limit A, and after point B material starts deforming plastically. The maximum stress a material can endure while being stretched or pulled before breaking or fracturing is termed tensile strength. Some materials break without plastic deformation, which are termed brittle material.

Toughness: It measures to what extent a material can absorb energy before fracture takes place.

Nanomaterials typically possess high strength due to the presence of a large volume fraction of grain boundaries in nano-crystalline materials and reduced defect sizes. However, their toughness can vary depending on factors such as grain size, grain boundary characteristics, and the presence of defects.

Hardness: This is the property of a material by virtue of its being able to resist abrasion, indentation, and scratching by relatively harder bodies. It is the resistance of a material to permanent deformation of the surface.

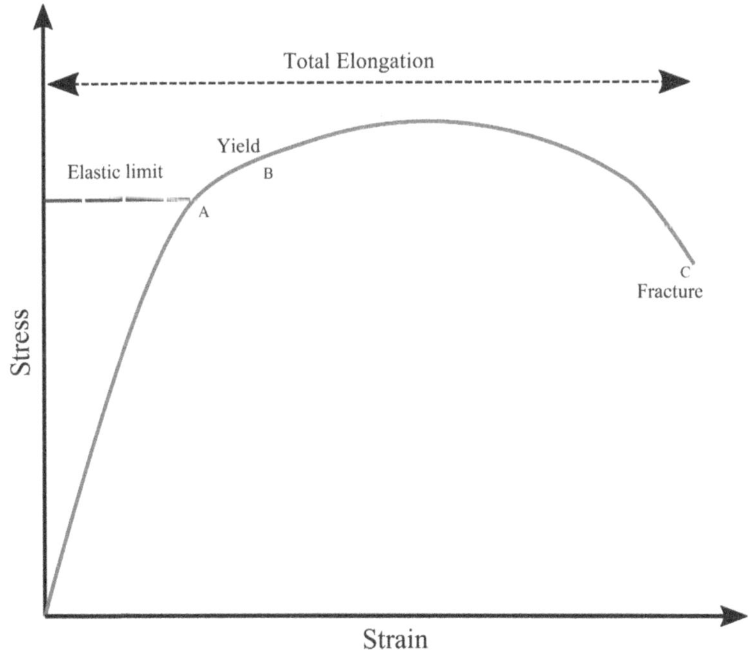

FIGURE 2.3 Stress and strain curve.

Creep: This is essentially a slow manifestation of plastic deformation under the action of stresses below the yield strength of the material. Creeps are typically caused by thermal activation of the deformation of crystal. In some studies, it has been observed that in certain nanomaterials, deformation-induced grain growth occurs and thereby resists creep more efficiently.[10]

The grain size plays an important role for resisting the plastic deformation of materials. Therefore, the yield stress (σ) and grain diameter (d) followthe Hall-Petch relation:[10]

$$\sigma = \sigma_0 + c\frac{1}{\sqrt{d}}, \tag{2.21}$$

where σ_0 is the resistance of the lattice to movement of dislocations, and c is the constant.

A similar type of relation is also applicable for hardness of the material. Experiments on nanomaterial show that the hardness increases with the reduction of grain size in nano materials. In some nanomaterials, different trends are also observed, which do not obey Hall-Petch behavior.[10]

Generally, crystal imperfections (point defects, line defects, and surface defects) play a major role in the mechanical behavior of materials.

2.6 THERMAL PROPERTIES

Thermal properties of materials for micro- or nanoscale devices are critical, as they influence heat dissipation, thermal stability, and thermal management in these miniature technologies.[17,18] Understanding and engineering these properties is essential for ensuring the reliable operation of micro- and nanoscale devices. Here are some key thermal properties and their significance:

i. **Thermal Conductivity (K):** This property measures a material's ability to conduct heat. High thermal conductivity is essential for efficient heat dissipation in micro/nanoscale devices, preventing overheating and maintaining device reliability. Materials with high thermal conductivity are used for heat spreaders, thermal interfaces, and substrates. Thermal conductivity[2] is defined as the time rate of heat *(Q/t)* transfer through a unit thickness *(d)* of a material per unit area *(A)* per unit temperature difference (" T)

$$K = \frac{Q/t}{Ad\Delta T}. \tag{2.22}$$

In an insulator, heat is primarily conducted by phonons, while in metal, it is carried by both phonons and electrons. For an insulator, K can be modeled as[2,19]

$$K = \frac{1}{3} C_v v l, \tag{2.23}$$

where C_v is the specific heat, v the velocity of the carriers involved, and l is the mean free path of the carrier. However, in the case of metallic heat conduction via electrons, it can be described by the electronic properties of the substance and it becomes:

$$K = \frac{\pi^2 N k^2 T \tau}{3m^*}, \tag{2.24}$$

where N is the number of carriers/unit volume; T is the absolute temperature; k is the Boltzman constant; τ is the collision time for the charge carriers at the Fermi surface; and m^* be is the effective mass of the carrier. It is clear that the nanostructure has an exotic band structure, and we therefore modulate the thermal conductivity according to our requirements. The electronic heat conductivity and electrical conductivity at a particular temperature bear a constant ratio (Lorenz number) as $\frac{1}{3}\left(\frac{\pi k}{e}\right)^2$.

ii. **Specific Heat Capacity (C):** It is the amount of heat energy required to raise the temperature of unit mass of the material by one degree. According to Debye's theory of lattice vibration or phonon, the specific heat is proportional to the T^3. Specific heat depends only on temperature for bulk material. For nanomaterials, specific heat also depends on the diameter of the nanoparticle. Debye specific heat becomes[20]

$$C_n = C_b \left(1 - \frac{2d}{D}\right)^{-1}, \tag{2.25}$$

where C_n is the specific heat of the nanoparticle; C_b is the specific heat of the bulk; d is the atomic diameter of the material; and D is the diameter of the nanoparticle. It is clear that specific heat increases due to the size reduction of the nanoparticles. The main contribution comes from the presence of significant surface atoms in the nanoparticles.

iii. **Melting point**: The melting point of a material is the temperature at which the solid-to-liquid phase transition occurs. The melting point of nanomaterial (T_n) also depends on the size of nanoparticles as[20]

$$T_n = T_b \left(1 - \frac{2d}{D}\right). \tag{2.26}$$

Therefore, the melting point of a nanomaterial decreases with respect to the corresponding bulk material.

2.7 CHALLENGES AND FUTURE PERSPECTIVES

Now, we look into the challenges and future perspectives of different properties of materials for micro/nanoscale devices in more detail.

In terms of mechanical properties, developing robust materials at the micro/nanoscale can be challenging due to their unique structure and size, and so maintaining mechanical stability and reliability is a major challenge. The future perspectives of this property is the integration of novel nanocomposites and advanced manufacturing techniques, such as 3D printing at the nanoscale, will lead to stronger and more durable micro/nanoscale devices. Researchers are also exploring the use of biomimetic materials inspired by nature for improved mechanical properties.

At the nanoscale, quantum effects can result in challenges like quantum tunneling and charge confinement, affecting the precise control of electrical properties. The future perspectives of this property is quantum engineering and the use of new materials, such as 2D materials like graphene and transition metal dichalcogenides, offer promise for overcoming these challenges. These materials enable the development of more efficient transistors, sensors, and optoelectronic devices.

Achieving precise control over optical properties, especially at the nanoscale, can be complex. Nano-structuring and manufacturing challenges may limit the full exploitation of desired optical effects. Advancements in materials, plasmonics, and nano-photonics will continue to unlock unprecedented control over light–matter interactions. These materials will enable revolutionary technologies such as ultra-high-resolution imaging, cloaking devices, and advanced photo-detectors.

Efficient heat management in micro/nanoscale devices is challenging due to reduced heat dissipation and the prevalence of thermal stresses. Nanomaterials, such as carbon nanotubes and graphene, hold promise for improving thermal properties. The development of nanoscale heat pipes and thermoelectric materials will enhance heat dissipation and energy conversion in micro/nanoscale devices.

Ensuring the safety and ethical use of nanomaterials and micro/nanoscale technologies is crucial. Concerns related to toxicity, privacy, and ethical use must be addressed. Ethical guidelines, regulations, and safety protocols are essential for responsible development and deployment. Ongoing research into the potential health and environmental impacts of nanomaterials will guide these efforts.

Addressing these challenges while exploring future perspectives will drive innovations in micro/nanoscale materials and devices. These innovations have the potential to revolutionize various fields, including electronics, healthcare, energy, and environmental monitoring, with more efficient, reliable, and sustainable technologies.

2.8 CHRONICLES OF GRAPHENE

Graphene is a remarkable two-dimensional carbon allotrope with a single layer of carbon atoms arranged in a hexagonal lattice. It possesses a wide range of extraordinary properties that have garnered significant attention in various scientific and technological fields.[13–16] Some key features of graphene include exceptional intrinsic tensile strength, high electrical conductivity, high thermal conductivity (~5,300 W/mK), transparency, flexibility and lightweightness, impermeability, and unique optical properties.

In graphene, sp^2 hybridization occurs between one s-orbital and two in-plane p-orbitals. Each carbon atom in graphene is surrounded by three other carbon atoms, generating three in-plane bonds per atom, resulting in a trigonal profile for the hybridized orbital. Because these bonds are much stronger than the pi bond, graphene has an indestructible hexagonal structure. The last p-orbital crosswise creates pi-bonds above and below each atomic layer by overlapping with neighboring carbon atoms. Because the electrons are tightly bound to the atom, they are not responsible for the conduction phenomenon; nevertheless, the valence and conduction bands of the ∏ and ∏*orbitals become valence and conduction bands in enormous sheets, resulting in planar conduction phenomena. The hexagonal carbon lattice can be divided into basic identical trigonal sub-lattices. Three sub-lattice B atoms are connected to a sub-lattice an atom, and vice versa. The armchair or zigzag graphene layer is determined by the arrangement of carbon atoms on the graphene layer's edges. The lattice of the hexagonal graphene shows both armchair and zigzag edges are made up of two similar sub-lattices, A and B (Figure 2.4(a)). The zigzag pattern adds to a substance's metallic property, whereas the armchair pattern contributes to its semiconducting behavior.

Electronic Properties of Graphene

Because graphene is distinguishable from other 3D materials, its electrical attribute can be mirrored in its electronic structure. Electrons produce an operating magnetic field proportional to their momentum when they hop between sub-lattices. At Dirac points, when the conduction and valence bands create tapering curved valleys that meet like the hexagon at the Brillouin zone, the momentum-dependent field vanishes, creating graphene a zero-gap material in those directions (Figure 2.4 (b) and (c)). Depending on the definition of momentum vector, the effective mass (m) of an electron is the second derivative of energy with respect to the electron's momentum vector.

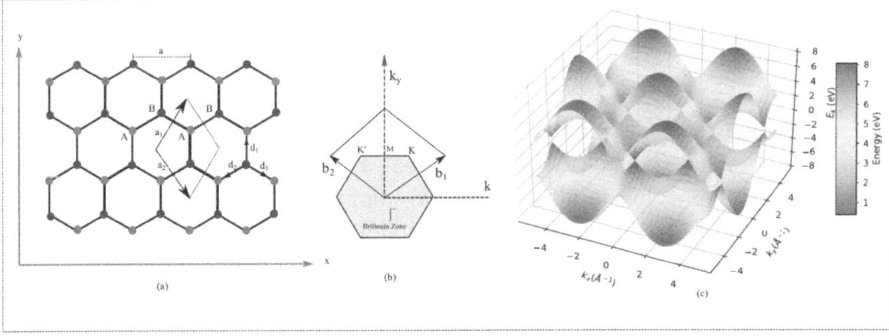

FIGURE 2.4 (a) Lattice of the hexagonal graphene, (b) Reciprocal lattice space and the corresponding Brillouin zone, (c) Energy dispersion diagram (E-k) of graphene.

Because the second derivative of any linear function is zero, electrons in graphene have no feasible mass in the region of the Dirac points. Graphene is a zero-gap material in which massless electrons flow at a constant rate along lattice directions with great symmetry. The electron effective mass, on the other hand, is a measure of how electrons move in a given field. On its approach to the Fermi level, the Fermi level travels via the K or Dirac points. At the graphene/oxide edge, the exact location of the Dirac point and the Fermi level has yet to be determined. The Fermi level is located at the converging points of these cones in a pure graphene sheet. The electrical conduction of pure graphene is relatively poor due to the zero density of states at that converging point. External energy affects the Fermi level's site, which becomes either n-doped (with electrons) or p-doped (without electrons), depending on whether it is n-doped (with electrons) or p-doped (without electrons) (with holes). Doping can be done in three different ways. The initial step is heteroatom doping, which includes arc discharge, chemical vapor deposition (CVD), and electrothermal reaction; the second step is chemical adjustment; and the third step is electrostatic field regulation. Doped graphene has a high electrical conductivity, possibly even greater than copper.

Mechanical Properties of Graphene

Graphene strength is over 100 times greater than a steel film, so it is one of the strongest materials. The mechanical properties of graphite should be estimated since they may differ from what is known about graphite. In graphene, the carbon–carbon bond length is 0.142nm. We can precisely measure the dimensions of the graphene sheet, such as length, width, and thickness, as well as the spring constant and other properties, using atomic force microscopy (AFM). The Van der Waals forces are primarily accountable for firmly binding carbon atoms together to form a graphene sheet, which was probed with an AFM tip to assess its mechanical properties. A monolayer graphene sheet has a spring constant of 1–5 N/m and a Young's modulus of 1.0 TPa for a perfect graphene sheet, which can vary up to 0.5 TPa with some interstitial defect. Researchers used mathematical modeling and numerical simulations to study the mechanical characteristics of monolayer graphene. The Young's modulus of a nanometer-thick suspended graphite sheet was investigated using AFM on a strip of suspended graphene. Force–volume measurements in AFM have also been used to investigate circular membranes of few-layer graphene (FLG). AFM has recently been used to investigate the fracture strength of single monolayer graphene 130 GPa is the fracture strength.

Optical Properties of Graphene

When black graphite is peeled away to reveal a graphene monolayer, it becomes extremely transparent. In the ultraviolet (UV) range, electrons absorb light, but not in the visible region. To the naked eye, pure monolayer graphene appears translucent. Because it absorbs around 2.3% of white light, this single-atom thick crystal may be viewed with the naked eye. The absorbance value of each layer is 2.3%. By utilizing the band-gap energy, graphene can also become photo luminescent. In graphene, there are two ways to participate in the photo luminescent property. One method involves shaping graphene into nanoribbons and quantum dots, while the other method involves chemical treatment with various gases to enhance orbital electron connection.

Electronic Transport Properties under External Fields

"At room temperature, suspended monolayer graphene has very high electron mobility, reported to be 200,000 cm^2/V/s."[21] In the temperature range of 10 to 100 K, mobility is self-regulating, implying that defect scattering is the dominant scattering mode. The carrier mobility is proportional to the relaxation time of the carrier, but not to the true mass of the carrier. Electron mobility is usually greater than hole mobility because electron effective mass is smaller than hole effective mass. Because the magnitude of electron and hole relaxation durations are similar, they do not transform. Electrons and holes have less collision experience with crystal defects, impurities, and other materials. Charge carriers (i.e., electrons and holes) scattering on imperfections in a crystal can decrease the mean free path and flexible scattering time, both of which control mobility; as a result, mobility is reduced. At a carrier density of 10^{12}cm-^{2}, carrier mobility of graphene is 200,000 $cm^2/V/s$ at room temperature. The resistivity of a graphene sheet is $10^{-6}\Omega$cm, which is lower than that of silver, the famous substance with the lowermost resistivity at ambient temperature. Schedin et al.[22] observed that gradual heating of the graphene sheet in a vacuum can improve the initial pure graphene assembly of doped graphene with various species (some donors, acceptors). They also found that when chemical dopant concentrations exceed 10^{12}cm-2, Dirac points indicate a near-zero concentration of carriers at that place, meaning that conductivity is also insignificant. However, due to the peculiarity of graphene's electrons, this will not occur. Rather than acting like a particle, it behaves like a wave.

Anomalous Quantum Hall Effect

In the presence of a transverse magnetic field, an electric current traveling through a rectangular metallic conductor produces a voltage difference (Hall voltage) among the two opposing faces of the conductor. The Hall resistivity ratio, which is proportional to the applied magnetic field, is directly proportional to the potential difference with respect to the current. For the measurement of this magnetic field, the Hall effect is commonly used. At temperatures close to absolute zero, Hall resistivity quantizes in a 2D structure, with an estimated value of h/ne2 (where h = Planck's constant, n = positive integer number, and e= electric charge).Only for the odd value of n was Hall resistivity discovered, which is a result of a quantum mechanical phenomenon known as Berry's phase. The quantum Hall effect is the name for this phenomenon. When compared to other 2D electron systems, graphene has a unique way of describing the quantum Hall effect. If the filling factor is a half-integer, however, the Hall conductivity is quantized for SLG. The quantification of Hall resistance and conductivity requires the development of discrete Landau levels. The massless Dirac's equation of the pseudo-relativistic principle can explain this unusual phenomenon, which is caused by quasi-particle excitation in graphene. As a result, a zero energy Landau level will develop, resulting in a 12-shift in the filling factor in regard to conductivity quantization. This is most likely due to the fact that electrons in graphene have 1,000 times the magnetic energy of electrons in conventional materials. Graphene electrons can pass through any potential barrier at 100% efficiency. The quantum mechanical explanation of this reflection is that the barrier can prevent electrons from passing through a certain valley by changing its alignment, but it violates the chirality conservation laws.

Magnetic Properties of Graphene

At low temperatures, vacancies (missing atoms in the hexagonal structure) function like little magnets (they have a magnetic properties) and have an unlimited impact on the electrons in graphene that convey electrical currents, resulting in a massive increase in electrical resistance. This strong technique has the potential to be used in nanoscale magnetic field sensors as well as spintronics. Encroachment of impurities into the graphitic layer by physisorption mechanically lowers the inter-graphene layer distance, disrupting magnetic moment arrangement and lowering the net magnetic moment.

Thermal Properties of Graphene

Non-doped graphene has a low carrier density, and the electronic influence on heat conductivity is insignificant. The phonon transport phenomenon is responsible for graphene's thermal conductivity. When the temperature is high, conduction occurs through diffusion; when the temperature is low, ballistic conduction happens since there is no scattering of electrons. Graphene has a longer mean free path than other metals A suspended monolayer graphene has a thermal conductivity of 6000 W/m/K at normal temperature that is substantially higher than graphite. The graphene was suspended in a trench. The center of the suspended graphene was fixed using a laser beam. The heat moved outward from the graphene's center to the far end. The heat loss through air was found to be lower than that through graphene. Because of link weakening, the Raman G peak red shifts as the temperature of the heated graphene rises. Another work[23] used CVD-grown graphene on a thin silicon nitride membrane to achieve a thermal conductivity of 2500 W/m/K (at 350 K); the silicon nitride was covered with a thin coating of gold for increased thermal contact.

While graphene offers a wide array of remarkable properties, its commercial and industrial applications are still evolving. Researchers continue to explore and develop new uses for this material in fields such as electronics, energy, materials science, and biotechnology. Its potential for revolutionary advancements in various industries makes it a subject of ongoing research and innovation.

2.9 CONCLUSIONS

In conclusion, the properties of materials for micro- and nanoscale devices are instrumental in shaping the design, performance, reliability, and functionality of these miniature technologies. Some key conclusions regarding the properties of materials for micro/nanoscale devices are Size-Dependent Phenomena, Mechanical Properties, Electrical Properties, Thermal Properties, Optical Properties, Biocompatibility, Environmental Considerations, and so on. Materials at the micro/nanoscale exhibit unique structural behaviors, including the emergence of quantum effects, size-dependent properties. Precise control over the structural properties of materials is essential for tailoring them to specific applications in micro/nanoscale devices. The electrical properties also play a fundamental role in micro/nanoscale electronics, influencing device operation, energy efficiency, and performance. Quantum effects at the nanoscale provide opportunities for advanced electronic devices, while challenges like quantum tunneling must be addressed. Efficient thermal management is another crucial property for micro/nanoscale devices to prevent overheating and maintain reliability. Advances in nanomaterials and manufacturing techniques offer promise for enhancing heat dissipation and energy conversion. Precise control over optical properties, including wavelength selectivity and nonlinear optical effects, is vital for various applications. Meta-materials, plasmonics, and nanophotonics enable unprecedented control over light–matter interactions, opening up transformative technologies. In the concept of morphological properties, size, shape, surface area, and roughness significantly influence materials' interaction with their surroundings. Nanostructuring and tailored morphological properties are crucial for optimizing material behavior in micro/nanoscale devices. The mechanical properties, such as stiffness, strength, and resilience, are essential for ensuring the structural integrity and reliability of micro/nanoscale devices. Advanced materials and manufacturing techniques, along with biomimetic approaches, offer opportunities for stronger and more durable micro/nanoscale components.

In summary, the properties of materials at the nanoscale are not merely scaled-down versions of macroscopic properties; they exhibit unique behaviors and characteristics that can be harnessed for specific applications. Addressing the challenges associated with these properties and capitalizing on future perspectives in materials science and engineering will drive innovations in nanotechnology, leading to more efficient, reliable, and sustainable devices with applications across a wide range of fields. Advances in materials science and engineering continue to drive innovation in micro/nanoscale technology, enabling a wide range of applications in fields such as electronics, optics, medicine, and more.

REFERENCES

1. Charles, K. (2004). *Introduction to Solid State Physics*. Wiley.
2. Omar, M.A. (2008). *Elementary Solid State Physics: Principles and Applications*. Pearson.
3. Wolf, E.L. (2004). *Nanophysics and Nanotechnology: An Introduction to Modern Concepts in Nanoscience*. Wiley-VCH.
4. Kuno, M. (2011). *Introductory Nanoscience: Physical and Chemical Concepts*. Garland Science.
5. Vladimir, V.M., DmitryI., Sementsov, I., Nizami, K., Vagidov, Z. (2010). *Quantum Mechanics for Nanostructures*. Cambridge University Press.
6. Zhang, B. (2018). *Physical Fundamentals of Nanomaterials*. Elsevier.
7. Kaur, D., Bharti, A., Sharma, T., Madhu, C. (2021). Dielectric Properties of ZnO-Based Nanocomposites and Their Potential Applications, *International Journal of Optics*, 9950202, 20. https://doi.org/10.1155/2021/9950202
8. Wang, H.W., Chi, L., Al-Hashimi, M., Marks, T.J., Facchetti, A. (2018). High-k Gate Dielectrics for Emerging Flexible and Stretchable Electronics, *Chemical Reviews*, 118 (11), 5690–5754. https://doi.org/10.1021/acs.chemrev.8b00045
9. Kumbhakar, P., Ray, S.S., Stepanov, A.L. (2014). Optical Properties of Nanoparticles and Nanocomposites. http://dx.doi.org/10.1155/2014/1813655
10. Bhushan, B., Scott, D. L., Schricker, R., Sigmund, W., Zauscher, S. (Eds). (2014). *Handbook of Nanomaterials Properties*. Springer Science & Business Media.
11. Cotta, M.A. (2020). Quantum Dots and Their Applications: What Lies Ahead? *ACS Applied Nano Materials*, 3 , 4920–4924.

12. Zhang, J.Z. (2009).*Optical Properties and Spectroscopy of Nanomaterials.* Word Scientific. https://doi.org/10.1142/7093
13. Isono, Y., Namazu, T., Terayama, N. (2006). Development of AFM Tensile Test Technique for Evaluating Mechanical Properties of Sub-Micron Thick DLC Films, *Journal of Micro-electromechanical Systems*, 15(1), 169–180.
14. Tsuchiya, T., Tabata, O., Sakata, J., Taga, Y. (1998). Specimen Size Effect on Tensile Strength of Electromechanical Systems, *Journal of Microelectromechanical Systems*, **7**(1), 106–113.
15. Pandey, V., Kumar, M. (2023). Development of a Simple Model for Size and Shape-Dependent Young's Modulus for Nanomaterials, *Pramana – Journal of Physics*, 97, 88. https://doi.org/10.1007/s12043-023-02552-x
16. Liang, L., Hansong, M., Yueguang, W. (2011). Size-Dependent Elastic Modulus and Vibration Frequency of Nanocrystals, *Journal of Nanomaterials*, 2011, 670857. https://doi.org/10.1155/2011/670857
17. Porter, A.L., Jan, Y. (2009). Where does Nanotechnology Belong in the Map of Science? *Nature Nanotechnology*, 4, 534–536. http://dx.doi.org/10.1038/nnano.-2009.207
18. Chen, Y.J., Groves, B., Muscat, R.A. (2015). DNA Nanotechnology from the Test Tube to the Cell, *Nature Nanotechnology*, 10, 748–760. http://dx.doi.org/10.1038/nnano.2015.195
19. Kalantar, M.H., Zhang, X. (2023).Thermal Transport in 2D Materials, *Nanomaterials,* 13, 117.https://doi.org/10.3390/nano13010117
20. Rawat, K., Goyal, M. (2021). Theoretical Analysis of Thermo-physical Properties of Nanomaterials, *Materials Today: Proceedings,* 47(P16), 1210–1214. https://doi.org/10.1016/j.matpr.2020.08.7544
21. Bolotin, K.I., Sikes, K.J., Jiang, Z., Klima, M., Fudenberg, G., Hone, J., Kim, P., Stormer, H.L. (2008). Ultrahigh electron mobility in suspended graphene. Solid State Communications, 146(9–10), 351–355. https://doi.org/10.1016/j.ssc.2008.02.024
22. Schedin, F., Geim, A.K., Morozov, S.V., Hill, E.W., Blake, P., Katsnelson, M.I., Novoselov, K.S. (2007). Detection of individual gas molecules adsorbed on graphene. *Nature Materials*, 6, 652–655. https://doi.org/10.1038/nmat1967
23. Yasaei, P., Kumar, B., Foroozan, T., Wang, C., Foster, C.W., Asadi, M., ... Salehi-Khojin, A. (2017). In-plane thermal conductivity of polycrystalline chemical vapor deposition graphene. *ACS Nano*, 11(1), 1250–1258. https://doi.org/10.1021/acsnano.6b06836

3 Nanotechnology Tools for Diagnosis and Therapy

Arnab De, Santanu Ghosh, Sourav De, Sabyasachi Banerjee, Indranil Banerjee, Amit Kundu, Soumik Podder, and Biplab Debnath

3.1 INTRODUCTION

Nanotechnology is now regarded as the most promising technology of the twenty-first century based on the exploitation of the unique properties of quasi-atomic particles at a size scale ranging from 1 to 100 nm.[1] The concept of nanotechnology was first introduced in 1959 by Richard Feynman, an American physicist. Nanotechnology has been widely used in the field of drug delivery systems, tissue-engineered constructs, implants, and pharmaceutical therapeutics. Nanocarriers can be used for targeted drug delivery, which can improve the dosage and accuracy of drug release, thereby reducing the toxic and side effects of drugs on normal tissues. Additionally, nanoparticles (NPs) can be functionalized based on the surface chemistry of the particles and conjugation of ligands (small molecules, antibodies, and peptides) to precisely target the expressed receptors.[2,3]

Managing and treating a disease requires an understanding of its underlying causes, which includes information about the patient's medical background and physical examination, as well as applying diagnostic methods to identify and characterize symptoms.[4] Nanodiagnostics has emerged as a modality that utilizes nanoscale properties, such as unique physiochemical and optical properties to meet the demand for early detection, high sensitivity, and point of-care testing, particularly for community clinics in developing countries.[5,6]

Recent advancements in nanotechnology have enabled an opportunity to combine diagnosis and therapy in a single platform or carrier.[7] Theranostic nanosystems have been recognized as a potential breakthrough to revolutionize future disease management. The term 'theranostics' integrates the technology with concurrent and complementary diagnostic and therapeutic capabilities to achieve more specific and personalized disease management.[8,9] Some common nanotechnology tools employed in the development of theranostic formulations are shown in Figure 3.1.

DOI: 10.1201/9781003453635-3

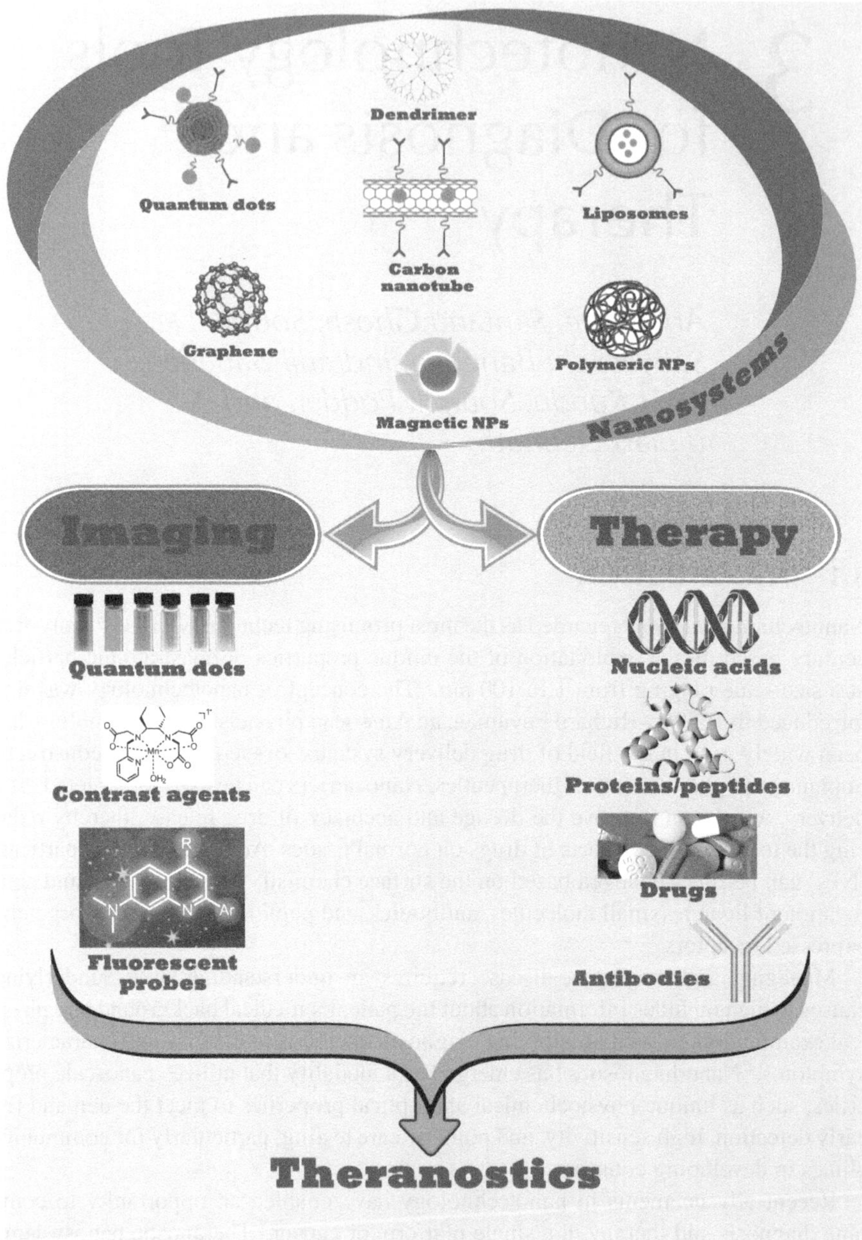

FIGURE 3.1 Nanotechnology-based tools used for diagnosis, therapy, and theranostic purposes.

3.2 PROPERTIES OF NANOTECHNOLOGY-BASED TOOLS/ PROPERTIES OF NPS

NPs are distinct and promising candidates for many uses because of their many desirable physicochemical features, such as their high surface area, optical activity, mechanical strength, and chemical reactivity. The following sections summarize the key features at a nanoscale level.

3.2.1 Electronic and Optical Properties

Quantum confinement and the Localized Surface Plasmon Resonance (LSPR) phenomena provide metallic and semiconductor NPs with unique linear photoluminescence emissions, absorption, and non-linear optical characteristics.[10,11] Experimental studies have demonstrated that the optical characteristics of AgNPs exhibit variations that are dependent on their size. It was found that the larger sphere of 60 nm corresponds to the quadrupole peak of 358 nm light, whereas the 30 nm sphere represents the major extinction peak of 369 nm light for this size.[12] The optical properties of NPs are affected by the solvent and substrate's dielectric environment. Numerous studies, encompassing both practical and theoretical approaches, have demonstrated that the plasmon wavelength of AgNPs exhibits a linear relationship with the refractive index of the surrounding solvent.[12,13] Furthermore, it was observed that the wavelength of LSPR shifted toward the red region by 10 nm when AgNPs were deposited on mica surfaces, as compared to unsupported NPs.[14]

3.2.2 Magnetic Properties

At ambient temperatures, magnetic compounds are characterized by the inclusion of a specific magnetic element, namely Fe, Co, or Ni, within their chemical formula. Only $ZrZn_2$, Sc_3In, and $TiBe_{2-x}Cu_x$ are the three exceptions formed of mixed diamagnetic elements.[15–18] According to the literature, the predicted critical radius for Co is 8 nm, Ni is 35 nm, and Fe is 1 nm.[19] Beyond this threshold, a phenomenon known as multidomain magnetism occurs, which necessitates a lower magnetic field for achieving a state of zero net magnetization.[20] These nanoparticles display superparamagnetic characteristics, with no coercivity observed at room temperature.[21] The magnetism of NPs is substantially altered by the incorporation of magnetic dopants, as has been widely observed.[22]

3.2.3 Mechanical Properties

Nanomaterials exhibit a range of mechanical properties that can be categorized into ten distinct attributes, namely toughness, brittleness, strength, hardness, fatigue strength, elasticity, plasticity, ductility, yield stress, and rigidity.[23] Conventional FeAl powder consisting of microparticles bigger than 4 μm is brittle, but ultrafine FeAl alloy powder has better ductility, strength, and plasticity.[24] The van der Waals forces including three parts, namely, Debye, Keesom, and London forces, which are the main interaction forces in the particles' mechanical properties.[25] Some other forces such as electrostatic double layer forces, solvation, structural, or hydration forces also affect interactions.[25] The JKR (Johnson-Kendall-Roberts) theory, the DLVO (Derjaguin-Landau-Verwey-Overbeek) theory, and the DMT (Derjaguin-Muller-Toporov) theory are among the various explanations proposed to elucidate the mechanisms underlying the emergence of unique mechanical properties resulting from the interaction forces between NPs.[25] The DLVO theory, which considers both van der Waals attraction and electrostatic repulsion, provides a description of the stability of colloidal dispersions.[25] The JKR theory can be employed to analyze the prevalence of powerful, limited-range adhesion forces on the surfaces of deformable, large objects with elevated surface energy.

3.2.4 Thermal Properties

NPs transmit heat through electrons, photons, and scattering.[26] Thermoelectric power, thermal conductivity, thermal stability, and heat capacity make up a material's thermal properties.[27,28] An increase in thermal conductivity of up to 40% is achieved, for instance, by mixing Cu NPs with ethylene glycol.[29] A material's electrical conductivity and its Seebeck coefficient determine its thermoelectric power ($P = S^2\sigma$, where P is the material's thermoelectric power, S is its Seebeck coefficient, and σ is its electrical conductivity). Thermoelectric power factor is increased by the scattering of NPs in bulk materials (doping).[30] This improvement can result from a rise in electrical conductivity or the Seebeck coefficient. Because of electron energy filtering, the embedding of size-controlled NPs in bulk thermoelectric materials lowers the lattice thermal conductivity and increases the Seebeck coefficient.[31,32] When the particle size reduces, because of the increased surface-to-volume ratio and increased free energy at the particle surface, the melting temperature decreases.[33] For example, the AuNPs of 3 nm size showed a melting temperature that is 300° lower than the bulk gold.[34] Furthermore, thermal stability is also affected by NP composition.

3.2.5 Catalytic Properties

Compared to bulk catalysts, NP catalysts have superior reactivity and selectivity.[35] The catalytic characteristics of NPs are influenced by various factors such as NP size, shape, oxidation state, composition, and interparticle spacing. Cuenya et al. reported the highest normalized current densities in the smallest NPs.[36] The research findings indicated that hemispherical AuNPs exhibited greater efficacy in the oxidation of CO compared to their round counterparts.[37] It was investigated how different shapes of silver NPs affected the oxidation rate of styrene. The results indicated that the reaction rate of nanocubes were 14 times and 4 times higher than nanoplates and near-spherical NPs, respectively.[38] The interparticle spacing considerably affected the catalytic activity and stability of dispersed Au NPs on polycrystalline titanium carbide (TiC) films. The study found that samples with larger interparticle distances exhibited longer lifetimes, while samples with smaller interparticle distances showed shorter lifetimes.[39]

To date, several nanomaterials have been investigated and developed. Some common types are discussed below.

3.3 POLYMERIC NANOPARTICLES

Different types of medical imaging modalities are essential for the diagnosis of various diseases including cardiovascular diseases, cancer, and infections. Apart from diagnosis, noninvasive medical imaging techniques such as positron emission tomography (PET), single photon emission computed tomography (SPECT), and so forth can also be used to specify disease progression through the visualization of organs and biological compartments.[40,41] However, each imaging technique has some advantages as well as disadvantages, for example, in its spatial resolution, sensitivity, probing depth, and so on. Therefore, it is often necessary to combine data obtained from more than one imaging modality to improve the overall diagnostic efficacy. To overcome disadvantages associated with one imaging technique, the unique construction and multifunctionality (e.g., various imaging probes, targeting ligands, etc. that are attached throughout the polymer structure) of polymeric nanoparticles (PNPs) can be leveraged. Thus, PNPs represent a practical approach to the development of multimodal imaging that can bypass limitations related to conventional imaging techniques.[42] A lot of studies have been performed to develop PNPs as contrast agents for multimodal medical imaging.[40,42] As mentioned earlier, the ability of surface modification as well as cargoes of different structures enable polymeric nanoparticles as versatile platforms for various magnetic resonance imaging (MRI) and optical imaging probes. Researchers have reported an MRI/optical probe prepared from poly(glycidyl methacrylate) (PGMA) conjugated with Rhodamine B (as a fluorophore) and superparamagnetic iron oxide (SPIO) nanoparticles (as an MRI agent).[43] Other researchers have also reported another similarly functional probe made from poly(lactic acid)-D-α-tocopheryl poly(ethylene glycol)$_{1000}$ succinate (PLA-TPGS), that encapsulated quantum dots (QDs) for fluorescence imaging and SPIO nanoparticles for MRI.[44] PLA-TPGS nanoparticles can be used for the controlled delivery of imaging agents to tumor tissues by passive targeting.

PNPs are suitable delivery vehicles for different groups of therapeutics because they are biocompatible, water-soluble, and stable during storage. Their surfaces can be modified with different ligands; thus, additional targeting can be achieved to deliver drugs to various organs. The limitations of PNPs include particle aggregation and toxicity.[45] However, the US Food and Drug Administration (FDA) approved some PNPs for the treatment of various diseases including cancer, multiple sclerosis, infection, hemophilia, and so forth. It is also encouraging to note that a lot of clinical trials are going on with PNPs to develop them as therapeutics.[45]

PNPs have been used to improve the delivery efficacy of anticancer drugs to the tumor site. For example, Le et al. speculated a PNP formulation of paclitaxel to get an increased efficacy after oral chemotherapy.[46] Other routes of administration (e.g., intravenous, intratumoral, etc.) were also used to deliver anticancer drugs through PNPs. Cisplatin and its prodrugs have been delivered through PNPs to avoid systemic side effects of these hydrophilic and cytotoxic small molecular drugs.[47,48] Apart from cancer, PNPs have shown superior delivery efficacy of antimicrobial agents to the infection site as compared to raw drugs alone. PNPs could improve the pharmacokinetics of anti-infective agents, provide sustained release of drugs at the infection sites, and overcome the drug resistance exhibited by microbes.[49] It is intriguing to note that PNPs have also been investigated for the treatment of cardiovascular diseases, for instance, hypertension, atherosclerosis, and so forth.[50,51] PNPs could be useful for cell therapy and provide appropriate materials for stents. Although a lot of polymeric materials have been developed for cardiovascular applications, there is still a need to develop novel polymeric structures for effective clinical utilization. Apart from the above-mentioned disease conditions, PNPs have also been investigated in delivering insulin,[52] immunotherapy for cancer,[53] small interfering RNA (siRNA) for treating inflammation,[54]and so on. Biopolymers like gum Arabic and gellan gum also have the potential for use in nanomedicine due to the emergence of nanotechnology, which necessitates the use of renewable resources.[55–57] However, more studies are needed to get those therapies into the clinic.

Theranostic PNPs combine diagnosis and therapy in a single platform. The development of multifunctional PNPs that can co-deliver multiple therapeutics and imaging agents to diseased organs is of paramount importance. Multifunctional PNPs can be synthesized by several strategies, for example, by combining different nanostructures into a single platform. Hierarchically assembled theranostic (HAT) nanoparticles have been developed to deliver siRNA alone as well as a combination of paclitaxel and siRNA.[58,59] It was observed that the combination of paclitaxel and siRNA in nanoparticles induces an enhanced amount of cancer cell death as compared to siRNA loaded in lipofectamine or to HAT assemblies having individual drugs. Moreover, different methods were used to assess the toxicity profile and intracellular delivery mechanisms of HAT nanoparticles. Combining cylindrical and spherical nanoparticles of the opposite charges by controlling polymer chemistry has resulted in decreased immunotoxicity of both individual PNPs, whereas the combination maintained prolonged circulation, active targeting ability, and retained both diagnostic and therapeutic agents with tunable control over the ratio of incorporated drugs and nanostructure. Agyare et al. have developed PNPs for the targeted treatment (targeting by using a fragment of anti-amyloid antibody and cyclophosphamide loaded in PNPs for therapy) and early diagnosis (by using MRI) of cerebral amyloid angiopathy.[60] Guo et al. showed that the use of theranostic PNPs loaded with the diacylglycerol kinase-α gene (a suppressor of immune function) reduced the immune response in the allogenic heart transplant model generated in rats, and the treatment course could be monitored by MRI.[61] In a recent study, Vaughan et al. reported the development of theranostic poly(beta-amino-ester) (PBAE) nanoparticles loaded with genes for the treatment and imaging of hepatocellular carcinoma.[62] However, several technical and biological limitations still exist to develop efficient theranostic nanoparticles. It is anticipated that with the advent of advanced technology and extensive research work, the usefulness of theranostic PNPs will be implemented in the clinic.

3.4 LIPOSOMES

Liposomes are spherical vesicles of sizes ranging from 10 nm to several microns, and are composed of phospholipid bilayers with an aqueous volume entirely enclosed by a membrane. Liposomes can be classified into small unilamellar vesicles (SUVs), large unilamellar vesicles (LUVs), giant unilamellar vesicles (GUVs), multilamellar vesicles (MLVs), and multivesicular vesicles (MVVs).[63] Liposomes have been suggested as a suitable candidate for drug delivery, due to their biocompatibility, biodegradability, and also to being able to encapsulate both hydrophilic and hydrophobic molecules, biologics, and macromolecules.[64,65] Liposomes can be engineered with specific ligands to identify chemical and biological targets with high specificity.[66,67] Liposomes have been widely explored, and a number of liposomal formulations have been approved for clinical use in different fields.[68] In 1995, the FDA authorized the intravenous Doxil®, which is a PEG-coated liposome containing doxorubicin hydrochloride. Doxil is indicated for the treatment of Kaposi's sarcoma, multiple myeloma, ovarian cancer, and breast cancer.[69] Onivyde® is the trade name for Irinotecan (a topoisomerase I inhibitor) encapsulated in a pegylated liposome, approved by the FDA in 2015 and by the European Medicines Agency (EMA) in 2016. Intravenous administration of Onivyde in combination with 5-fluorouracil and Leucovorin has been approved as a second-line treatments of metastatic pancreatic cancer.[70] Because of the EPR effect, Onivyde causes increased accumulation of Irinotecan in the tumor tissue. Furthermore, compared to the nonliposomal form, it has a longer half-life due to its pegylation. Onivyde, on the other hand, is detected in higher concentrations in the intravascular fluid because of its significantly smaller volume of distribution (2.61 L/m2) compared to nonliposomal Irinotecan (138 L/m2) (European Medicines Agency. Onivyde (irinotecan liposome injection): EU summary of product characteristics. www.ema.europa.eu/en/.) Another example is the liposomal formulation containing Daunorubicin (DaunoXome®), which was developed to improve the in vivo stability and specificity resulting in higher cumulative chemotherapeutic doses without significant cardiotoxicity. DaunoXome is indicated for the treatment of Kaposi's sarcoma, acute myeloid leukemia, and a wide variety of solid tumors.[71] Vyxeos® liposomal is used to treat the adults with newly diagnosed acute myeloid leukemia. The active substances in Vyxeos liposomal are daunorubicin and cytarabine.[72] Vyxeos liposomal has been shown to remain in the patient's body for longer than conventional daunorubicin and cytarabine. Moreover, Vyxeos is expected to be very stable, to protect the drugs from being broken down early, and to enhance their effect in cancer cells.

Liposomes have the potential to be utilized as a diagnostic tool for a range of biomolecular targets, such as proteins, DNA, and small organic molecules.[73] It has also been exploited as nanocarriers for a variety of imaging agents, such as Gd^{3+}, ^{64}Cu, and ^{18}F; for magnetic resonance imaging (MRI); and for nuclear imaging.[74] Due to excellent carrier properties and maneuverability, multifunctional liposomes can be employed as a "smart" theranostics system for simultaneous delivery of therapeutic and diagnostic agents.[75] For instance, Xiao et al. showed that sorafenib (SF) and gadolinium (Gd) co-loaded liposomes (SF/Gd-liposomes) prepared by the thin-film hydration method could be utilized as promising nano-carriers for the MRI-guided in vivo visualization of the delivery and biodistribution of liposomes and to achieve the therapy for hepatocellular carcinoma (HCC).[76]

Xu et al. synthesized a theranostic liposome (QSC-Lip) containing cilengitide (CGT), quantum dots (QDs), and SPIONs to target gliomas via magnetic targeting (MT).[77] QSC-Lip displayed a biphasic release pattern. The QSC-Lip produced a negative-contrast enhancing impact on gliomas by MRI. The precise resection of glioma by surgery was tracked by dual imaging of QSC-Lip. In another study, the theranostic efficacy of tumor-specific, pH-responsive, stabilized liposome-containing paclitaxel (PTX) and SPIONs modified with peptide $H_7K(R_2)_2$ was reported by

Zheng et al.[78] A thin-film hydration approach was employed to prepare the formulation (PTX/SPIO-SSL-$H_7K(R_2)_2$), with 50 $H_7K(R_2)_2$ serving as the targeted ligand. The theranostic impacts (i.e., targeted impact, MRI, and anticancer activity) of PTX/SPIO-SSL-$H_7K(R_2)_2$ were assessed and confirmed in MDA-MB-231 cell line in vitro and in vivo. In another study, Chang et al. investigated the tumor targeting and localization of 188Re-*N*,*N*-bis (2-mercaptoethyl)-*N*′,*N*′-diethylethylenediamine (BMEDA)-labeled pegylated liposomal doxorubicin (DXR) (188Re-DXRliposome) in a C26 murine colon carcinoma solid tumor model by means of biodistribution, pharmacokinetics and in vivo micro-SPECT/CT imaging.[79]

3.5 DENDRIMERS

Dendrimers are precisely structured nanoscale molecules with a symmetrical configuration consisting of a core, an inner shell, and an outer shell. They possess a well-defined, uniform, and consistent structure. Dendrimers exhibit a diverse array of morphologies, each with distinct biological attributes such as solubility, chemical stability, low cytotoxicity, self-assembly, and electrostatic interactions. Dendrimers possess a wide range of features, making them very suitable for applications in the field of therapeutics.[80]

To enhance the quality of MRI imagery, contrast agents are used, and the administration of contrast agents may be achieved by the use of dendrimers.[81,82] The contrast agents often used in MRI are paramagnetic metal cations of gadolinium, namely Gd(III)-N,N′,N″,N‴-tetracarboxymethyl-1,4,7,10-tetraazacyclododecane (Gd(III)-DOTA)and Gd(III)-diethylenetriaminepentaacetic acid (Gd(III)-DTPA), as well as their related derivatives. Nevertheless, because of their limited molecular size, these substances are capable of spreading via blood arteries and have a brief duration in the body's circulation. Moreover, their ability to selectively target sick and normal tissues is ineffective. To mitigate this drawback, one may bind gadolinium cations to the exteriors of dendrimers.[83,84]

Dendrimers may effectively serve as drug carriers in cancer therapy by covalent conjugation or physical encapsulation.[85] Dendrimers can effectively address the challenges that arise while administering anti-cancer treatments. Presently, a multitude of research studies are underway to investigate their potential as vehicles for medicinal compounds. There is a wide range of potential modifications due to the specificity of its construction. Hence, it is feasible to use them in focused therapeutic approaches, since they may directly adhere to the receptors in cancerous tumors by means of ligand attachment.[86] This enables the drug to infiltrate the neoplastic cells, thereby circumventing the growing resistance to treatment drugs.[87] The emergence of carriers that can modulate the expression of oncogenes and cancer-related genes is of utmost significance in cancer therapy. The efficiency and biocompatibility of drugs are significantly influenced by the dendrimer design and the core.[88]

Thakur et al. evaluated the possibility of employing poly (propylene-imine) dendrimers (PPI) as a carrier for targeting docetaxel (DTX) to MCF-7 breast cancer.[89] Researchers have used PPI dendrimers to enhance the transportation of paclitaxel (PTX) to the brain.[90] Prior research has used a dendrimer that contains transferrin to specifically treat the brain.[91] The study used PPI dendrimers of the 4th generation that were partly modified with maltose. These dendrimers were employed as carriers for the CCRF-1301 lymphoid leukemia cell lines and the HL-60 myeloid cell line.[92] In addition, Franiak-Pietryga conducted in vivo experiments to investigate the possibility of maltotriose-modified PPI dendrimers as a novel therapy approach for chronic lymphocytic leukemia.[93]

Polyethylene glycol (PEG)-coated polyamidoamine (PAMAM) dendrimers have been used to attach antibodies, folic acid, or biotin molecules.[94,95] In addition, the outer layer of the dendrimer may be coated with DNA/RNA or biotinylated antibodies for the purpose of detecting cancer antigens.[94,96] Hence, dendrimers have versatile applications in cancer diagnostics, allowing for the integration of several functionalities of these large molecules in a single study. For instance, dendrimers may be used to target tumors and enhance the quality of MRI images. Dendrimers with a ferrocene core, or dendrimers coated with cysteamine-modified gold electrodes, were used in another study that aimed at developing a complicated immunodiagnostic device based on dendrimers.[97] The molecules acquired had the capability to provide the analytical redox signal produced by the ferrocene fragments and to anchor the prostate-specific antibody (PSA) by using the major amino group on the dendrimer's surface. Optimal outcomes were achieved by using first-generation dendrimers, whereby these compounds exhibited the ability to detect PSA within a concentration range of 10 pg/mL to 100 ng/mL.[97]

3.6 MAGNETIC NANOPARTICLES (MNPS)

There is a growing trend in using magnetic nanoparticles (MNPs) for theranostic purposes due to their small sizes, bio-compatibility, capacity to be functionalized with biological molecules, and their responsiveness to external magnetic fields.[98] Due to their stability and functionality, they have been primarily utilized to enhance the contrast and therapeutic capabilities of MRI,[99,100] as well as for targeted drug delivery, hyperthermia applications,[101,102] and the accurate capture of specific nucleic acids.[103] Iron oxide nanoparticles (IONPs) have been conjugated with antibodies, peptides, proteins, and small molecules, and are used for both diagnostic and therapeutic purposes.[104] Composite/hybrid core/shell systems, formed by combining MNPs with polymers, provide multi-stimuli-responsive platforms that serve both diagnostic and therapeutic purposes.[105] Below is a concise summary of the MNPs often used in traditional biomedical applications.

Biomedical imaging with MRI is used to capture the contrast signal generated by the tissue in response to an external magnetic field, which is influenced by the proton density and magnetic relaxation time of a proton. The researchers investigated the use of MNPs to enhance the signal contrast and improve the resolution of soft tissue images, as shown in Figure 3.2 for both in vitro and in vivo MRI studies.

Currently, contrast agents have been examined in more than 30% of MRI investigations. The excellent spatial resolution of MRI and the accuracy of MNP enable the imaging of tumors as small as 2–3 mm in clinical applications. Superparamagnetic iron oxide nanoparticles (SPIONs) were coupled with monoclonal antibody C595, resulting in SPIONs-C595. These nanoprobes, with a hydrodynamic size of 87 nm, were successfully used for MRI. The SPIONs-C595 nanoprobe has the potential to be used as a T_2-weighted MRI contrast agent for detecting breast cancer MCF-7 cells.[107] MNPs have also been used in ultrasonic CT (UCT) imaging conducted on a

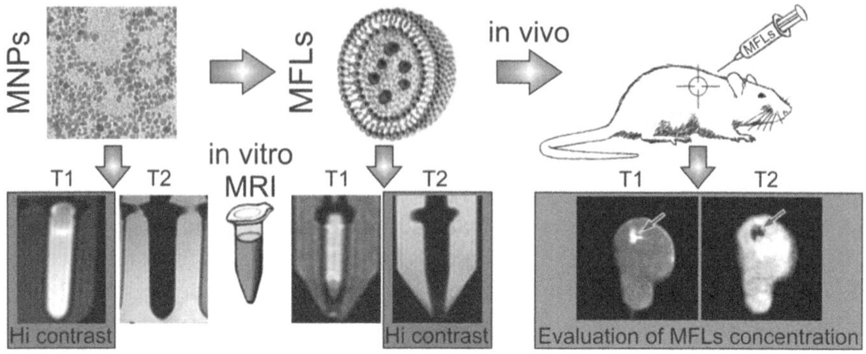

FIGURE 3.2 Improved T1 and T2 signal strength by MNP contrast agent for in vivo and in vitro MRI (Reprinted with permission from German et al. 2015).[106]

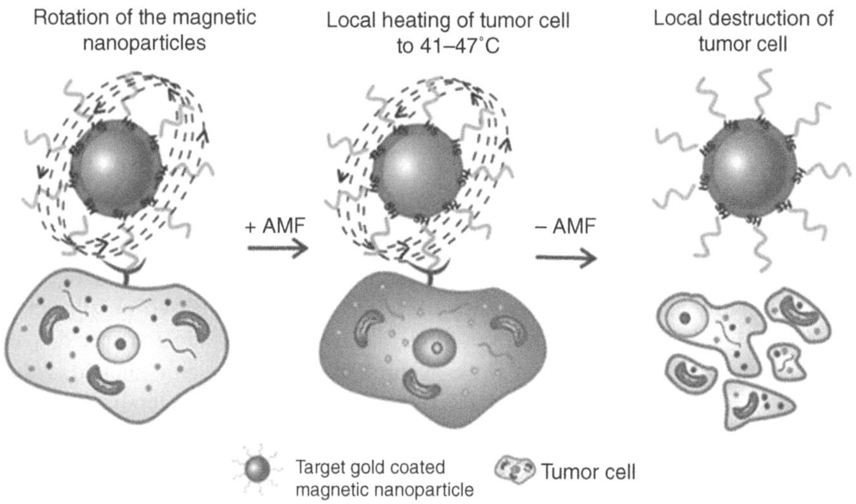

FIGURE 3.3 Principle of magnetic hyperthermia. Cancer cells in an AMF are the intended targets of MNPs. Subsequently, cancer cells are heated to a narrow range of 41–47°C by the heat produced by the MNPs (Reprinted from Belyanina et al. 2017, distributed under CC-BY 4.0 license, which permits unlimited distribution in any medium, https://creativecommons.org/licenses/by/4.0/).[111]

breast-mimicking phantom and on an ex vivo tissue model.[108] A multimodal imaging-guided cancer theranostic agent called Au/PPY@Fe_3O_4 was created by combining gold nanorods and superparamagnetic iron oxide (Fe_3O_4) with polypyrrole. They generate a significant contrast for MR and X-ray CT imaging. The photothermal effects of Au/PPY@Fe_3O_4 nanocomposites can effectively destroy cancer cells.[109]

Due to great effectiveness and, to lesser negative effects, magnetic fluid hyperthermia utilizing biocompatible magnetic nanoparticles as heat mediators for cancer therapy has been thoroughly studied recently. As shown in Figure 3.3, the heat produced by MNPs in a non-intrusive applied alternating magnetic field (AMF) leads to cell death via magnetic hyperthermia, which are irreversible physiological alterations.[110]

By functionalizing the silica coating with anti-αvβ6 antibodies, the nanoparticles could be directly targeted to cells overexpressing αvβ6, and the use of thermal therapy resulted in a considerable increase in the death of the targeted cancer cells.[112]

The IONPs-based delivery system has great promise for the effective treatment of ovarian cancer based on chemotherapy and mild hyperthermia. The engineered nanocarrier using IONPs served as an effective vehicle for doxorubicin and could remotely heat cancer cells in an alternating magnetic field (AMF). The nanocarrier, coated with PEG and functionalized with a luteinizing hormone-releasing hormone (LHRH) peptide, enhances its stability in physiological fluids and enables it specifically to target cancer cells.[113] A multifunctional DOX-loaded magnetic alginate-chitosan microsphere with SPIONs was developed for the treatment of breast cancer. In vitro results on MCF-7 breast cancer cells revealed significant cell death based on a combinatorial strategy.[114]

3.7 QUANTUM DOTS

Quantum dots are zero-dimensional, semiconductor nanomaterials. Various types of quantum dots have been developed until now, majorly based on group I–III–VI and II-VI compounds, along with carbon, transition-metal dichalcogenides and so forth. Owing to their quantum size effect, they exhibit excellent optical, electronic, and photoluminescence properties, contributing their applications in diverse scientific fields including the fields of biomedical sciences.[115,116]

Apart from suitable treatment strategies, proper diagnosis of diseases, preferably in their early stage, is of paramount importance. There are situations when a colonoscopy cannot provide a quick and precise diagnosis, particularly when dealing with colon tumors associated with persistent inflammation. A colonoscopy significantly relies on visual observation, and the diagnosis is primarily based on morphology-specific information, i.e., pit pattern. Numerous endoscopic procedures have been created to help with the endoscopic diagnostic more effectively, like chromoendoscopy, narrow-band imaging (NBI), magnifying endoscopy, and so on. Chromoendoscopy utilizes dye molecules to improve visualization of the epithelial lesion's surface architecture, while vasculature of the mucosal layer can be visualized using a narrow-band filter in narrow-band imaging. On the other hand, the mucosal crypt pattern can be visualized in detail by using the magnifying endoscopy technique. Regrettably, improvements in colonoscopic procedures have only slightly enhanced the ability to diagnose adenomas, flat adenomas, and colon polyps. Moreover, biopsy is not a rapid process, even failing to show the details of the entire colon. In this context, fluorescence imaging (a molecular imaging technique) appears as an alternative, where the targeted areas can be selectively visualized by color emission, and even the signals can be quantified. Owing to the excellent photoluminescence, photostability, and symmetric and narrow emission profiles, quantum dots are frequently studied as the bioimaging probes for disease diagnosis. For example, a group of researchers developed antibody-coated CdSe/CdS/ZnS (Core/Shell/Shell) quantum dots for the detection of colorectal cancer. The developed probes not only detect cancer successfully but also detect the adenoma and hyperplasia regions. The authors claimed that such a probe can be useful to detect small tumors or flat tumors, which may not be done by conventional endoscopic processes.[117] As the technology matures day by day, new tools are being explored for disease diagnosis. For example, a research team developed a straightforward and inexpensive wireless, chip-based, multiplex diagnostic modality by fusing the quantum dot barcode technology with smartphones and isothermal amplification. In order to diagnose individuals with HIV or hepatitis B, the researchers characterized the analytical performance of this instrument and showed that it is proficient in detecting viral genetic copies as low as 1,000 per milliliter. More crucially, the barcoding made it possible to identify numerous pathogenic diseases in a single test within less than an hour. Due to common symptoms like a fever or rash, infections that are challenging to distinguish clinically can now be diagnosed due to the device's multiplexing capabilities. The combination of quantum dot-based barcoding technology and a smartphone reader enables real-time knowledge transfer of emerging or urgent disease threats with health-care and military organizations, along with the capacity for global infectious disease surveillance.[118] Moreover, researchers have checked the clinical viability of the new quantum dot barcoding technology to support its feasibility in real-world applications in the future.[119] Moreover, other life-threatening diseases like tuberculosis can be diagnosed using the modality developed by quantum dot nanobeacon and multicomponent nucleic acid enzyme, which enable detection by the naked eye using a portable and cheap UV flashlight.[120]

Quantum dots are explored for their therapeutic potential against various disease conditions. Quantum dots exert such benefits majorly through photodynamic and photothermal therapy. Briefly, a photosensitizer molecule absorbs the incident radiation, and thereby transits from the ground state to a short-lived singlet excited state. From the singlet excited state, it reaches more stable triplet excited state through intersystem crossing. From there, the molecule comes back to the ground state through different photodynamic reactions (type I and type II). In type I, the molecule transfers an electron to the substrate, which generates various ROS, like $O_2^{\bullet-}$, $HO^{\bullet}_2$,while in type II, the energy is shifted to ground-state molecular oxygen, resulting in the formation of singlet oxygen, $^1O_2$. Such highly reactive species trigger many biological processes, such as cytotoxicity, inflammation, and so forth. A study demonstrated that the aqueous soluble CdSe quantum dot conjugated with leukemia cell-specific antibodies effectively served as a photosensitizer and promoted the activity of classical photosensitizers, like sulfonated aluminum phthalocyanine and trifluoperazine in photodynamic therapy. The study demonstrated that the detrimental effect on cancer cell viability was due to the generation of free radicals upon UV irradiation,[121] while photothermal therapy typically uses radiation of the near-infrared region, which raises the tissue temperature, thereby resulting in local photocoagulation. Photothermal therapy requires light energy to raise temperature to attain coagulative (55–100°C) or sub-coagulative (43–55°C) temperatures. Such temperatures bring rapid cell death through the denaturation of protein and by damaging the cell membranes. However, moderate heating (41–43°C) is referred to as hyperthermia. Hyperthermia is not tumoricidal unless used for a prolonged period of time (more than1h).[122] For instance, polyethylene glycol-coated copper/carbon dot nanosheets were generated and explored for image-guided photothermal therapy. The study claimed that as-prepared nanosheets had a 41.3% photothermal conversion efficiency with good photothermal stability.[123]

On a different note, biomass-derived carbon dots were utilized to prepare a green synthesis of silver nanoparticles. As-developed nanoconjugates demonstrated significant antibacterial activity against *E. coli*. Moreover, synergistic anticancer activity against breast cancer cells was observed, which was attributed to the enhancement in intracellular ROS generation.[124]

Quantum dots are also explored for theranostic applications. For instance, a quantum dot (QD)-based micelle loaded with aminoflavone, an anticancer drug, has been coupled with an anti-epidermal growth factor receptor (EGFR) nanobody for the purpose of EGFR-overexpressing cancer theranostics.[125] In another study, Pei and colleagues synthesized a fluorescent hyper-cross-linked β-cyclodextrin-carbon quantum dot (β-CD-CQD) hybrid nanosponge loaded with doxorubicin, as theranostic nanomedicine, which could be internalized into HepG2 cells, and showed effective antitumor efficacy.[126] In another study, a doxorubicin-loaded folic acid-derived carbon dot-functionalized amphiphilic nanomicelle was developed. Such a nanomicelle showed targeted delivery of doxorubicin and imaging capability for triple negative breast cancer cell lines.[127] Moreover, nanoplatforms were developed using quantum dots for bioimaging and ion detection, along with having therapeutic benefits. For example, Ghosh et al. developed a dendrimer-functionalized biomass-derived carbon quantum dot for theranostic application.[128] The study revealed that the as-prepared nanoplatform can selectively detect intracellular copper ions, able to image cells and deliver genes for cancer theranostic.[128]

3.8 CARBON NANOTUBES

Carbon nanotubes (CNTs) are cylindrical nanostructures consisting solely of carbon atoms arranged in a hexagonal lattice. They have been widely explored for diverse applications in various scientific fields, due to their excellent properties like high thermal conductivity, electrical conductivity, exceptional strength, chemical stability, high surface-to-volume ratio, and so forth. Single-walled carbon nanotubes (SWCNTs) are hexagonal-packed bundles made up of a single graphene layer with a diameter ranging from 0.4 to 2 nm. Multi-walled carbon nanotubes (MWCNTs) are made up of two or more cylinders formed of graphene sheets. The diameter ranges between 1 and 3 nm.[129] Moreover, owing to their unique features, such as tunable surface chemistry, outstanding biocompatibility, and high surface area (when appropriately functionalized and purified), CNTs have shown significant promise in a variety of biomedical applications, among which are its application in diagnosis, therapy and theranostic purposes, briefly discussed below.

CNTs have found application as electrochemical biosensors. CNT-based biosensors may detect target biomolecules such as proteins, DNA, or small molecules with high sensitivity and selectivity when functionalized with specific recognition components (e.g., antibodies, aptamers, or enzymes). The target molecule's interaction changes the electrical characteristics of the CNTs, allowing for quantitative measurement. In the amperometric sensing of hydrogen peroxide, networks of single-walled carbon nanotubes (SWCNTs) decorated with Au-coated Pd (Au/Pd) nanocubes are used as electrochemical biosensors with outstanding sensitivity and a low estimated detection limit (2.3 nM).[130] The selective immobilization of fluorescently labeled streptavidin on the nanocube surfaces by thiol linkage demonstrates biofunctionalization of the Au/Pd nanocube-SWCNT biosensor. Similarly, for amperometric glucose sensing, glucose oxidase is connected to the surface of the nanocubes. The nanocube-SWCNT biosensor is particularly useful for glucose detection in diabetic patients and is well adapted to an array of amperometric detection techniques for clinically significant biomarkers.[130]

Like other carbon nanomaterials, CNTs have grown in popularity for photothermal therapy. It was observed from the previous literature that SWCNTs can deliver therapeutic medicines and diagnostic compounds into cells. Through absorbing near-infrared light, SWCNTs can result in cell death through localized photothermal effects. The use of SWCNTs will aid in the development of novel multimodal therapies combining PTT with PDT. Being a non-invasive phototherapy, PDT is utilized in clinical practices; thus, combining PTT with PDT is possible and clinically beneficial.[131] Such multimodal treatment techniques with synergistic effects is a potential approach, improving effectiveness and optimizing cancer treatment. Zhang and co-researchers developed Ru(II) complex-functionalized SWCNT nanotemplates and explored their application in photothermal and two-photon photodynamic therapy.[131] The study revealed that the as-developed nanotemplates have superior anticancer efficacies to Ru(II) complex-induced PDT or SWCNT-induced PTT. Ru@SWCNTs have greater anticancer efficacies than either PDT using Ru(II) complexes or PTT using SWCNTs in two-dimensional (2D) cancer cell and three-dimensional (3D) multicellular tumor spheroid (MCTS) models. Moreover, the simultaneous use of PTT with gene therapy has the potential to create synergistic anticancer efficacy. Nevertheless, the release of genes from carriers in a controlled fashion remains a significant impediment. To overcome such an issue, carbon nanotubes are coated with temperature-sensitive lipids for delivery of siRNA and PTT for synergistic cancer therapy.[132] On the other hand, depending on the disease status, on-demand release of payloads is often required, which could be possible using an external stimulus. For instance, a group of researchers showed carbon nanotube-liposome complexes in hydrogels for controlled drug delivery via near-infrared laser stimulation.[133]

CNTs and their conjugates were assessed for cancer theranostics. For example, on the oxidized MWCNTs, branched gold nanoparticles were grown, while doxorubicin (Dox) can be adsorbed in high quantity on both inner and outer surfaces of oxidized carbon nanotubes by π–π stacking interactions between doxorubicin aromatic groups and carbon nanotube (CNT) backbone. Carbon nanotube/gold hybrids display a broad absorption band in the red and near-infrared regions allowing their use for imaging applications. In vitro cellular tests showed that the nanostructures can efficiently transport and deliver doxorubicin inside the cells.[134] Such a nanohybrid was explored for the delivery of the anticancer drug into the lung cancer cell line (A549). Moreover, the nanohybrid showed a wide absorption band in the infrared region, demonstrating its applicability for imaging applications. The research demonstrated that the CNTs' conjugates can be efficiently used in cancer theranostics. On the other hand, Das et al. synthesized a theranostic MWCNT by coupling acid-oxidized MWCNTs with four moieties: methotrexate, a radionuclide (Technetium-99m), a fluorochrome (Alexa-fluor, AF488/647), and folic acid. This theranostic MWCNT is selectively internalized by human cancer cells (A549) and MCF-7 cells that express the folate receptor.[135]

3.9 GRAPHENE

The thinnest, yet still strongest material, graphene, is the allotrope of the carbon family, having a two-dimensional honeycomb structure arranged in a single monolayer comprising sp^2 hybridized carbon atoms.[136] Owing to its extraordinary physicochemical and mechanical properties, it has been widely explored in various scientific fields like electronics, sensing, optoelectronics, energy, and even biomedical sciences.[136,137] Notably, graphene and its derivatives (i.e., graphene oxides, reduced graphene oxides) of various forms (e.g., nanosheets, quantum dots) were explored for therapy, diagnosis, and theranostic applications, which are briefly discussed here.

Early detection of any disease condition is immensely critical as it can effectively extend human life. Graphene-derived biosensors were researched for diagnostic applications owing to their abilities to monitor and/or detect biomarkers. Such capabilities of graphene and its derivatives provide a better understanding of altered physiological conditions. For example, the graphene field-effect transistor with a wide planar graphene surface allows it to study cell membrane-related protein interactions. When the graphene field-effect transistor was modified with a ganglioside GM1-enriched lipid bilayer, it efficiently monitored the aggregation of amyloid-beta 40.[138] Notably, it was able to detect the aggregation of amyloid-beta 40 in the early nucleation stage, which was undetectable through conventional thioflavin T assay. The study, thereby, depicted the significant potential of a lipid bilayer-supported graphene field-effect transistor as an efficient biomimetic sensor for studies of membrane-related protein activities and interaction-kinetics.[138] Exosomes are nanoscale (100–200 nm) cell-derived extracellular vesicles that circulate in the body's fluids. Exosomes have demonstrated their significance in a number of biological processes and their potential for early cancer detection. It was found that breast cancer cell-derived exosomes can be isolated and detected quantitatively by the hydroxylated graphene as a biosensor.[139] Another study reported that a biofunctionalized graphene quantum dot was used to develop a fluorescent turn-on biosensor. Such a biosensor was helpful in the highly sensitive detection of the biomarker specific to small cell lung cancer.[140] Moreover, the biological alteration of a porous three-dimensional (3D) graphene holds significant importance in expanding its applicability as a precise biosensor. Through controllable chitosan electrodeposition technology, an enzymatic biosensor was developed using 3D graphene foam. As-developed reagentless biosensors precisely detected glucose in a phosphate buffer solution.[141]

Graphene derivatives, i.e., graphene oxide and reduced graphene oxide, showed therapeutic potential through photothermal therapy along with the capability to deliver drugs. Several research studies have been done since nearly the last two decades. For example, a graphene oxide and silver nanoparticle-based, hyaluronidase-triggered photothermal framework for antibacterial therapy was developed. When exposed to near infrared light, the developed graphene oxide-based nanomaterials elevated the local temperature, resulting in substantial bacterial death. The study showed that the nanoplatform was highly effective against *S. aureus*, while it was non-lethal for normal cells, and also demonstrated its efficacy in an in vivo wound disinfection model.[142] On the other hand, glutaraldehyde-functionalized magnetic reduced graphene oxide showed effective killing of both *S. aureus* and *E. coli* upon the exposure of near-infrared light.[143] Through the hydrophobic interaction and π-π stacking, graphene derivatives can interact with drugs, particularly hydrophobic ones, and are therefore used as drug delivery platforms. Several research studies were done to demonstrate the ability of graphene-based nanoplatforms for drug delivery, along with photothermal therapy. For instance, a research group developed bovine serum albumin-functionalized reduced graphene oxide nanosheets for near infrared responsive photothermal therapy and controlled release of doxorubicin.[144] Such synergistic activity is always beneficial to mitigate cancer by targeting multiple pathways simultaneously. In another study, anti-PDL1-conjugated doxorubicin loaded tea polyphenol-reduced graphene oxide was developed, which showed pH-responsive doxorubicin release, and near-infrared mediated photothermal effect along with target specificity toward human tongue squamous cancer cells, while safeguarding the normal cells.[145] Therefore, graphene derivative-based nanoplatforms showed great promise in chemo-photothermal-based cancer therapies.

In cancer nanomedicine, selective tissue visualization and localized tumor regression without damaging the healthy tissues are significant considerations. In this context, a pressing requirement for targeted combination therapies is the development of a safe and biodegradable nanohybrid with significant theranostic capacity. Toward this aim, a folic acid-conjugated graphene oxide decorated liposome was fabricated. The as-developed nanohybrid showed photo-triggered tumor imaging and tumor regression, thereby showcasing its theranostic applications.[146] There is a growing interest in developing physiologically robust substances to avoid instability and dissociation in biological environments while designing theranostic agents for integrating diagnosis and diverse modalities of therapy. To achieve stability, a covalently linked silicon phthalocyanine and graphene oxide composite was synthesized by conjugation reaction.[147] When compared to pure graphene oxide, this unique highly water-soluble substance exhibits intrinsic fluorescence and simultaneous PTT-PDT action, as well as threefold enhanced near-infrared absorbance. In vitro studies depicted that as-prepared nanohybrid could simultaneously generate reactive oxygen species, fluorescence intracellularly, and photothermal effects. This study also demonstrated the efficient photoablation of tumor cells, which could be induced by both laser (671 and 808 nm)-mediated synergistic PTT/PDT and near infrared photothermal effects.[147]

3.10 CONCLUSION

Nanotechnology has emerged as a groundbreaking tool in the fields of diagnosis and therapy, revolutionizing the landscape of medical research and healthcare. Nanotechnological tools have already been investigated for a wide range of applications both in vitro and in vivo. Nanoscale materials, such as nanoparticles and nanocarriers, offer unique properties that enable precise imaging and targeted drug delivery. In diagnostics, nanotechnology enhances imaging modalities, providing unprecedented resolution and sensitivity for early disease detection. Concurrently, in therapeutic interventions, nanocarriers enable site-specific drug delivery, minimizing systemic side effects and maximizing therapeutic efficacy. In summary, a number of research studies have demonstrated the potential growth of nanotheranostics, which has attracted considerable attention in clinics due to its diagnostic potential in multimodal imaging and therapy platforms. The integration of nanotechnology into medical practice holds immense promise for improving patient outcomes, introducing a new era of precision medicine and transformative health-care solutions.

REFERENCES

1. Schmidt, C., & Storsberg, J. (2015). Nanomaterials-Tools, technology and methodology of nanotechnology based biomedical systems for diagnostics and therapy. *Biomedicines*, *3*(3), 203–223. https://doi.org/10.3390/biomedicines3030203
2. Shen, Q., Peng, L., Zhang, Y., & Wang, R. (2023). Editorial: Application of nanotechnology in diagnosis and/or therapy of non-small cell lung cancer. *Frontiers in Oncology*, *13*, 1234727. https://doi.org/10.3389/fonc.2023.1234727
3. Bag, J., Banerjee, S., De, A., Manna, S., Banerje, S., Ashok Kumar., S. K., & De, S. (2023). Nanoengineered approaches to improve the efficacy of targeted drug delivery for the treatment of malignancy: A comprehensive review. *Future Journal of Pharmaceutical Science*, *9*, https://doi.org/10.1186/s43094-023-00541-w
4. Thwala, L. N., Ndlovu, S. C., Mpofu, K. T., Lugongolo, M. Y., & Mthunzi-Kufa, P. (2023). Nanotechnology-based diagnostics for diseases prevalent in developing countries: Current advances in point-of-care tests. *Nanomaterials (Basel, Switzerland)*, *13*(7), 1247. https://doi.org/10.3390/nano13071247
5. Wang, Y., Yu, L., Kong, X., & Sun, L. (2017). Application of nanodiagnostics in point-of-care tests for infectious diseases. *International journal of nanomedicine*, *12*, 4789–4803. https://doi.org/10.2147/IJN.S137338
6. Asdaq, S. M. B., Ikbal, A. M. A., Sahu, R. K., Bhattacharjee, B., Paul, T., Deka, B., Fattepur, S., Widyowati, R., Vijaya, J., Al Mohaini, M., Alsalman, A. J., Imran, M., Nagaraja, S., Nair, A. B., Attimarad, M., & Venugopala, K. N. (2021). Nanotechnology integration for SARS-CoV-2 diagnosis and treatment: An approach to preventing pandemic. *Nanomaterials (Basel, Switzerland)*, *11*(7), 1841.https://doi.org/10.3390/nano11071841
7. Indoria, S., Singh, V., & Hsieh, M. F. (2020). Recent advances in theranostic polymeric nanoparticles for cancer treatment: A review. *International Journal of Pharmaceutics*, *582*, 119314. https://doi.org/10.1016/j.ijpharm.2020.119314
8. Jokerst, J. V., & Gambhir, S. S. (2011). Molecular imaging with theranostic nanoparticles. *Accounts of Chemical Research*, *44*(10), 1050–1060. https://doi.org/10.1021/ar200106e
9. De, A., Das, S., Ghosh, S., Das, B., Samanta, S., Bhattacharya, B., & Samanta, A. (2022). Theranostic nanovesicles. In A. K. Nayak, M. S. Hasnain, T. M. Aminabhavi, V. P. Torchilin (Eds.), *Applications of Nanovesicular Drug Delivery* (pp. 39–51). Academic Press: An imprint of Elsevier.
10. Kumbhakar, P., Ray, S. S., & Stepanov, A. (2014). Optical properties of nanoparticles and nanocomposites. http://hdl.handle.net/10204/7669
11. Khlebtsov, N. G., & Dykman, L. A. (2010). Optical properties and biomedical applications of plasmonic nanoparticles. *Journal of Quantitative Spectroscopy and Radiative Transfer*, *111*(1), 1–35.

12. Kelly, K. L., Coronado, E., Zhao, L. L., & Schatz, G. C. (2003). The optical properties of metal nanoparticles: The influence of size, shape, and dielectric environment. *Journal of Physical Chemistry B*, *107*, 668–677.
13. Malinsky, M. D., Kelly, K. L., Schatz, G. C., & Van Duyne, R. P. (2001). Nanosphere lithography: Effect of substrate on the localized surface plasmon resonance spectrum of silver nanoparticles. *Journal of Physical Chemistry B, 105*(12), 2343–2350.
14. Jensen, T. R., Duval, M. L., Kelly, K. L., Lazarides, A. A., Schatz, G. C., & Van Duyne, R. P. (1999). Nanosphere lithography: Effect of the external dielectric medium on the surface plasmon resonance spectrum of a periodic array of silver nanoparticles. *Journal of Physical Chemistry B, 103*(45), 9846–9853.
15. Nealon, G. L., Donnio, B., Greget, R., Kappler, J. P., Terazzi, E., & Gallani, J. L. (2012). Magnetism in gold nanoparticles. *Nanoscale*, *4*(17), 5244–5258. https://doi.org/10.1039/c2nr30640a
16. Matthias, B. T., Clogston, A. M., Williams, H. J., Corenzwit, E., & Sherwood, R. C. (1961). Ferromagnetism in solid solutions of scandium and indium. *Physical ReviewLetters, 7*(1), 7.
17. Matthias, B. T., & Bozorth, R. M. (1958). Ferromagnetism of a zirconium–zinc compound. *Physical Review*, *109*(2), 604.
18. Acker, F., Fisk, Z., Smith, J. L., & Huang, C. Y. (1981). Enhanced paramagnetism of TiBe2 and ferromagnetic transitions in TiBe2-xCux. *Journal of Magnetism and Magnetic Materials, 22*(3), 250–256.
19. Podaru, G., & Chikán, V. (2017). Magnetism in nanomaterials: Heat and force from colloidal magnetic particles. In S. H. Bossmann, H. Wang (Eds.), *Magnetic Nanomaterials: Applications in Catalysis and Life Sciences* (pp. 1–24). The Royal Society of Chemistry.
20. Jun, Y. W., Seo, J. W., & Cheon, J. (2008). Nanoscaling laws of magnetic nanoparticles and their applicabilities in biomedical sciences. *Accounts of Chemical Research*, *41*(2), 179–189. https://doi.org/10.1021/ar700121f
21. Park, J. I., & Cheon, J. (2001). Synthesis of "solid solution" and "core-shell" type cobalt--platinum magnetic nanoparticles via transmetalation reactions. *Journal of the American Chemical Society*, *123*(24), 5743–5746. https://doi.org/10.1021/ja0156340
22. Lee, J. H., Huh, Y. M., Jun, Y. W., Seo, J. W., Jang, J. T., Song, H. T., Kim, S., Cho, E. J., Yoon, H. G., Suh, J. S., & Cheon, J. (2007). Artificially engineered magnetic nanoparticles for ultra-sensitive molecular imaging. *Nature Medicine*, *13*(1), 95–99. https://doi.org/10.1038/nm1467
23. Wu, Q., Miao, W., Zhang, Y., Gao, H., & Hui, D. (2020). Mechanical properties of nanomaterials: A review. *Nanotechnology Reviews*, *9*(1), 259–273. https://doi.org/10.1515/ntrev-2020-0021
24. Pithawalla, Y. B., El-Shall, M. S., Deevi, S. C., Ström, V., & Rao, K. V. (2001). Synthesis of magnetic intermetallic FeAl nanoparticles from a non-magnetic bulk alloy. *Journal of Physical Chemistry B, 105*(11), 2085–2090.
25. Guo, D., Xie, G., & Luo, J. (2013). Mechanical properties of nanoparticles: Basics and applications. *Journal of Physics D: Applied Physics*, *47*(1), 13001.
26. Fu, L., & Fu, Z. (2015). *Plectranthusamboinicus* leaf extract-assisted biosynthesis of ZnO nanoparticles and their photocatalytic activity. *Ceramics International*, *41*(2), 2492–2496.
27. Andrievski, R. A. (2014). Review of thermal stability of nanomaterials. *Journal of Materials Science, 49*, 1449–1460.
28. Qiu, L., Zhu, N., Feng, Y., Michaelides, E. E., Żyła, G., Jing, D., Zhang, X., Norris, P. M., Markides, C. N., & Mahian, O. (2020). A review of recent advances in thermophysical

properties at the nanoscale: From solid state to colloids. *Physics Reports, 843*, 1–81. doi: 10.1016/j.physrep.2019.12.001.

29. Eastman, J. A., Choi, S. U. S., Li, S., Yu, W., & Thompson, L. J. (2001). Anomalously increased effective thermal conductivities of ethylene glycol-based nanofluids containing copper nanoparticles. *Applied Physics Letters*, *78*(6), 718–720.
30. Zebarjadi, M., Esfarjani, K., Shakouri, A., Bahk, J. H., Bian, Z., Zeng, G., ... & Gossard, A. (2009). Effect of nanoparticle scattering on thermoelectric power factor. *Applied Physics Letters*, *94*(20), 202105.
31. Zeng, G., Zide, J. M., Kim, W., Bowers, J. E., Gossard, A. C., Bian, Z., ... & Majumdar, A. (2007). Cross-plane seebeck coefficient of ErAs: InGaAs/InGaAlAs superlattices. *Journal of Applied Physics*, *101*(3), 34502.
32. Kim, W., Singer, S. L., Majumdar, A., Vashaee, D., Bian, Z., Shakouri, A., Zeng, G., Bowers, J. E., Zide, J. M. O.,& Gossard, A. C. (2006). Cross-plane lattice and electronic thermal conductivities of ErAs:InGaAs/InGaAlAs superlattices. *Applied Physics Letters, 88*(24), 242107.
33. Shim, J-H., Lee, B-J., & Cho, Y. W. (2002). Thermal stability of unsupported gold nanoparticle: A molecular dynamics study. *Surface Science*. *512*(3), 262–268.
34. Roduner, E. (2006). Size matters: Why nanomaterials are different. *Chemical Society Reviews, 35*(7), 583–592. https://doi.org/10.1039/B502142C
35. Cuenya, B. R. (2010). Synthesis and catalytic properties of metal nanoparticles: Size, shape, support, composition, and oxidation state effects. *Thin Solid Films, 518*(12), 3127–3150.
36. Cuenya, B. R., Baeck, S. H., Jaramillo, T. F., & McFarland, E. W. (2003). Size- and support-dependent electronic and catalytic properties of Au0/Au3+ nanoparticles synthesized from block copolymer micelles. *Journal of the American Chemical Society*, *125*(42), 12928–12934. https://doi.org/10.1021/ja036468u
37. Haruta,M. (2004). Nanoparticulate gold catalysts for low-temperature CO oxidation. *ChemInform*, *35*(48), 163–172.
38. Xu, R., Wang, D., Zhang, J., & Li, Y. (2006). Shape-dependent catalytic activity of silver nanoparticles for the oxidation of styrene. *Chemistry, An Asian Journal*, *1*(6), 888–893. https://doi.org/10.1002/asia.200600260
39. Ono, L. K., & Roldán-Cuenya, B. (2007). Effect of interparticle interaction on the low temperature oxidation of CO over size-selected Au nanocatalysts supported on ultrathin TiC films. *Catalysis Letters, 113,* 86–94. https://doi.org/10.1007/s10562-007-9027-7
40. Elsabahy, M., Heo, G. S., Lim, S. M., Sun, G., & Wooley, K. L. (2015). Polymeric nanostructures for imaging and therapy. *Chemical Reviews*, *115*(19), 10967–11011. https://doi.org/10.1021/acs.chemrev.5b00135
41. Behera, A., Banerjee, I., De, K., Munda, R. N., Chattopadhayay, S., Samanta, A., Sarkar, B., Ganguly, S., & Misra, M. (2013). Synthesis, characterization, conformational analysis of a cyclic conjugated octreotate peptide and biological evaluation of (99m)Tc-HYNIC-His(3)-Octreotate as novel tracer for the imaging of somatostatin receptor-positive tumors. *Amino Acids*, *44*(3), 933–946. https://doi.org/10.1007/s00726-012-1423-7
42. Yao, J., Yang, M., & Duan, Y. (2014). Chemistry, biology, and medicine of fluorescent nanomaterials and related systems: New insights into biosensing, bioimaging, genomics, diagnostics, and therapy. *Chemical Reviews*, *114*(12), 6130–6178. https://doi.org/10.1021/cr200359p
43. Harrison, J., Bartlett, C. A., Cowin, G., Nicholls, P. K., Evans, C. W., Clemons, T. D., Zdyrko, B., Luzinov, I. A., Harvey, A. R., Iyer, K. S., Dunlop, S. A., & Fitzgerald, M. (2012). In vivo imaging and biodistribution of multimodal polymeric nanoparticles

delivered to the optic nerve. *Small (Weinheim an der Bergstrasse, Germany)*, *8*(10), 1579–1589. https://doi.org/10.1002/smll.201102648

44. Tan, Y. F., Chandrasekharan, P., Maity, D., Yong, C. X., Chuang, K. H., Zhao, Y., Wang, S., Ding, J., & Feng, S. S. (2011). Multimodal tumor imaging by iron oxides and quantum dots formulated in poly (lactic acid)-D-alpha-tocopheryl polyethylene glycol 1000 succinate nanoparticles. *Biomaterials*, *32*(11), 2969–2978. https://doi.org/10.1016/j.biomaterials.2010.12.055
45. Mitchell, M. J., Billingsley, M. M., Haley, R. M., Wechsler, M. E., Peppas, N. A., & Langer, R. (2021). Engineering precision nanoparticles for drug delivery. *Nature Reviews. Drug Discovery*, *20*(2), 101–124. https://doi.org/10.1038/s41573-020-0090-8
46. Le, Z., Chen, Y., Han, H., Tian, H., Zhao, P., Yang, C., He, Z., Liu, L., Leong, K. W., Mao, H. Q., Liu, Z., & Chen, Y. (2018). Hydrogen-bonded Tannic acid-based anticancer nanoparticle for enhancement of oral chemotherapy. *ACS Applied Materials & Interfaces*, *10*(49), 42186–42197. https://doi.org/10.1021/acsami.8b18979
47. Dhar, S., Kolishetti, N., Lippard, S. J., & Farokhzad, O. C. (2011). Targeted delivery of a cisplatin prodrug for safer and more effective prostate cancer therapy in vivo. *Proceedings of the National Academy of Sciences of the United States of America*, *108*(5), 1850–1855. https://doi.org/10.1073/pnas.1011379108
48. Graf, N., Bielenberg, D. R., Kolishetti, N., Muus, C., Banyard, J., Farokhzad, O. C., & Lippard, S. J. (2012). α(V)β(3) integrin-targeted PLGA-PEG nanoparticles for enhanced anti-tumor efficacy of a Pt(IV) prodrug. *ACS Nano*, *6*(5), 4530–4539. https://doi.org/10.1021/nn301148e
49. Andrade, F., Rafael, D., Videira, M., Ferreira, D., Sosnik, A., & Sarmento, B. (2013). Nanotechnology and pulmonary delivery to overcome resistance in infectious diseases. *Advanced Drug Delivery Reviews*, *65*(13–14), 1816–1827. https://doi.org/10.1016/j.addr.2013.07.020
50. Mulder, W. J., Jaffer, F. A., Fayad, Z. A., & Nahrendorf, M. (2014). Imaging and nanomedicine in inflammatory atherosclerosis. *Science Translational Medicine*, *6*(239), 239sr1. https://doi.org/10.1126/scitranslmed.3005101
51. Wang, Y. Q., Wang, F., Deng, X. Q., Sheng, J., Chen, S. Y., & Su, J. (2013). Delivery of therapeutic AGT shRNA by PEG-Bu for hypertension therapy. *PloSOne*, *8*(7), e68651. https://doi.org/10.1371/journal.pone.0068651
52. Volpatti, L. R., Matranga, M. A., Cortinas, A. B., Delcassian, D., Daniel, K. B., Langer, R., & Anderson, D. G. (2020). Glucose-responsive nanoparticles for rapid and extended self-regulated insulin delivery. *ACS Nano*, *14*(1), 488–497. https://doi.org/10.1021/acsnano.9b06395
53. Shae, D., Becker, K. W., Christov, P., Yun, D. S., Lytton-Jean, A. K. R., Sevimli, S., Ascano, M., Kelley, M., Johnson, D. B., Balko, J. M., & Wilson, J. T. (2019). Endosomolyticpolymersomes increase the activity of cyclic dinucleotide STING agonists to enhance cancer immunotherapy. *Nature Nanotechnology*, *14*(3), 269–278. https://doi.org/10.1038/s41565-018-0342-5
54. He, C., Yue, H., Xu, L., Liu, Y., Song, Y., Tang, C., & Yin, C. (2020). siRNA release kinetics from polymeric nanoparticles correlate with RNAi efficiency and inflammation therapy via oral delivery. *Acta Biomaterialia, 103,* 213–222. www.ncbi.nlm.nih.gov/pubmed/31812844
55. De, A., Nayak, A. K., Kundu, A., Das, B., & Samanta, A. (2021). Gum arabic-based nanomaterials in drug delivery and biomedical applications. In H. Bera, C. M.Hossain, S. Saha(Eds.), *Biopolymer-Based Nanomaterials in Drug Delivery and Biomedical Applications* (pp. 165–182). Academic Press: An imprint of Elsevier.

56. Nayak, A., Bera, H., Hasnain, M., De, A., Pal, D., & Samanta, A. (2021). Gellan gum-based nanomaterials in drug delivery applications. In H. Bera, C. M. Hossain, S. Saha(Eds.), *Biopolymer-Based Nanomaterials in Drug Delivery and Biomedical Applications* (pp. 313–336). Academic Press: An imprint of Elsevier.
57. Kundu, A., Dey, P., Banerjee, S., Sarkar, P., De, A., Maity, A., & Kim, H. S. (2018). Heparin and its derivatives-based nanoparticles for cancer therapy. In T. K. Giri & B. Ghosh (Eds.), *Polysaccharide Based Nano-Biocarrier in Drug Delivery* (pp. 81–97). CRC Press.
58. Shrestha, R., Elsabahy, M., Luehmann, H., Samarajeewa, S., Florez-Malaver, S., Lee, N. S., Welch, M. J., Liu, Y., & Wooley, K. L. (2012). Hierarchically assembled theranostic nanostructures for siRNA delivery and imaging applications. *Journal of the American Chemical Society*, *134*(42), 17362–17365. https://doi.org/10.1021/ja306616n
59. Elsabahy, M., Shrestha, R., Clark, C., Taylor, S., Leonard, J., & Wooley, K. L. (2013). Multifunctional hierarchically assembled nanostructures as complex stage-wise dual-delivery systems for coincidental yet differential trafficking of siRNA and paclitaxel. *Nano Letters*, *13*(5), 2172–2181. https://doi.org/10.1021/nl4006645
60. Agyare, E. K., Jaruszewski, K. M., Curran, G. L., Rosenberg, J. T., Grant, S. C., Lowe, V. J., Ramakrishnan, S., Paravastu, A. K., Poduslo, J. F., & Kandimalla, K. K. (2014). Engineering theranosticnanovehicles capable of targeting cerebrovascular amyloid deposits. *Journal of Controlled Release: Official Journal of the Controlled Release Society*, *185*, 121–129. https://doi.org/10.1016/j.jconrel.2014.04.010
61. Guo, Y., Chen, W., Wang, W., Shen, J., Guo, R., Gong, F., Lin, S., Cheng, D., Chen, G., & Shuai, X. (2012). Simultaneous diagnosis and gene therapy of immuno-rejection in rat allogeneic heart transplantation model using a T-cell-targeted theranosticnanosystem. *ACS Nano*, *6*(12), 10646–10657. https://doi.org/10.1021/nn3037573
62. Vaughan, H. J., Zamboni, C. G., Hassan, L. F., Radant, N. P., Jacob, D., Mease, R. C., Minn, I., Tzeng, S. Y., Gabrielson, K. L., Bhardwaj, P., Guo, X., Francisco, D., Pomper, M. G., & Green, J. J. (2022). Polymeric nanoparticles for dual-targeted theranostic gene delivery to hepatocellular carcinoma. *Science Advances*, *8*(29), eabo6406. https://doi.org/10.1126/sciadv.abo6406
63. Li, S., Goins, B., Zhang, L., & Bao, A. (2012). Novel multifunctional theranostic liposome drug delivery system: Construction, characterization, and multimodality MR, near-infrared fluorescent, and nuclear imaging. *Bioconjugate Chemistry*, *23*(6), 1322–1332. https://doi.org/10.1021/bc300175d
64. Gonzalez Gomez, A., Syed, S., Marshall, K., & Hosseinidoust, Z. (2019). Liposomal nanovesicles for efficient encapsulation of staphylococcal antibiotics. *ACS Omega*, *4*(6), 10866–10876.
65. Janjic, J.M.,& Bai, M. (2014). Design and development of theranostic nanomedicines. In M. Y. Berezin (Ed.), *Nanotechnology for Biomedical Imaging and Diagnostics*. https://doi.org/10.1002/9781118873151.ch15
66. Bally, M., Bailey, K., Sugihara, K., Grieshaber, D., Vörös, J., & Städler, B. (2010). Liposome and lipid bilayer arrays towards biosensing applications. *Small (Weinheim an der Bergstrasse, Germany)*, *6*(22), 2481–2497. https://doi.org/10.1002/smll.201000644
67. Bally, M., & Vörös, J. (2009). Nanoscale labels: Nanoparticles and liposomes in the development of high-performance biosensors. *Nanomedicine (London, England)*, *4*(4), 447–467. https://doi.org/10.2217/nnm.09.16
68. Tenchov, R., Bird, R., Curtze, A. E., & Zhou, Q. (2021). Lipid Nanoparticles–from liposomes to mRNA vaccine delivery, a Landscape of Research Diversity and Advancement. *ACS Nano*, *15*(11), 16982–17015. https://doi.org/10.1021/acsnano.1c04996

69. Barenholz Y. (2012). Doxil®--the first FDA-approved nano-drug: lessons learned. *Journal of Controlled Release: Official Journal of the Controlled Release Society, 160*(2), 117–134. https://doi.org/10.1016/j.jconrel.2012.03.020
70. Lamb, Y. N., & Scott, L. J. (2017). Liposomal Irinotecan: A review in metastatic pancreatic adenocarcinoma. *Drugs, 77*(7), 785–792. https://doi.org/10.1007/s40265-017-0741-1
71. Petre, C. E., & Dittmer, D. P. (2007). Liposomal daunorubicin as treatment for Kaposi's sarcoma. *International Journal of Nanomedicine, 2*(3), 277–288.
72. Mayer, L. D., Tardi, P., & Louie, A. C. (2019). CPX-351: A nanoscale liposomal co-formulation of daunorubicin and cytarabine with unique biodistribution and tumor cell uptake properties. *International Journal of Nanomedicine, 14*, 3819–3830. https://doi.org/10.2147/IJN.S139450
73. Liu, Q., & Boyd, B. J. (2013). Liposomes in biosensors. *Analyst, 138*(2), 391–409.
74. Fatima, A., Ahmad, M. W., Al Saidi, A. K. A., Choudhury, A., Chang, Y., & Lee, G. H. (2021). Recent advances in Gadolinium based contrast agents for bioimaging applications. *Nanomaterials (Basel, Switzerland), 11*(9), 2449. https://doi.org/10.3390/nano11092449
75. Li, Y., Zhang, R., Xu, Z., & Wang, Z. (2022). Advances in nanoliposomes for the diagnosis and treatment of liver cancer. *International Journal of Nanomedicine, 17*, 909–925. https://doi.org/10.2147/IJN.S349426
76. Xiao, Y., Liu, Y., Yang, S., Zhang, B., Wang, T., Jiang, D., Zhang, J., Yu, D., & Zhang, N. (2016). Sorafenib and gadolinium co-loaded liposomes for drug delivery and MRI-guided HCC treatment. *Colloids and Surfaces. B, Biointerfaces, 141*, 83–92. https://doi.org/10.1016/j.colsurfb.2016.01.016
77. Xu, H. L., Yang, J. J., ZhuGe, D. L., Lin, M. T., Zhu, Q. Y., Jin, B. H., Tong, M. Q., Shen, B. X., Xiao, J., & Zhao, Y. Z. (2018). Glioma-targeted delivery of a Theranostic Liposome integrated with Quantum Dots, Superparamagnetic Iron Oxide, and Cilengitide for dual-imaging guiding cancer surgery. *Advanced Healthcare Materials, 7*(9), e1701130. https://doi.org/10.1002/adhm.201701130
78. Zheng, X. C., Ren, W., Zhang, S., Zhong, T., Duan, X. C., Yin, Y. F., Xu, M. Q., Hao, Y. L., Li, Z. T., Li, H., Liu, M., Li, Z. Y., & Zhang, X. (2018). The theranostic efficiency of tumor-specific, pH-responsive, peptide-modified, liposome-containing paclitaxel and superparamagnetic iron oxide nanoparticles. *International Journal of Nanomedicine, 13*, 1495–1504. https://doi.org/10.2147/IJN.S157082
79. Chang, Y. J., Chang, C. H., Yu, C. Y., Chang, T. J., Chen, L. C., Chen, M. H., Lee, T. W., & Ting, G. (2010). Therapeutic efficacy and microSPECT/CT imaging of 188Re-DXR-liposome in a C26 murine colon carcinoma solid tumor model. *Nuclear Medicine and Biology, 37*(1), 95–104. https://doi.org/10.1016/j.nucmedbio.2009.08.006
80. Abbasi, E., Aval, S. F., Akbarzadeh, A., Milani, M., Nasrabadi, H. T., Joo, S. W., Hanifehpour, Y., Nejati-Koshki, K., & Pashaei-Asl, R. (2014). Dendrimers: Synthesis, applications, and properties. *Nanoscale Research Letters, 9*(1), 247. https://doi.org/10.1186/1556-276X-9-247
81. Bober, Z., Bartusik-Aebisher, D., & Aebisher, D. (2022). Application of Dendrimers in Anticancer Diagnostics and Therapy. *Molecules (Basel, Switzerland)*, 27(10), 3237.https://doi.org/10.3390/molecules27103237
82. Longmire, M. R., Ogawa, M., Choyke, P. L., & Kobayashi, H. (2014). Dendrimers as high relaxivity MR contrast agents. *Wiley Interdisciplinary Reviews. Nanomedicine and Nanobiotechnology, 6*(2), 155–162. https://doi.org/10.1002/wnan.1250
83. Svenson, S., & Tomalia, D. A. (2012). Dendrimers in biomedical applications-reflections on the field. *Advanced Drug Delivery Reviews, 64*, 102–115. https://doi.org/10.1016/j.addr.2012.09.030

84. Wahsner, J., Gale, E. M., Rodríguez-Rodríguez, A., & Caravan, P. (2019). Chemistry of MRI contrast agents: Current challenges and new frontiers. *Chemical Reviews*, *119*(2), 957–1057. https://doi.org/10.1021/acs.chemrev.8b00363
85. Xiong, Z., Shen, M., & Shi, X. (2018). Dendrimer-based strategies for cancer therapy: Recent advances and future perspectives. *Science China Materials, 61,* 1387–1403. https://doi.org/10.1007/s40843-018-9271-4
86. Tekade, R. K., Kumar, P. V., & Jain, N. K. (2009). Dendrimers in oncology: An expanding horizon. *Chemical Reviews*, *109*(1), 49–87. https://doi.org/10.1021/cr068212n
87. Larsen, A. K., Escargueil, A. E., & Skladanowski, A. (2000). Resistance mechanisms associated with altered intracellular distribution of anticancer agents. *Pharmacology & Therapeutics*, *85*(3), 217–229. https://doi.org/10.1016/s0163-7258(99)00073-x
88. Hu, J., Hu, K., & Cheng, Y. (2016). Tailoring the dendrimer core for efficient gene delivery. *Acta Biomaterialia*, *35*, 1–11. https://doi.org/10.1016/j.actbio.2016.02.031
89. Thakur, S., Tekade, R. K., Kesharwani, P. & Jain, N. K. (2013). The effect of polyethylene glycol spacer chain length on the tumor-targeting potential of folate-modified PPI dendrimers. *Journal of Nanoparticle Research, 15,* 1625. https://doi.org/10.1007/s11051-013-1625-2
90. Patel, S. K., Gajbhiye, V., & Jain, N. K. (2012). Synthesis, characterization and brain targeting potential of paclitaxel loaded thiamine-PPI nanoconjugates. *Journal of Drug Targeting*, *20*(10), 841–849. https://doi.org/10.3109/1061186X.2012.719231
91. Somani, S., Blatchford, D. R., Millington, O., Stevenson, M. L., & Dufès, C. (2014). Transferrin-bearing polypropylenimine dendrimer for targeted gene delivery to the brain. *Journal of Controlled Release: Official Journal of The Controlled Release Society*, *188*, 78–86. https://doi.org/10.1016/j.jconrel.2014.06.006
92. Studzian, M., Szulc, A., Janaszewska, A., Appelhans, D., Pułaski, Ł., & Klajnert-Maculewicz, B. (2017). Mechanisms of internalization of Maltose-Modified Poly(propyleneimine) Glycodendrimers into Leukemic Cell Lines. *Biomacromolecules*, *18*(5), 1509–1520. https://doi.org/10.1021/acs.biomac.7b00046
93. Franiak-Pietryga, I., Ziemba, B., Sikorska, H., Jander, M., Kuncman, W., Danilewicz, M., Appelhans, D., Lewkowicz, P., Ostrowska, K., Bryszewska, M., & Borowiec, M. (2020). Maltotriose-modified poly(propylene imine) Glycodendrimers as a potential novel platform in the treatment of chronic lymphocytic Leukemia. A proof-of-concept pilot study in the animal model of CLL. *Toxicology and Applied Pharmacology*, *403*, 115139. https://doi.org/10.1016/j.taap.2020.115139
94. Cai, X., Hu, J., Xiao, J., & Cheng, Y. (2013). Dendrimer and cancer: A patent review (2006-present). *Expert Opinion on Therapeutic Patents*, *23*(4), 515–529. https://doi.org/10.1517/13543776.2013.761207
95. Zhao, Y., Liu, S., Li, Y., Jiang, W., Chang, Y., Pan, S., Fang, X., Wang, Y. A., & Wang, J. (2010). Synthesis and grafting of folate-PEG-PAMAM conjugates onto quantum dots for selective targeting of folate-receptor-positive tumor cells. *Journal of Colloid and Interface Science*, *350*(1), 44–50. https://doi.org/10.1016/j.jcis.2010.05.035
96. Palmerston Mendes, L., Pan, J., & Torchilin, V. P. (2017). Dendrimers as nanocarriers for nucleic acid and drug delivery in cancer therapy. *Molecules (Basel, Switzerland)*, *22*(9), 1401.https://doi.org/10.3390/molecules22091401
97. Çevik, E., Bahar, Ö., Şenel, M., & Abasıyanık, M. F. (2016). Construction of novel electrochemical immunosensor for detection of prostate specific antigen using ferrocene-PAMAM dendrimers. *Biosensors & Bioelectronics*, *86*, 1074–1079. https://doi.org/10.1016/j.bios.2016.07.064
98. Gul, S., Khan, S. B., Rehman, I. U., Khan, M. A, Khan, M. I. (2019). A comprehensive review of magnetic nanomaterials modern day theranostics. *Frontiers in Materials, 6*, https://doi.org/10.3389/fmats.2019.00179

99. Stephen, Z. R., Kievit, F. M., & Zhang, M. (2011). Magnetite nanoparticles for medical MR imaging. *Materials Today (Kidlington, England)*, *14*(7–8), 330–338. https://doi.org/10.1016/S1369-7021(11)70163-8
100. Estelrich, J., Sánchez-Martín, M. J., & Busquets, M. A. (2015). Nanoparticles in magnetic resonance imaging: From simple to dual contrast agents. *International Journal of Nanomedicine*, *10*, 1727–1741. https://doi.org/10.2147/IJN.S76501
101. Johannsen, M., Gneveckow, U., Thiesen, B., Taymoorian, K., Cho, C. H., Waldöfner, N., Scholz, R., Jordan, A., Loening, S. A., & Wust, P. (2007). Thermotherapy of prostate cancer using magnetic nanoparticles: Feasibility, imaging, and three-dimensional temperature distribution. *European Urology*, *52*(6), 1653–1661. https://doi.org/10.1016/j.eururo.2006.11.023
102. Ito, A., Tanaka, K., Honda, H., Abe, S., Yamaguchi, H., & Kobayashi, T. (2003). Complete regression of mouse mammary carcinoma with a size greater than 15 mm by frequent repeated hyperthermia using magnetite nanoparticles. *Journal of Bioscience and Bioengineering*, *96*(4), 364–369. https://doi.org/10.1016/S1389-1723(03)90138-1
103. Veyret, R., Delair, T., & Elaissari, A. (2005). Preparation and biomedical application of layer-by-layer encapsulated oil in water magnetic emulsion. *Journal of Magnetism and Magnetic Materials, 293*(1), 171–176.
104. Lima-Tenório, M. K., Pineda, E. A., Ahmad, N. M., Fessi, H., & Elaissari, A. (2015). Magnetic nanoparticles: In vivo cancer diagnosis and therapy. *International Journal of Pharmaceutics*, *493*(1–2), 313–327. https://doi.org/10.1016/j.ijpharm.2015.07.059
105. Vega-Chacón, J., Tarhini, M., Lebaz, N., Jafelicci, M., Zine, N., Errachid, A., & Elaissari, A. (2020). Protein-silica hybrid submicron particles: Preparation and characterization. *Chemistry Africa*, *3*, 793–801.
106. German, S. V., Navolokin, N. A., Kuznetsova, N. R., Zuev, V. V., Inozemtseva, O. A., Anis'kov, A. A., Volkova, E. K., Bucharskaya, A. B., Maslyakova, G. N., Fakhrullin, R. F., Terentyuk, G. S., Vodovozova, E. L., & Gorin, D. A. (2015). Liposomes loaded with hydrophilic magnetite nanoparticles: Preparation and application as contrast agents for magnetic resonance imaging. *Colloids and Surfaces. B, Biointerfaces*, *135*, 109–115. https://doi.org/10.1016/j.colsurfb.2015.07.042
107. Moradi Khaniabadi, P., Shahbazi-Gahrouei, D., Malik Shah Abdul Majid, A., Suhaimi Jaafar, M., Moradi Khaniabadi, B., & Shahbazi-Gahrouei, S. (2017). In vitro study of SPIONs-C595 as molecular imaging probe for specific breast cancer (MCF-7) cells detection. *Iranian Biomedical Journal*, *21*(6), 360–368. https://doi.org/10.18869/acadpub.ibj.21.6.360
108. Perlman, O., & Azhari, H. (2017). Ultrasonic computed tomography imaging of iron oxide nanoparticles. *Physics in Medicine and Biology*, *62*(3), 825–842. https://doi.org/10.1088/1361-6560/aa51ab
109. Feng, W., Zhou, X., Nie, W., Chen, L., Qiu, K., Zhang, Y., & He, C. (2015). Au/polypyrrole@Fe3O4 nanocomposites for MR/CT dual-modal imaging guided-photothermal therapy: An in vitro study. *ACS Applied Materials & Interfaces*, *7*(7), 4354–4367. https://doi.org/10.1021/am508837v
110. Das, P., Colombo, M., & Prosperi, D. (2019). Recent advances in magnetic fluid hyperthermia for cancer therapy. *Colloids and Surfaces. B, Biointerfaces*, *174*, 42–55. https://doi.org/10.1016/j.colsurfb.2018.10.051
111. Belyanina, I., Kolovskaya, O., Zamay, S., Gargaun, A., Zamay, T., & Kichkailo, A. (2017). Targeted magnetic nanotheranostics of cancer. *Molecules (Basel, Switzerland)*, *22*(6), 975. https://doi.org/10.3390/molecules22060975
112. Legge, C. J., Colley, H. E., Lawson, M. A., & Rawlings, A. E. (2019). Targeted magnetic nanoparticle hyperthermia for the treatment of oral cancer. *Journal of Oral*

Pathology & Medicine: Official Publication of the International Association of Oral Pathologists and the American Academy of Oral Pathology, *48*(9), 803–809. https://doi.org/10.1111/jop.12921

113. Taratula, O., Dani, R. K., Schumann, C., Xu, H., Wang, A., Song, H., Dhagat, P., & Taratula, O. (2013). Multifunctional nanomedicine platform for concurrent delivery of chemotherapeutic drugs and mild hyperthermia to ovarian cancer cells. *International Journal of Pharmaceutics*, *458*(1), 169–180. https://doi.org/10.1016/j.ijpharm.2013.09.032
114. Xue, W., Liu, X. L., Ma, H., Xie, W., Huang, S., Wen, H., Jing, G., Zhao, L., Liang, X.J., & Fan, H. M. (2018). AMF responsive DOX-loaded magnetic microspheres: Transmembrane drug release mechanism and multimodality postsurgical treatment of breast cancer. *Journal of Materials Chemistry B, 6*(15), 2289–2303.
115. Cotta, M. A. (2020). Quantum dots and their applications: What lies ahead? *ACS Applied Nano Materials, 3,* 4920–4924.
116. Manna, S., Banerjee, S., De, A., Banerjee, S., Das, S., Rakshit, P., Kumar, S. K. A., Sen, K. K., & De, S. (2024). Therapeutic and diagnostic implications of carbon dot: An advancement in the avenue towards cancer, diabetes and neurodegenerative disorders. *Pharmaceutical Nanotechnology*, 10.2174/0122117385314533240824090949. Advance online publication. https://doi.org/10.2174/0122117385314533240824090949
117. Park, Y., Ryu, Y. M., Jung, Y., Wang, T., Baek, Y., Yoon, Y., Bae, S. M., Park, J., Hwang, S., Kim, J., Do, E. J., Kim, S. Y., Chung, E., Kim, K. H., Kim, S., & Myung, S. J. (2014). Spraying quantum dot conjugates in the colon of live animals enabled rapid and multiplex cancer diagnosis using endoscopy. *ACS Nano*, *8*(9), 8896–8910. https://doi.org/10.1021/nn5009269
118. Ming, K., Kim, J., Biondi, M. J., Syed, A., Chen, K., Lam, A., Ostrowski, M., Rebbapragada, A., Feld, J. J., & Chan, W. C. (2015). Integrated quantum dot barcode smartphone optical device for wireless multiplexed diagnosis of infected patients. *ACS Nano*, *9*(3), 3060–3074. https://doi.org/10.1021/nn5072792
119. Kim, J., Biondi, M. J., Feld, J. J., & Chan, W. C. (2016). Clinical validation of quantum dot barcode diagnostic technology. *ACS Nano*, *10*(4), 4742–4753. https://doi.org/10.1021/acsnano.6b01254
120. Hu, O., Li, Z., Wu, J., Tan, Y., Chen, Z., & Tong, Y. (2023). A multicomponent nucleic acid Enzyme-Cleavable quantum dot Nanobeacon for highly sensitive diagnosis of tuberculosis with the naked eye. *ACS Sensors*, *8*(1), 254–262. https://doi.org/10.1021/acssensors.2c02114
121. Bakalova, R., Ohba, H., Zhelev, Z., Nagase, T., Jose, R., Ishikawa, M., & Baba, Y. (2004). Quantum dot anti-CD conjugates: are they potential photosensitizers or potentiators of classical photosensitizing agents in photodynamic therapy of cancer? *Nano Letters, 4*(9), 1567–1573.
122. Overchuk, M., Weersink, R. A., Wilson, B. C., & Zheng, G. (2023). Photodynamic and photothermal therapies: Synergy opportunities for nanomedicine. *ACS Nano*, *17*(9), 7979–8003. https://doi.org/10.1021/acsnano.3c00891
123. Bao, Y. W., Hua, X. W., Li, Y. H., Jia, H. R., & Wu, F. G. (2018). Hyperthemia-promoted cytosolic and nuclear delivery of copper/carbon quantum dot-crosslinked nanosheets: Multimodal imaging-guided photothermal cancer therapy. *ACS Applied Materials & Interfaces*, *10*(2), 1544–1555.
124. Ghosal, K., Ghosh, S., Ghosh, D., & Sarkar, K. (2020). Natural polysaccharide derived carbon dot based in situ facile green synthesis of silver nanoparticles: Synergistic

effect on breast cancer. *International Journal of Biological Macromolecules, 162*, 1605–1615.

125. Wang, Y., Wang, Y., Chen, G., Li, Y., Xu, W., & Gong, S. (2017). Quantum-Dot-Based Theranostic Micelles conjugated with an Anti-EGFR nanobody for Triple-Negative Breast Cancer Therapy. *ACS Applied Materials & Interfaces*, *9*(36), 30297–30305. https://doi.org/10.1021/acsami.7b05654
126. Pei, M., Pai, J. Y., Du, P., & Liu, P. (2018). Facile synthesis of fluorescent Hyper-Cross-Linked β-Cyclodextrin-Carbon quantum dot hybrid nanosponges for tumor theranosticapplication with enhanced antitumor efficacy. *Molecular Pharmaceutics*, *15*(9), 4084–4091. https://doi.org/10.1021/acs.molpharmaceut.8b00508
127. Sarkar, P., Ghosh, S., & Sarkar, K. (2021). Folic acid based carbon dot functionalized stearic acid-g-polyethyleneimine amphiphilic nanomicelle: Targeted drug delivery and imaging for triple negative breast cancer. *Colloids and Surfaces. B, Biointerfaces*, *197*, 111382. https://doi.org/10.1016/j.colsurfb.2020.111382
128. Ghosh, S., Ghosal, K., Mohammad, S. A., & Sarkar, K. (2019). Dendrimer functionalized carbon quantum dot for selective detection of breast cancer and gene therapy. *Chemical Engineering Journal*, *373*, 468–484.
129. Anzar, N., Hasan, R., Tyagi, M., Yadav, N., & Narang, J. (2020). Carbon nanotube-A review on synthesis, properties and plethora of applications in the field of biomedical science. *Sensors International*, *1*, 100003.
130. Claussen, J. C., Franklin, A. D., Ul Haque, A., Porterfield, D. M., & Fisher, T. S. (2009). Electrochemical biosensor of nanocube-augmented carbon nanotube networks. *ACS Nano*, *3*(1), 37–44. https://doi.org/10.1021/nn800682m
131. Zhang, P., Huang, H., Huang, J., Chen, H., Wang, J., Qiu, K., Zhao, D., Ji, L., & Chao, H. (2015). Noncovalent Ruthenium(II) Complexes-Single-Walled carbon nanotube composites for bimodal photothermal and photodynamic therapy with near-infrared irradiation. *ACS Applied Materials & Interfaces*, *7*(41), 23278–23290. https://doi.org/10.1021/acsami.5b07510
132. Zhao, Y., Zhao, T., Cao, Y., Sun, J., Zhou, Q., Chen, H., Guo, S., Wang, Y., Zhen, Y., Liang, X. J., & Zhang, S. (2021). Temperature-sensitive lipid-coated carbon nanotubes for synergistic photothermal therapy and gene therapy. *ACS Nano, 15*(4), 6517–6529.
133. Madani, S. Z. M., Safaee, M. M., Gravely, M., Silva, C., Kennedy, S., Bothun, G. D., & Roxbury, D. (2021). Carbon nanotube–liposome complexes in hydrogels for controlled drug delivery via near-infrared laser stimulation. *ACS Applied Nano Materials, 4*(1), 331–342.
134. Minati, L., Antonini, V., Dalla Serra, M., & Speranza, G. (2012). Multifunctional branched gold-carbon nanotube hybrid for cell imaging and drug delivery. *Langmuir: The ACS Journal of Surfaces and Colloids*, *28*(45), 15900–15906. https://doi.org/10.1021/la303298u
135. Das, M., Datir, S. R., Singh, R. P., & Jain, S. (2013). Augmented anticancer activity of a targeted, intracellularly activatable, theranostic nanomedicine based on fluorescent and radiolabeled, methotrexate-folic Acid-multiwalled carbon nanotube conjugate. *Molecular Pharmaceutics*, *10*(7), 2543–2557. https://doi.org/10.1021/mp300701e
136. Ghosal, K., Mondal, P., Bera, S., & Ghosh, S. (2021). Graphene family nanomaterials-opportunities and challenges in tissue engineering applications. *FlatChem*, *30*, 100315.
137. Ghosh, S., & Chatterjee, K. (2020). Poly(Ethylene Glycol) functionalized Graphene Oxide in tissue engineering: A review on recent advances. *International Journal of Nanomedicine*, *15*, 5991–6006. https://doi.org/10.2147/IJN.S249717

138. Kuo, C. J., Chiang, H. C., Tseng, C. A., Chang, C. F., Ulaganathan, R. K., Ling, T. T., Chang, Y. J., Chen, C. C., Chen, Y. R., & Chen, Y. T. (2018). Lipid-Modified Graphene-Transistor biosensor for monitoring Amyloid-βaggregation. *ACS Applied Materials & Interfaces*, *10*(15), 12311–12316. https://doi.org/10.1021/acsami.8b01917
139. Huang, J., Yan, H., Ma, M., Tian, N., Ni, H., Xia, M., Pei, K., & Ye, P. (2022). Hydroxylated Graphene Porous membrane-based biosensor for exosome isolation and detection. *ACS Applied Nano Materials, 5*(5), 6115–6124.
140. Kalkal, A., Pradhan, R., Kadian, S., Manik, G., & Packirisamy, G. (2020). Biofunctionalized Graphene Quantum Dots based fluorescent biosensor toward efficient detection of small cell lung cancer. *ACS Applied Biomaterials*, *3*(8), 4922–4932. https://doi.org/10.1021/acsabm.0c00427
141. Liu, J., Wang, X., Wang, T., Li, D., Xi, F., Wang, J., & Wang, E. (2014). Functionalization of monolithic and porous three-dimensional graphene by one-step chitosan electrodeposition for enzymatic biosensor. *ACS Applied Materials & Interfaces*, *6*(22), 19997–20002. https://doi.org/10.1021/am505547f
142. Ran, X., Du, Y., Wang, Z., Wang, H., Pu, F., Ren, J., & Qu, X. (2017). Hyaluronic acid-templated Ag Nanoparticles/Graphene Oxide composites for synergistic therapy of bacteria infection. *ACS Applied Materials & Interfaces*, *9*(23), 19717–19724. https://doi.org/10.1021/acsami.7b05584
143. Wu, M. C., Deokar, A. R., Liao, J. H., Shih, P. Y., & Ling, Y. C. (2013). Graphene-based photothermal agent for rapid and effective killing of bacteria. *ACS Nano*, *7*(2), 1281–1290. https://doi.org/10.1021/nn304782d
144. Cheon, Y. A., Bae, J. H., & Chung, B. G. (2016). Reduced Graphene Oxide Nanosheet for Chemo-photothermal Therapy. *Langmuir: The ACS Journal of Surfaces and Colloids*, *32*(11), 2731–2736. https://doi.org/10.1021/acs.langmuir.6b00315
145. Hao, L., Song, H., Zhan, Z., & Lv, Y. (2020). Multifunctional reduced Graphene Oxide-Based nanoplatform for synergistic targeted Chemo-Photothermal Therapy. *ACS Applied Bio Materials*, *3*(8), 5213–5222. https://doi.org/10.1021/acsabm.0c00614
146. Prasad, R., Yadav, A. S., Gorain, M., Chauhan, D. S., Kundu, G. C., Srivastava, R., & Selvaraj, K. (2019). Graphene Oxide supported Liposomes as Red Emissive Theranostics for Phototriggeredtissue visualization and tumor regression. *ACS Applied Bio Materials*, *2*(8), 3312–3320. https://doi.org/10.1021/acsabm.9b00335
147. Pan, J., Yang, Y., Fang, W., Liu, W., Le, K., Xu, D., & Li, X. (2018). Fluorescent phthalocyanine–graphene conjugate with enhanced NIR absorbance for imaging and multi-modality therapy. *ACS Applied Nano Materials, 1*(6), 2785–2795.

4 Binary/Ternary, or beyond Oxide-Based Sensors for Disease Diagnosis

Suman Kumar and Basanta Bhowmik

4.1 INTRODUCTION

An individualized approach has been more popular in recent years in both health care and medicine. Providing reasonable, low-cost, noninvasive, improved, and earlier detection of illnesses and various diseases, including slow and invasive blood tests for diagnosing diseases, is a challenging task.[1–3] To lower mortality and medical expenses, it is crucial to monitor physical conditions and diagnose diseases early. Therefore, in order to accomplish real-time and on-site diagnostics, there is a lot of focus on developing portable and affordable diagnostic platforms. It is necessary to identify certain biomarkers with high sensitivity and selectivity that are connected to a given physical condition in order to create a viable diagnostic platform. The blood typically contains these indicators in rather high concentrations. However, because it is an invasive diagnostic technique, blood analysis inevitably causes pain. An additional, noninvasive, and painless method is to identify biomarkers that are produced by the metabolism of breath, perspiration, and urine as well as by blood.[4] An appealing option with real-time, quick, and reliable disease diagnosis is gas sensors made exclusively for human breath analysis.[1] The use of human exhaled breath for disease diagnosis provides a quick, low-cost, noninvasive, and affordable method of diagnosis.[2] Exhaled human breath, which is a combination of CO_2, nitrogen, water, oxygen and other gases such as NO_2, H_2S, acetone, NH_3 and different VOCs, are the key biomarker gases to diagnose several diseases as cancer, diseases due to halitosis, asthma, diabetes, and so on.[1,2,5] A specific disease may be indicated by variations in these gases' concentrations. Diagnosing diabetes, halitosis, kidney dysfunction, asthma, and lung cancer can be done using variations in the concentration of various acetone, H_2S, NH_3, NO, and toluene species respectively.[4] Given that breath analysis is completely painless and hence acceptable to individuals, it is a very appealing diagnostic technique for this reason.

4.2 BINARY/TERNARY NANOCOMPOSITE THROUGH SURFACE FUNCTIONALIZATION AND DOPING

For an understanding of an alloy's physicochemical qualities, its structure and surface features are essential. Since one alloy component may have surface segregated, these

 DOI: 10.1201/9781003453635-4

properties frequently deviate greatly from their bulk equivalents. Because of their versatile chemical structures and the interplay of the individual metals, bimetallic and trimetallic alloys demonstrate higher catalytic activity than their monometallic counterparts.

Nanocomposite are substances that mix nanoparticles with metals/semiconductors, to produce a structure with unique physical and chemical characteristics. Particularly in the context of biomolecular and cellular systems, these characteristics are frequently quite beneficial for a variety of medicinal and diagnostic applications.[6] Size compatibility, enhanced surface area (high surface-to-volume ratio), improved sensitivity and selectivity, recognition of specific cell type and tissues, imaging and diagnostics, and so forth are some features of nanocomposites that make them suitable for a wide range of therapeutic and diagnostic applications in health care and biotechnology. Due to their size compatibility, nanocomposites can interact with biomolecules. Additionally, due to their larger surface area, they also permit effective interactions with biological targets. By adjusting the type, size, and concentration of nanoparticles in nanocomposites, it is possible to tune the nanocomposite properties. These features are achieved through surface functionalization and doping with binary or ternary composites. Selective bio-sensing can be achieved by tuning the surface functionalization.[6] Doping is a popular technique to improve bio-sensing capabilities and electronic properties. Doping is the process of adding metal or non-metal ions to semiconductors' crystalline structure. This process can affect charge carrier production, mobility, and recombination, which in turn affects the ability of the carriers to sense gas.[7] Due to doping with binary or ternary elements, a higher surface area and higher concentration of charge carriers are achieved. They also help to lower the required operating temperature for the sensors while simultaneously improving stability, selectivity, response time, and recovery time due to their catalytic properties. Su et al.[8] modified the SnO_2 composite by adding RGO and Pd by the one-pot route, and they observed that sensing performance was much better compared to bared SnO_2. Even at a low concentration of 5 ppm, they observed high sensitivity, good linearity, and good reversibility. Zhang et al.[9] prepared the Na-doped ZnO and RGO and evaluated their sensing performance for lung cancer surveillance diagnosis, which was synthesized by using the simple solution route technique. Compared to pure ZnO, the Na-doped ZnO has a higher surface-to-volume ratio. Notably, the surface-to-volume ratio of Na-doped ZnO is increased with the inclusion of graphene. When compared to pristine ZnO, Na-doped ZnO nanostructures offer more active sites for chemical adsorption, which causes a greater change in sensor resistance and consequently sensor response. Zhou et al.[10] prepared SnO_2 nanomaterials doping with Ni and Zn to detect carbon monoxide gas and studied the change in sensing parameters. Variations in sensitivity are seen in the gas response of pure SnO_2, Zn-doped, and Ni-doped materials to 100 μL/L of CO gas at 280°C. When compared to the response of pure SnO_2, which is 7.29, Zn-doped and Ni-doped materials show greater sensitivity, with values of 17.53 and 25.08, respectively, which is clearly shown in Figure 4.1. Reduced crystal size, more surface area, and the creation of n-p junctions, which improve the doped materials' gas-sensing capabilities, are some of the causes of this enhanced sensitivity. Lowering the operating temperature is an additional benefit of the doping procedure. The working temperature of pure SnO_2 is 310°C, but it

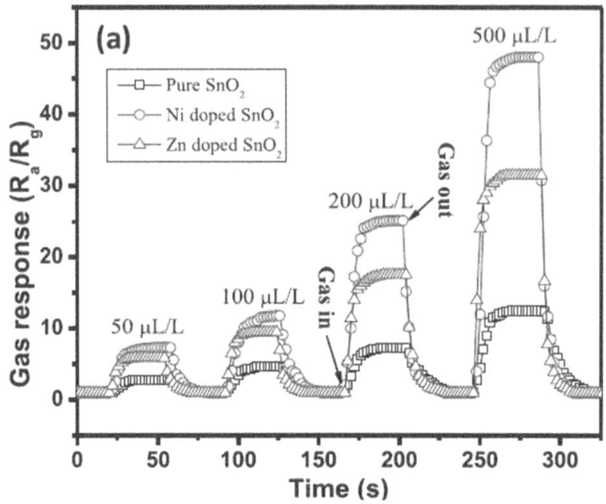

FIGURE 4.1 Gas responses for various concentrations of CO measured for gas sensors based on pure SnO_2, Ni-doped SnO_2, and Zn-doped SnO_2 nanomaterials. (Reproduced with permission[10]).

is lowered to 280°C for Zn- and Ni-doped materials. Compared to pure SnO_2, the response time and recovery time are improved for Ni-doped SnO_2, decreasing from 16 seconds and 28 seconds to 15 seconds and 16 seconds, respectively. And for Zn-doped, the response time and recovery time reduced to 11 s and 16 s respectively.

Wang et al.[11] fabricated a gas sensor by hydrothermal synthesis, which was used to create both undoped and tungsten (*W*)-doped hierarchical hollow spheres of NiO. The sensor responded (R_g/R_a) to 100 ppm acetone at 250°C at 4.0 atomic percent (at %) W-doping, whereas the undoped sensor's response was 139 times lower. The doped sensor's enhanced response can be ascribed to multiple factors, such as the W-doping-induced reduction in crystalline size, the larger surface area of 217.2 $m^2\ g^{-1}$, and the increased concentration of hole carriers that result in the introduction of additional holes within the NiO lattice.[11] Co-doped ZnO nanofibers (NFs) were synthesized by electrospinning and annealing. The 0.5 weight percent Co-doped ZnO NFs demonstrated a response (R_a/R_g) of roughly 16 to 100 ppm acetone at the ideal sensing temperature of 360°C, while the undoped sensor only displayed a response of 4.4. The response and recovery times of the Co-doped sensor were 4 and 6 seconds, respectively. The NFs' huge surface area and the numerous NF-NF junctions inside the netlike structure were blamed for the high acetone reaction. Furthermore, in comparison to the pristine ZnO NFs, the Co-doped sensor showed a reduced NF diameter. Acetone oxidation was made easier by Co_3O_4's catalytic action, which enhanced the sensor signal.[12]

4.3 ANALYSIS AND TECHNIQUES FOR COMPOSITE MATERIAL CHARACTERIZATIONS

There are several synthesis techniques for the fabrication of binary or ternary composite-based gas sensors for disease diagnosis like hydrothermal, atomic layer deposition, chemical vapor deposition, electrochemical anodization, and so forth. After fabrication of a gas sensor, the sensing of different gases is totally influenced by the chemical states of surface elements and surface activity of the sensing material. In this regard several characterization techniques are available to analyze their structural, morphological, compositional, optical, and electrical properties on these materials, both before and after exposure to gases. FESEM, TEM, scanning TEM (STEM), EDS, and x-ray photoelectron spectroscopy (XPS) are a few of the several characterization techniques that were used to analyze the binary or ternary composite-sensing layer.[13,14] A field emission scanning electron microscopy (FESEM) is an electron microscope that offers high-resolution surface imaging of a material. It works by focusing an electron beam using a field emission electron source, and then scanning the beam across the specimen's surface.

Transmission electron microscopy (TEM) offers high-resolution images that disclose detailed information regarding the size, shape, and dispersion of nanoparticles within sensor materials. This technique functions by transmitting an electron beam through a thin sample, generating images that reveal nanoscale particulars crucial for comprehending nanoparticle morphology. TEM plays a vital role in confirming the uniformity and structure of nanoparticles in gas sensors, directly influencing their sensitivity and performance. Furthermore, TEM enables investigation into the crystalline structure and composition of nanoparticles, providing insights into their chemical and physical properties pertinent to gas-sensing applications. There are four possible outcomes from an electron beam's interaction with a sample: transmission, scattering, backscattered, or diffractive.

X-ray photoelectron spectroscopy (XPS) is a surface-sensitive technique for determining the elemental composition, chemical states, and electronic states of materials. It involves irradiating the sample with X-rays and measuring the kinetic energy of the emitted electrons to generate spectra.

High-resolution transmission electron microscopy (HRTEM) is a useful tool for electron diffraction analysis and atomic-resolution characterization of sample crystallinity. The energy-dispersive X-ray spectroscopy (EDS) technique is employed for elemental evaluation or chemical characterization of a sample. It operates based on the interaction between X-rays and the material being analyzed. This method was used to determine the constituent elements. This was necessary in order to describe precisely the possible mechanism of interaction between the analyte and the sensing layer.

Kim et al.[4] fabricated highly porous WO_3 nanofibers (NFs) embedded with apoferritin-encapsulated Pt nanoparticles via the electrospinning technique. They also revealed numerous mesopores within the NFs' interior by the scanning transmission electron microscopy (STEM) technique. Further, they confirmed the extremely porous structure of the Apo-Pt WO_3 NFs by cross-sectional transmission electron microscopy (TEM) analysis. Measuring 23.3 ± 14.5 nm along the NFs, the distinct mesopores were ascribed to the protein shells breaking down during the calcination procedure. Zhang et al.[9] developed a gas sensor for lung cancer diagnosis based on a Na-doped ZnO/reduced graphene oxide (RGO) heterojunction using a simple solution route. They utilized several characterization techniques including X-ray photoelectron spectroscopy (XPS), scanning electron microscopy (SEM), and X-ray diffraction (XRD) to investigate the structure and chemical properties. XRD patterns were obtained at $\lambda = 1.54056$ Å, and XPS analysis revealed a peak at 284.6 eV. Surface morphology was examined using field-emission scanning electron microscopy (FESEM) at 10 kV.

Asal et al.[20] created a zinc oxide/tin dioxide nanocomposite for diabetes diagnosis using a hydrothermal method. Their XRD analysis identified characteristic peaks of the hexagonal wurtzite structure for ZnO and the tetragonal rutile structure for SnO_2 within the 2θ range of 15° to 80°. FESEM imaging revealed that the sample had a nanoplate morphology with irregular shapes and an average width of 30 to 50 nm. EDS spectrum analysis confirmed the presence of Zn, Sn, and O elements, with no impurities detected.

4.4 CARRIER TRANSPORT MECHANISMS

Depending on the type of sensing material, gas sensors utilize semiconductor materials that cause various phenomena. An electron depletion layer forms in n-type semiconductors, while a hole accumulation layer forms in p-type semiconductors. In n-type materials, the primary charge carriers are electrons; in p-type materials, the primary charge carriers are holes. As a result of adsorbed oxygen molecules, the resistance is higher in n-type materials and lower in p-type materials as compared to vacuum. Trapped electrons in n-type materials are released back into the sensing material when a gas sensor is exposed to a reducing gas, which lowers resistance. However, exposure to reducing gases causes additional holes to develop in p-type materials, which increases resistance. Once the exposure ends, the resistance in both types of materials typically returns to its initial value. On the other hand, because there are more trapped electrons in n-type materials, exposure to an oxidizing atmosphere causes an increase in resistance. As more holes grow in p-type materials, exposure to oxidizing agents reduces resistance, which is shown in Figure 4.2. Similar to

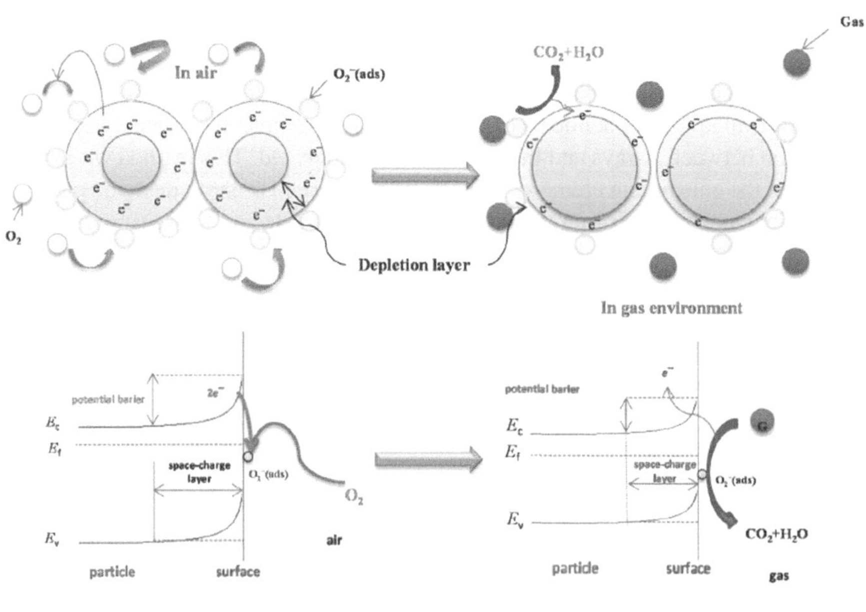

FIGURE 4.2 Pictorial representation showing electron transport mechanism of n-type semiconducting in reducing gas.[18]

reducing gases, once the exposure to the oxidizing gas ceases, the resistance generally returns to its original level. The target gases can be detected because of the variation in resistance brought about by these interactions. The measurement of the sensor response is commonly expressed as the inverse ratio (R_g/R_a) for p-type materials, and the ratio of resistance in the target gas (R_g) to resistance in air (R_a) for n-type materials. This value shows the gas sensor's selectivity and sensitivity to particular gases.[15,16] The types of adsorbed oxygen ions present on a sensitive material's surface are temperature-dependent. Specifically, when the temperature T is less than 100°C, the adsorbed oxygen ions primarily exist as O_2^-. At temperatures ranging from 100°C to 300°C, the dominant form of adsorbed oxygen ions shifts to O^-. For temperatures exceeding 300°C, the prevalent adsorbed oxygen ions are O_2^-, which is shown in Figure 4.4. Adsorbed oxygen ions play a crucial role by forming an electron depletion layer on the surface of the sensitive material. This layer increases the potential barrier at the grain boundaries and throughout the material, thereby elevating the material's resistance. This phenomenon is pivotal in the operation of gas sensors, influencing their sensitivity and response characteristics to different gases based on changes in electrical resistance.[17]

4.5 BINARY/TERNARY COMPOSITE-BASED SENSORS USED FOR DISEASE DIAGNOSIS

Diagnosing human diseases by analyzing the concentration of volatile organic compounds (VOCs), or specific gases in exhaled breath using sensor systems offers several advantages over traditional pathological procedures. These advantages include noninvasive diagnosis, portability and home care monitoring, cost-effectiveness, and safety. Research has shown that numerous VOCs can be detected in breath samples, with certain VOCs becoming more prominent in breath, or even in urine due to specific illnesses. This sensitivity makes VOC analysis a promising avenue for early and accurate disease detection. Research has shown that numerous VOCs can be detected in breath samples, with certain VOCs becoming more prominent in breath, or even in urine due to specific illnesses. This sensitivity makes VOC analysis a promising avenue for early and accurate disease detection. Many researchers have started preparing gas sensors based on binary and ternary composites, which are used to diagnose various diseases.

4.5.1 Diabetes

Diabetes is a condition characterized by the improper regulation of blood sugar levels in humans. It arises from a combination of genetic predisposition and environmental factors such as population growth, aging, urbanization, obesity, and a lack of physical activity. These factors contribute significantly to the prevalence of diabetes worldwide. Complications associated with diabetes include damage to blood vessels, which increases the risk of heart attacks and strokes. Additionally, diabetes can lead to complications affecting the kidneys, eyes, feet, and nerves. According to the World Health Organization (WHO), approximately 1.6 million deaths were attributed to diabetes globally in 2016. The prevalence of diabetes is projected to increase to 10.2% (578 million people) by 2030 and to 10.9% (around 700 million people) by 2045, compared to 9.3% (463 million people) in 2019. This upward trend underscores the growing impact of diabetes as a global health challenge.[13] Diabetes is typically diagnosed invasively by drawing blood samples, yet this approach is costly, uncomfortable, and problematic. The use of breath analysis is one noninvasive alternative method for the detection of diabetes. By examining volatile organic compounds in breath samples, breath analysis is utilized to gather data. The breath analysis tool known as electronic nose (E-nose) is capable of identifying, measuring, and analyzing a compound in order to obtain information.[19] It has been found that breath acetone concentrations in diabetic patients are greater than 1.8 ppm, but those in healthy individuals are less than 0.8 ppm.[5] Asal et al.[20] measured the low concentration of acetone gas in human exhale for diabetic diagnosis. To do this, they used a hydrothermal technique to create a porous zinc oxide/tin dioxide nanocomposite (ZS) with a specific surface area of 52.1 $m^2 \cdot g^{-1}$. The ZS had a nanoplate shape. Usman et al.[13] developed a ternary composite consisting of p-toluene sulfonic acid (PTSA) doped polyaniline (PANI), reduced graphene oxide (RGO), and chitosan to explore its capability in detecting low concentrations of acetone vapor. This research aims to facilitate noninvasive monitoring and screening of diabetes through analysis of exhaled breath. The study successfully detected acetone vapor at concentrations ranging from 1.8 ppm to 5 ppm. This detection range is crucial for potential applications in diabetes management, as elevated acetone levels in breath are indicative of certain diabetic conditions. This approach offers promise as a sensitive and noninvasive method for early detection and monitoring of diabetes using advanced sensor technology.

4.5.2 Chronic Kidney Disease

Chronic kidney disease, also known as kidney failure or renal insufficiency, occurs when the kidneys are unable to effectively filter and remove waste products (uremic toxins) from the blood.[21] In a healthy individual, the liver converts ammonia into urea. The urea is then transported via the bloodstream to the kidneys, which excrete it in the urine. For patients with chronic kidney disease (CKD), the excess urea is broken down by salivary urease into ammonia, which is subsequently expelled through respiration.[22] In stage 5 of chronic kidney disease, kidney function is severely impaired, with a filtration rate of less than 15 ml of plasma per minute, approximately representing 15% of normal function. At this stage, either dialysis treatment or a kidney transplant is required to preserve health.[21] Bevc et al.[22] designed an ammonia-based sensor that identified higher concentrations of ammonia in the breath of CKD patients (mean ppm 3.32 ± 2.19; range 1.26 to 6.33 ppm) compared to healthy volunteers (mean ppm 0.49 ± 0.08; range 0.38 to 0.64 ppm).

4.5.3 Blood Cholesterol

A vital component of mammalian cell membranes, cholesterol acts as a precursor of bile acids, vitamin D, steroid hormones, and other important compounds. Moreover, it is important for immune system performance, brain synapses, and cancer prevention. High blood cholesterol, especially low-density lipoprotein (LDL), can cause major health problems such hypertension, heart disease, stroke, and other ailments. Low cholesterol, on the other hand, can cause diseases like wasting syndrome and anemia. Consequently, it is crucial to regularly check blood cholesterol levels in order to diagnose and prevent a number of disorders.[23] Using β-CD, SPCE, and MWCNTs, Nawaz et al.[23] created a novel electrochemical sensor for the detection of cholesterol. The sensor responded to cholesterol electrochemically very well, with good sensitivity and repeatability in the range of 1 nM to 3 μM, and a detection limit of 0.5 nM.

4.5.4 Intestinal Disease

The intestine is essential for breaking down food, absorbing nutrients, and removing metabolic waste, which supports the body's normal metabolism. Issues in the intestine, such as ulcers, bleeding, inflammation, and tumors, can result in significant health problems and place a heavy burden on the health-care system and the economy.[24] Wilson[25] highlighted the use of E-nose technology for the early detection of gastrointestinal (GI) diseases. Detection methods for these diseases depend on the specific type and mechanism of each GI disease, the chemical classes of VOC-metabolites released from affected tissues, and the mobility of these metabolites within the body. Various VOC gas-sampling techniques, including analyses of human breath, urine, and fecal (stool) samples, have been employed to diagnose different types of GI tract diseases.

4.5.5 Halitosis

Halitosis, commonly known as 'bad breath' or 'oral malodor', affects approximately 50% of adults globally. This condition refers to the unpleasant odor emitted from an individual's exhaled breath (halitus), which can have notable social consequences for both the affected person and those around them.[26] Halitosis, a condition characterized by bad breath primarily due to volatile sulfur compounds (VSCs) such as methyl mercaptan (CH_3SH) and hydrogen sulfide (H_2S), is associated with the breakdown of sulfur-containing amino acids in the upper and lower respiratory tract, their exchange at the alveolar level with blood, and origin from the oral cavity.[5] Kim et al.[4] established an experimental setup to inject the captured breath into sensor arrays, which were composed of WO_3 NFs functionalized by Pt, Pd, Rh, and PtY that are shown in Figure 4.3 (a). In Figure 4.3 (b), they compared the response values of halitosis breath with healthy human. After performing PCA, they showed the simulated graph of halitosis breath and healthy human breath in Figure 4.3 (c).

4.5.6 Lung Cancer

One common and frequently deadly type of cancer is lung cancer (LC). Early signs and symptoms could be a mild fever and dry cough, which could easily be overlooked or mistaken for other common illnesses. This poses a serious obstacle to early diagnosis, since early identification is necessary for successful treatment and a better prognosis.[17] Still, it is difficult to conduct an accurate and convenient analysis of lung

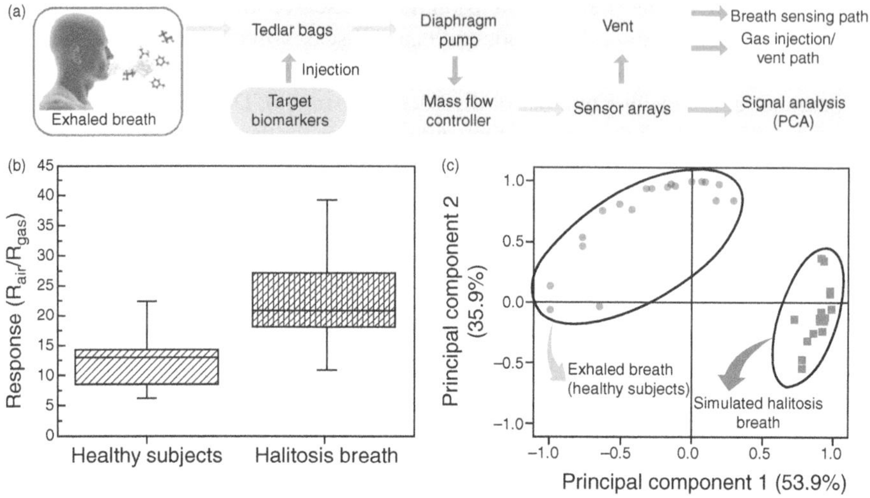

FIGURE 4.3 Experimental setup for real-time breath analysis aimed at diagnosing halitosis. (a) Overview of the breath analysis apparatus. (b) Response graph illustrating comparisons between simulated halitosis breath and breath samples from healthy individuals. (c) PCA-based classification demonstrating differentiation between simulated halitosis breath and healthy human breath. (Reproduced with permission[4]).

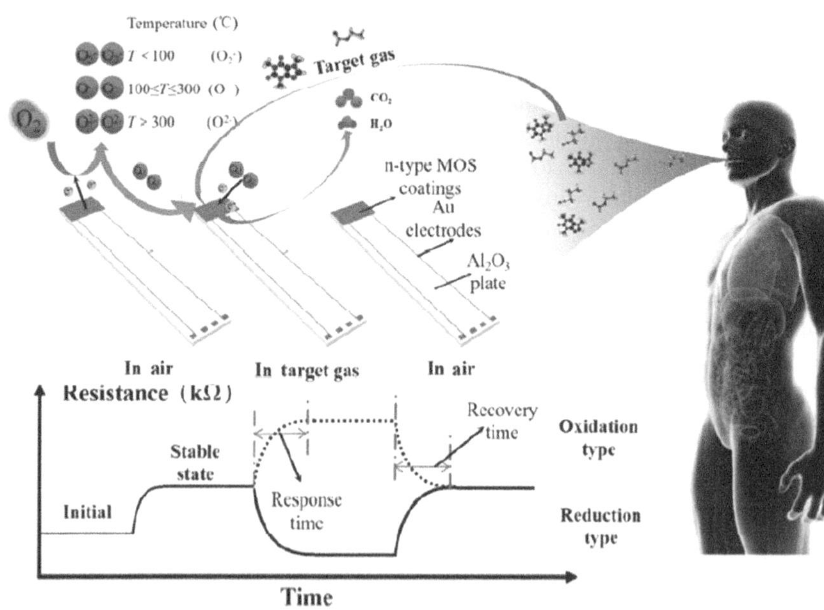

FIGURE 4.4 Working mechanism of n-type metal oxide gas sensors for LC detection. (open access license to reproduce[17]).

cancer in its early stages. The prognosis and treatment options for lung cancer patients are closely tied to the size of the tumor and its extent of spread at the time of diagnosis. Survival time decreases notably when the disease is more advanced at detection.[5,27] Traditional methods for diagnosing lung cancer, such as imaging examinations (like CT scans), histopathological examinations (biopsies), and detection of tumor markers in body fluids, do indeed have drawbacks that include high costs and potential health risks.[17] Xiaoxi HE et al.[17] clearly explained the working mechanism to detect lung cancer, which is shown in Figure 4.4. Hakim et al.[28] reported that there are two major forms of lung cancer: non-small-cell LC (NSCLC; ~85%) and small-cell LC (SCLC; ~15%). Diagnosis involves blood tests, chest X-ray, CT scans, MRI, and PET. VOC detection in breath offers a noninvasive, inexpensive diagnostic tool with the potential to complement or replace conventional methods. Marzorati et al.[29] reported the design and advancement of an E-nose employing metal oxide gas sensors for rapid analysis of exhaled breath, aimed at early detection of lung cancer. They achieved a sensitivity of 85.7% when testing breath samples from lung cancer patients.

4.6 FUTURE PERSPECTIVE

The collaborative efforts of experts in materials science, sensor engineering, information technology, and clinical medicine are essential to the interdisciplinary field of breath analysis for disease detection. While significant strides have been made in the creation of compact, portable gas sensors with high sensitivity, superior selectivity, and quick detection times, the majority of these instruments are still in the research and development stage and have not seen much application in clinical trials. To facilitate the adoption of breath-sensing technologies into mainstream clinical use, further advancements are necessary in several areas. These include improving production scalability, enhancing calibration accuracy, and miniaturizing sensors. These developments in manufacturing will enable widespread adoption of these advanced diagnostic tools, ensuring effective and noninvasive medical monitoring.

4.7 CONCLUSION

The chapter emphasizes the significant potential of oxide-based sensors, particularly binary and ternary composites, in noninvasive disease diagnosis through breath analysis. These sensors offer rapid, cost-effective, and portable diagnostic solutions, greatly improving patient comfort and accessibility compared to traditional invasive methods. By detecting specific volatile organic compounds (VOCs) in exhaled breath – such as acetone for diabetes and ammonia for chronic kidney disease – these sensors enable early diagnosis and effective monitoring of various health conditions. Current research is focused on optimizing these sensors to enhance their sensitivity, selectivity, and response times, while ensuring their reliability in clinical settings. The promising capabilities of these oxide-based systems could lead to significant advancements in personalized medicine, as early and accurate diagnosis is essential in managing chronic diseases effectively. Integrating interdisciplinary expertise across materials science, sensor technology, and clinical application is crucial to driving forward the research and development of these innovative diagnostic tools. Despite significant progress, further improvements in manufacturing processes and clinical validations are necessary to translate these technologies from laboratory settings to everyday medical practice. Therefore, the future of disease detection may increasingly rely on breath analysis, marking a crucial step toward less invasive and more efficient diagnostic methodologies.

REFERENCES

1. Nasiri N., & Clarke C. (2019). Nanostructured gas sensors for medical and health applications: Low to high dimensional materials. *Biosensors*, 9(1), 43. https://doi.org/10.3390/bios9010043
2. Drmosh Q. A., Olanrewaju Alade I., Qamar M., & Akbar S. (2021). Zinc oxide-based acetone gas sensors for breath analysis: A review. *Chemistry–An Asian Journal*, 16(12), 1519–1538. https://doi.org/10.1002/asia.202100303
3. Horne J., McLoughlin L., Bridgers B., & Wujcik E. K. (2020). Recent developments in nanofiber-based sensors for disease detection, immunosensing, and monitoring. *Sensors and Actuators Reports*, 2(1), 100005. https://doi.org/10.1016/j.snr.2020.100005
4. Kim S. J., Choi S. J., Jang J. S., Cho H. J., & Kim I. D. (2017). Innovative nanosensor for disease diagnosis. *Accounts of Chemical Research*, 50(7), 1587–1596. https://doi.org/10.1021/acs.accounts.7b00047
5. Yoon J. W., & Lee J. H. (2017). Toward breath analysis on a chip for disease diagnosis using semiconductor-based chemiresistors: Recent progress and future perspectives. *Lab on a Chip*, 17(21), 3537–3557. https://doi.org/10.1039/C7LC00810D
6. Mout R., Moyano D. F., Rana S., & Rotello V. M. (2012). Surface functionalization of nanoparticles for nanomedicine. *Chemical Society Reviews*, 41(7), 2539–2544. https://doi.org/10.1039/C2CS15294K
7. Zhang C., Luo Y., Xu J., & Debliquy M. (2019). Room temperature conductive type metal oxide semiconductor gas sensors for NO2 detection. *Sensors and Actuators A: Physical*, 289, 118–133. https://doi.org/10.1016/j.sna.2019.02.027
8. Su P. G., & Yang L. Y. (2016). NH3 gas sensor based on Pd/SnO2/RGO ternary composite operated at room-temperature. *Sensors and Actuators B: Chemical*, 223, 202–208. https://doi.org/10.1016/j.snb.2015.09.091
9. Zhang L., Guo Y. Y. H., Liu G., & Tan Q. (2020). Na-doped zno and rgo composite-based flexible acetone gas sensor operated in room temperature. *IEEE Access*, 8, 171568–171574. https://doi.org/10.1109/ACCESS.2020.3023028
10. Zhou Q., Chen W., Xu L., Kumar R., Gui Y., Zhao Z., ... & Zhu S. (2018). Highly sensitive carbon monoxide (CO) gas sensors based on Ni and Zn doped SnO2 nanomaterials. *Ceramics International*, 44(4), 4392–4399. https://doi.org/10.1016/j.ceramint.2017.12.038
11. Wang C., Liu J., Yang Q., Sun P., Gao Y., Liu F., ... & Lu G. (2015). Ultrasensitive and low detection limit of acetone gas sensor based on W-doped NiO hierarchical nanostructure. *Sensors and Actuators B: Chemical*, 220, 59–67. https://doi.org/10.1016/j.snb.2015.05.037
12. Amiri V., Roshan H., Mirzaei A., Neri G., & Ayesh A. I. (2020). Nanostructured metal oxide-based acetone gas sensors: A review. *Sensors*, 20(11), 3096. https://doi.org/10.3390/s20113096
13. Usman F., Dennis J. O., Mkawi E. M., Al-Hadeethi Y., Meriaudeau F., Ferrell T. L., ... & Sulieman A. (2020). Investigation of acetone vapour sensing properties of a ternary

composite of doped polyaniline, reduced graphene oxide and chitosan using surface plasmon resonance biosensor. *Polymers*, 12(11), 2750. https://doi.org/10.3390/polym12112750

14. Benelmekki M. (2015). Introduction to nanoparticles and nanotechnology. In *Designing Hybrid Nanoparticles* (Second Edition), IOP Publishing. https://doi.org/10.1088/978-1-6270-5469-0ch1
15. Uma S., & Shobana M. K. (2023). Metal oxide semiconductor gas sensors in clinical diagnosis and environmental monitoring. *Sensors and Actuators A: Physical*, 349, 114044. https://doi.org/10.1016/j.sna.2022.114044
16. Kumar S., & Bhowmik B. (2024). Frequency-tuned selectivity enhancement of Mg@ ZnO–TiO 2 nanoflake-based heterojunction sensor devices. *Analyst*, 149(6), 1895–1906. https://doi.org/10.1039/D3AN01562A
17. Xiaoxi H. E., Hongfeng C. H. A. I., Yifan L. U. O., Lingfeng M. I. N., Debliquy M., & Zhang C. (2023). Metal oxide semiconductor gas sensing materials for early lung cancer diagnosis. *Journal of Advanced Ceramics*, 12(2). https://doi.org/10.26599/JAC.2023.9220694
18. Nikolic M. V., Milovanovic V., Vasiljevic Z. Z., & Stamenkovic Z. (2020). Semiconductor gas sensors: Materials, technology, design, and application. *Sensors*, 20(22), 6694. https://doi.org/10.3390/s20226694
19. Sarno R., & Wijaya D. R. (2017, October). Detection of diabetes from gas analysis of human breath using e-Nose. In *2017 11th International Conference on Information & Communication Technology and System (ICTS)* (pp. 241–246). IEEE. https://doi.org/10.1109/ICTS.2017.8265677
20. Asal M., & Nasirian S. (2019). Acetone gas sensing features of zinc oxide/tin dioxide nanocomposite for diagnosis of diabetes. *Materials Research Express*, 6(9), 095093. https://doi.org/10.1088/2053-1591/ab332c
21. Davies S. J., Španěl P., & Smith D. (2014). Breath analysis of ammonia, volatile organic compounds and deuterated water vapor in chronic kidney disease and during dialysis. *Bioanalysis*, 6(6), 843–857. https://doi.org/10.4155/bio.14.26
22. Bevc S., Mohorko E., Kolar M., Brglez P., Holobar A., Kniepeiss D., ... & Hojs R. (2017). Measurement of breath ammonia for detection of patients with chronic kidney disease. *Clinical Nephrology*, 88(7), 14. https://doi.org/10.5414/cnp88fx04
23. Nawaz M. A. H., Majdinasab M., Latif U., Nasir M., Gokce G., Anwar M. W., & Hayat A. (2018). Development of a disposable electrochemical sensor for detection of cholesterol using differential pulse voltammetry. *Journal of Pharmaceutical and Biomedical Analysis*, 159, 398–405. https://doi.org/10.1016/j.jpba.2018.07.005
24. Zhao C., Zhang X., Chen G., & Shang L. (2022). Developing sensor materials for screening intestinal diseases. *Materials Futures*, 1(2), 022401. https://doi.org/10.1088/2752-5724/ac48a3
25. Wilson A. D. (2017). Recent applications of electronic-nose technologies for the non-invasive early diagnosis of gastrointestinal diseases. In *Proceedings* 2(3), p. 147). MDPI. https://doi.org/10.3390/ecsa-4-04918
26. Nakhleh M. K., Quatredeniers M., & Haick H. (2018). Detection of halitosis in breath: Between the past, present, and future. *Oral Diseases*, 24(5), 685–695. https://doi.org/10.1111/odi.12699
27. Nardi-Agmon I., & Peled N. (2017). Exhaled breath analysis for the early detection of lung cancer: Recent developments and future prospects. *Lung Cancer: Targets and Therapy*, 31–38. https://doi.org/10.2147/LCTT.S104205

28. Hakim M., Broza Y. Y., Barash O., Peled N., Phillips M., Amann A., & Haick H. (2012). Volatile organic compounds of lung cancer and possible biochemical pathways. *Chemical Reviews*, 112(11), 5949–5966. https://doi.org/10.1021/cr300174a
29. Marzorati D., Mainardi L., Sedda G., Gasparri R., Spaggiari L., & Cerveri P. (2019). A metal oxide gas sensors array for lung cancer diagnosis through exhaled breath analysis. In *2019 41st Annual International Conference of the IEEE Engineering in Medicine and Biology Society (EMBC)* (pp. 1584–1587). IEEE. https://doi.org/10.1109/EMBC.2019.8856750

5 Graphene-Based Sensor System for Noninvasive Diagnosis

Souvik Ghosh and Basanta Bhowmik

5.1 INTRODUCTION

Graphene-based transducers possess high sensitivity due to their unique honeycomb-like structure, formed with single thick planer SP^2 carbon sheet. Graphene's lattice structure is stable, which helps its durability as a sensing material. Owing to its lattice structure, it has high electron transfer rate and large surface area, which enhance its ability to immobilize molecules and increase its sensitivity. Naturally, graphene is easily available, cost effective, and falls into the category of semi-metals.

Graphene has been utilized for sensing applications in its nanomaterials-based synthesis form, usually referred to as GBN, i.e., graphene-based nanomaterial. The number of layers, the average lateral size, and the carbon-to-oxygen (C/O) ratio are key parameters to classify graphene in different synthesis methods. Broadly, it has been utilized as its pure form and its functionalized derivatives such as graphene quantum dots (GQDs), graphene oxide (GO), and reduced graphene oxide (rGO). Graphene oxide (GO) consists of various oxygen functional groups such as –OH, C–O–C, CO, and –COON, which enhance its bindings with strong reactivity properties. Similarly, removal of the oxygen-containing groups by chemical, electrochemical, or thermal methods, forms reduced graphene oxide (rGO), having superior thermal conductivity, chemical stability, and a large surface area. Graphene quantum dots (GQD) are derived from of carbon dots (CD) injected with graphene. They have extraordinary chemical, structural, electrical, and tunable optical properties of photoluminescence and electro-chemiluminescence. Due to their properties, GQDs forms highly sensitive sensors.

Graphene in its various synthesis forms possesses quantum hall effects, higher electron mobility and conductivity, and large surface-to-volume ratio. It possesses tunable optical properties and high mechanical strength, which make it an ideal nanomaterial for potential application in medical and industrial applications.

Pristine graphene is currently produced in limited quantities, primarily in research laboratories.[1] However, the challenges of large-scale production of graphene are difficult to overcome. In recent years, chemical methods such as Hummer's method and thermal exfoliation methods have been utilized for the large-scale production of graphene. These methods have resulted in the production of various graphene variants, which can be broadly classified into four categories: graphene nanoplatelets

DOI: 10.1201/9781003453635-5

(GNP), or flakes with minimal or no defects at the basal plane; graphene oxide (GO); reduced graphene oxide (rGO); and functionalized graphene (fGO). These different variants of graphene have been produced worldwide for specific applications.[1] GNP, or graphene flakes, can be directly produced from graphite using a mechanical exfoliation method, minimizing any functionalization at the C-C bond of the hexagonal structure.[2] Graphene oxide contains functional groups that can interact with other materials, making it suitable for applications in coatings, composites, water filtration, sensors, and more. On the other hand, reduced graphene oxide, which possesses good electrical properties, can be used in battery applications, electrochemical sensors, and other similar applications.

In addition to its production in small quantities and as variants, graphene can also be grown in the form of films, but only on certain metals like copper and nickel. To enable the development of applications using graphene films, a transfer process has been developed, allowing the transfer of graphene films from metal substrates to a variety of substrates such as silicon, PET, glass, and others. Efforts have also been made to develop a roll-to-roll growth and transfer process for graphene films. These films have the potential to be used in flexible displays and optoelectronics.[3]

5.2 PRODUCTION OF GRAPHENE

Carbon serves as the fundamental building block for all life forms in the universe. Two common crystalline forms of carbon are diamond and graphite. Diamond is known for its strength due to its ordered crystal structure in three-dimensional space through SP^3 carbon-carbon chemical bonds. On the other hand, graphite exhibits SP^2 chemical bonding between carbon atoms in a two-dimensional plane. Graphite is composed of stacked carbon planes held together by weak physical forces. Graphene, extracted from graphite, consists of these carbon planes. Graphite is commonly used in pencils due to the weak physical interaction, leaving thin layers of graphite on paper as we write.

There are three methods for extracting graphene from graphite:[1]

Chemical Exfoliation: Graphite is oxidized in the presence of strong acids and oxidizing agents in an aqueous solution. This oxidation process does not aim to produce CO or CO_2. Strong acids break some of the carbon-carbon bonds in graphite, leading to carbon bonding with -OH or =O functional groups. This increases the interplanar gap and weakens the Van der Waals force between layers. Mechanical agitation in aqueous media can separate carbon planes with oxygen functional groups, resulting in graphene oxide dispersion. Reduction with a reducing agent like hydrazine converts graphene oxide into graphene, forming reduced graphene oxide, which differs from pure graphene due to incomplete reduction.

Physical Exfoliation: Graphite particles are dispersed in a solvent with high wettability to the graphite surface. The dispersion is stirred at high RPM (10k–20k) to generate shear between graphite particles. Typically, 10–15 hours of shear mixing yield a significant amount of graphene dispersion.

Thermal Exfoliation: Graphite undergoes chemical treatment to intercalate sulphate ions between its planes, resulting in graphite intercalated compound. Heating this compound at approximately 800°C in air leads to thermal exfoliation. The sulphur located between the graphite planes reacts with oxygen to form SO2, causing a vigorous exfoliation of the graphene planes. This process can be further enhanced through physical exfoliation methods discussed in the previous section. These methods are commonly referred to as top-down processes, as they involve starting with graphite and thinning it down to graphene. Top-down processes can yield single-layer graphene or various numbers of layers. Graphitic flakes with up to ten layers can still exhibit graphene-like properties in certain applications.

Alternatively, graphene can be produced using a bottom-up approach, where the starting material is a carbon-containing precursor like methane. The potential routes for bottom-up production are as follows.[2-4]

Chemical vapor deposition (CVD): This method involves bombarding a copper substrate heated to 1000°C with gaseous methane (CH4) as a precursor. At such high temperatures, copper acts as a catalyst, breaking the carbon-hydrogen bonds. The free carbon then deposits on the copper substrate in the form of a graphene film. This process is widely utilized to create single layer to multilayer graphene films on copper and nickel substrates. To transfer the graphene to another substrate, the copper is etched using acids. In addition to gaseous precursors, liquid and solid precursors can also be employed. Sometimes shellac (a solid precursor) is used to produce graphene-like films on copper substrates. These films are not pure graphene but rather graphene with a few carbon atoms attached to functional groups (-OH, =O).

5.3 PROPERTIES OF GRAPHENE

Graphene's unique properties, such as its thinness, strength, stiffness, flexibility, and electrical conductivity, make it a highly versatile material with immense potential for various applications.

Graphene is incredibly thin, with a thickness of only 0.77mg/m^2. To put it into perspective, just 1 gram of graphene would be enough to cover an entire football field, which has a typical size of 7140 m^2. Graphene possesses exceptional tensile strength, measuring at 130 GPa. This is significantly higher than the tensile strength of steel, which is only 2 GPa. The in-plane carbon-carbon bonds in graphene are the strongest bonds known to science, contributing to its remarkable mechanical strength and flexibility. With a stiffness of 1000 GPa, graphene exhibits a high resistance to external forces while still being able to return to its original shape. The strong covalent bonds between carbon atoms allow graphene to stretch, bend, and deform without breaking.

Graphene can withstand up to 25% strain before reaching failure, making it comparable to rubber in terms of flexibility. The carbon atoms within the graphene structure can flex without breaking the atomic bonds, enabling it to maintain its integrity even under significant strain.[2]

Graphene demonstrates excellent electrical conductivity, with a resistivity of 1x10-8 ohm/m. This is slightly lower than copper, which has a resistivity of 1.68x10-8 ohm/m. Graphene's high conductivity can be attributed to its higher mobility compared to copper, as it has a higher carrier density and electron mobility. Graphene, having thermal conductivity 2000 w/mK (compared to copper = 400, Aluminum = 237), is widely recognized for its high intrinsic thermal conductivity, typically ranging from 3000 to 5000 W/mK. At room temperature, this surpasses the in-plane thermal conductivity of graphite, which is approximately 2000 W/mK. However, when graphene is placed on substrates, its thermal conductivity decreases to around 600 W/mK due to phonon scattering on the defects and interface of the substrate. Copper, on the other hand, has a relatively small contribution from phonons to its overall thermal conductivity at bulk values and room temperature, accounting for only about 2%. While the thermal conductivity of graphene is primarily phonon-based, its electronic-based thermal conductivity represents less than 1% of the total thermal conductivity at room temperature. Phonons are associated with the lattice vibrations of atoms in a crystal structure. Graphene, with its single atomic crystal structure, exhibits an impressive transparency of 98%. However, with the addition of each layer, its transparency decreases by 2%. This exceptional thinness is responsible for graphene's well-known optical transparency.

The angstrom-scale hexagonal voids in a graphene sheet resemble the openings in chicken wire. Despite this appearance, studies have shown that defect-free graphene is nearly impermeable, blocking all liquids and gases except for protons. The permeation of hydrogen atoms is attributed to the presence of nanometer-sized ripples on graphene membranes, which act as catalytically active regions and dissociate absorbed molecular hydrogen into two hydrogen atoms. Graphene oxide membranes allow for the lateral flow of water through the interplanar gaps of multilayer graphene oxide. The permeation of water through nanometer-pore materials has garnered significant interest due to the fundamental differences in governing mechanisms at macroscopic and molecular scales, as well as the importance of water permeation in living systems.

Graphene, with its exceptional properties, holds immense potential for various applications across different fields. Therefore, it is crucial to have a comprehensive understanding of its structural integrity, layer thickness, defects, and interactions with other materials through characterization techniques. This knowledge plays a vital role in ensuring the quality of synthesized graphene, optimizing fabrication processes, and tailoring its properties for specific purposes. Additionally, characterization provides valuable insights into the behavior of graphene in devices, composites, and functionalized forms, enabling the development of innovative technologies. Ultimately, a thorough characterization process drives advancements in graphene-based materials and their practical implementations. There are several technical methods available for fully characterizing graphene and similar materials. Some of the most important and widely used techniques are discussed below:[1,5]

Raman Spectroscopy: Raman scattering is utilized to examine the electronic structure of graphene materials, providing valuable information about molecular vibrations, bonds, compounds, and crystalline structures. The G and 2D peaks are fundamental features in Raman spectroscopy. The G peak, observed at around 1580 cm−1, confirms the presence of graphitic material, representing carbon-carbon bonds in the hexagonal lattice. The 2D peak, observed at around 2700 cm−1, relates to the layer count and stacking arrangement of graphene. The intensity ratio of the 2D to G peak provides insights into the layer number and quality of graphene. The intensity ratio of the D to G peak can offer valuable information about the degree of structural disorder in graphene. Raman spectroscopy allows for a thorough evaluation of the number of layers in graphene grown in a film-like structure on any substrate.

Transmission Electron Microscopy (TEM): This is a highly valuable method used to investigate the properties and features of materials at the nanoscale, providing detailed insights into atomic intricacies. In the case of graphene, TEM enables the visualization and comprehensive analysis of its atomic arrangement, imperfections, layer alignment, and interconnections.[1] TEM allows for the capture of high-resolution images of individual graphene sheets, enabling the study of the precise arrangement of carbon atoms within the hexagonal lattice. Through TEM, various flaws within graphene, including vacancies, grain boundaries, and edge dislocations, can be revealed and scrutinized. TEM has the capability to differentiate between different graphene layers, ranging from monolayer to multilayer configurations. This ability aids researchers in understanding the organization of layer stacking, which is crucial for producing high-quality graphene suitable for diverse applications. The determination of the graphene layer is achieved by quantifying the number of lines present at the edge.

Selected Area Electron Diffraction (SAED): SAED is a significant tool for characterizing the quality and functionality of graphene in various applications. It is a technique that can unveil the crystalline structure and orientation of materials like graphene. By analyzing the SAED pattern of graphene, one can determine the degree of graphitization, the number of layers, and the presence of defects or strains in the graphene lattice. These factors can impact the electronic and optical properties of graphene, such as its conductivity, bandgap, and photoluminescence.[1]

X-Ray Diffraction (XRD) Spectroscopy: This is a powerful technique used to analyze the crystalline structure and thickness of graphene samples. By studying the XRD pattern of graphene, valuable information can be obtained, such as the size of the crystallites, the spacing between layers, and the number of layers per domain. These factors have a significant impact on the physical and chemical properties of graphene, including its mechanical strength, thermal conductivity, and surface area. Therefore, XRD plays a crucial role in characterizing the morphology and quality of graphene for a wide range of applications.[4,1]

Analysis of Graphene Oxide (GO) and Reduced Graphene Oxide (rGO): XRD enables the examination of structural changes in GO and rGO compared to pristine graphene, providing insights into the modification processes and their corresponding applications.

In cases where graphene is incorporated into composite materials, XRD allows for the analysis of alterations in the diffraction patterns of the composite, enabling comparisons with its constituent elements. XRD provides valuable information about the arrangement and alignment of graphene layers grown on substrate surfaces. XRD is a widely used X-ray-based method for deciphering crystal structures and determining the composition and purity of graphene crystallinity. The XRD pattern of graphene exhibits strong peaks within the 20–27° (2θ) range.

Scanning Electron Microscope (SEM): This is a powerful imaging technique that is widely used in the fields of materials science and nanotechnology. It plays a crucial role in characterizing the surface morphology, impurities, and material structures of graphene. With SEM, researchers can study the size and distribution of graphene flakes at a microscopic level. SEM enables researchers to obtain high-resolution images that reveal the surface features of graphene. This allows them to analyze the arrangement of graphene layers; identify imperfections, creases, folds; and examine the sizes and shapes of graphene flakes, along with other surface attributes. SEM provides detailed insights into the topography and texture of graphene samples. It allows for the observation of variations in elevation, surface roughness, and other factors that can potentially influence the characteristics of graphene.

SEM enables meticulous scrutiny and analysis of the physical structure of graphene-based devices. This ensures their precise construction and optimal functionality. SEM is particularly valuable in visualizing the distribution and alignment of functional groups or particles that are decorated on the surface of graphene, even at lower magnifications.

Atomic Force Microscopy (AFM): It operates by utilizing a pointed and probing tip attached to a spring-like cantilever. The interaction forces between the tip and the sample cause deflections in the cantilever, which are then detected using a deflection sensor. Through the scanning of the sample, a feedback mechanism can maintain a consistent deflection. The AFM technique allows for the measurement of the topography and physical properties of graphene samples. By analyzing the AFM images of graphene, one can determine the number of layers, thickness, roughness, and adhesion of the material. These factors have an impact on the optical, electrical, and mechanical properties of graphene, including transparency, conductivity, and flexibility.

Specific Surface Area using Brunauer Emmett teller (BET) method: The Brunauer-Emmett-Teller (BET) analysis, also known as BET analysis, is a method used to calculate the specific surface area of various materials, including graphene. This technique involves the adsorption of gas molecules onto the material's surface and the subsequent measurement of the resulting adsorption isotherms. By analyzing these isotherms, it becomes possible to determine the surface area, which provides valuable insights into the material's reactivity, porosity, and potential applications. In the case of graphene, BET analysis allows for the evaluation of its available surface area, which holds significance in applications such as catalysis, sensors, filters, and energy storage.

However, it is important to note that the theoretical surface area of graphene, which is approximately 2600 m2/g, is not achievable in practical devices due to factors like aggregation, wrinkling, and stacking of graphene sheets. These factors lead to a reduction in the exposed surface area. The actual surface area of graphene depends on various factors, including the synthesis method, processing conditions, and measurement techniques employed.

5.4 USE OF GRAPHENE IN BIOSENSING

Graphene, as a biosensor (both invasive and noninvasive, see Figure 5.1) is suitable and applicable due to its excellent property as a recognition element that binds with targeted receptors. Biosensors can be classified based on the receptor elements, transducer, or signal amplification used. As graphene is a medium of recognition, it is suitable for enzymatic sensors, immunosensors, DNA-based sensors, aptamer-based sensors, or whole-cell microbial biosensors. Graphene's mechanical flexibility is notably essential for invasive- or noninvasive-based health monitoring biosensors.[3,5]

In invasive health monitoring sensors, highly cross-linked GBNs are being developed as molecularly imprinted polymers (MIPs) having high selectivity and affinity toward a targeted molecular analyte. Many studies have shown that GBN has good biocompatibility and low toxicity with the human body. The main aim is

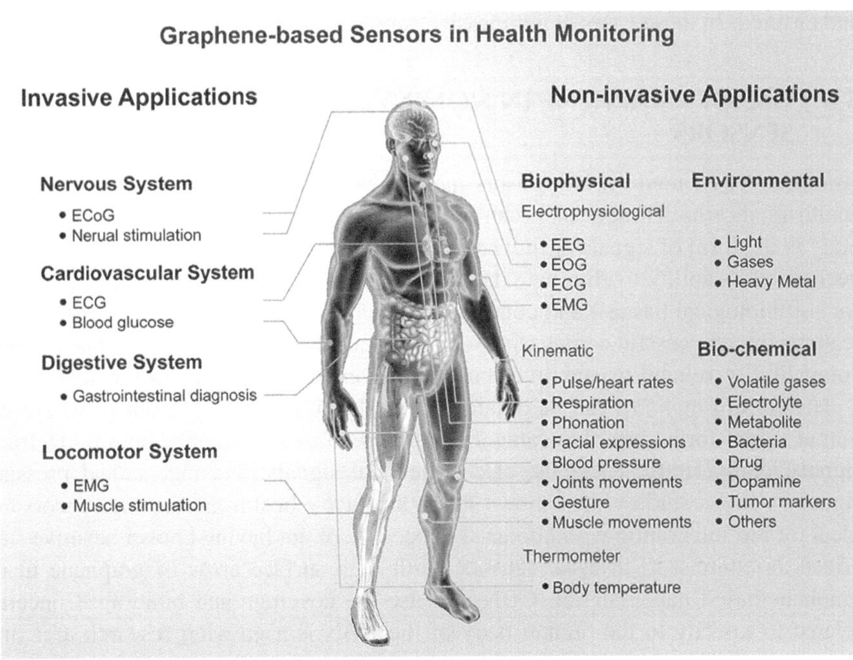

FIGURE 5.1 Graphene-based sensors in health monitoring. Courtesy: www.frontiersin.org/journals/chemistry/articles/10.3389/fchem.2019.00399/full

to evaluate whether graphene-based implants are suitable for human health. This is why they are widely used in neuroscience, which shows the effects of GBN on the central nervous system. Graphene is an attractive material for the realization of many implantable neural devices due to its unique electrical properties, such as flexibility, high transparency, and electrical conductivity, through which brain cells and neuron circuits are directly exposed to graphene-based implants.

The negative effects of graphene-based invasive sensors are less obvious, but their long duration causes problems in cell functioning. Graphene oxide and GO can be adsorbed by bacterial cells, and the adsorption of GO leads to sharp edges of bacteria, which can damage the cell membrane. GO also replaces some of the oxygen in cells, causing oxidative damage to the body's materials, damage to cells, and eventually cell death. To overcome these limitations and expand the biological applications of GBN, many graphene-biomacromolecule hybrid materials have been developed, such as graphene-biopolymer nanohybrid materials (made by combining GBN with biocompatible polymers), and graphene-polysaccharide nanohybrid materials (made by combining GBN with biocompatible materials). GBN was developed to meet the needs of biomedical and pharmaceutical applications with improved biocompatibility, reduced low toxicity, improvement of solubility and stability, and even support of cell proliferation. Additionally, many green methods have been proposed to reduce the toxicity in rGO production using bacterial and plant extracts.[4] It is also a good way to avoid direct contact between GBN and the human body or even biological fluids. The biodegradability of the materials in case of peeling or tearing must be taken into account and ensured. In these cases, a noninvasive sensor plays a crucial role.

5.5 USE OF GRAPHENE IN NONINVASIVE HEALTH-MONITORING SENSORS

Graphene-based noninvasive sensors include wearable and implantable sensors for health monitoring. These sensors can detect and measure various changes in the human body in the form of signals or movements, with higher sensitivity. Sensitivity is high because of its ability to eliminate the mechanical mismatch between the human skin (or soft biological tissues) and conventional rigid diaphragm-based sensors. Notable features for a successful noninvasive sensor include the zero constraints related to biocompatibility, reliability, stability, comfort, convenience, miniaturization, and cost.[6]

Health information t can be obtained from the skin by using a sensor where the output signal forms into five categories: thermo-electrical signals, neural electrical signals, photoelectrical signals, electrochemical signals, and mechanical pressure signals. Various studies have shown that noninvasive health-monitoring sensors are ideal for bio-integration with human skin because of not having known negative side effects in contrast to invasive sensors. With high surface areas of graphene in its graphene-based nanomaterial, GBN increase the coverage and bonding. Concerns related to toxicity to the human body of the GBN is a growing research area and shows a promising outcome in mitigating the negative effects. Skin irritation and allergies are still the most dominant negative effects of GBN-based noninvasive sensors. GBN's reactivity to various micronutrients of the human body like protein, vitamin and starch is also an area of concern. Yet other factors like high concentrations of graphene in base substrate or a long exposure time to human skin can accelerate the penetration in the human body.

Recent automation in signal processing has increased the efficiency and affectivity of noninvasively collecting patient vitals. This could have the potential to significantly improve the efficiency of modern health care. Combining the collection of these vitals with a system that can present a summary of possible diagnostic conclusions and the time spent in a waiting room can be better utilized to aid both the patient and the medical professional. These systems can measure the electro-ocular movement, temperature, blood oxygen saturation, arterial blood pressure, heart rate variability, breathing rate, and so forth. The measurements are combined using an embedded microcontroller that is wirelessly linked to a base station processor via Bluetooth or other peer-to-peer communication protocols.[2,4]

Although graphene-based sensors have received great attention in health and biomedical applications, the health impact of graphene for humans (e.g., safety biocompatibility, toxicity, and potential hazard to the environment) should be considered before graphene is used. Physicochemical properties of GBNs, such as dose, shape, purity, surface chemistry, number of layers, thickness, and lateral dimensions, depend mainly on the degree of toxicity and are often affected by their biodistribution to other organs. The original properties and biological behavior of GBNs change dynamically upon exposure to the immune system or biomolecules in the biological environment, which can lead to degradation or biotransformation. Additionally, when these GBNs are transferred to other biological sites over time, their properties will change. Therefore, field evaluation is very important for future applications. In recent years, the toxicity of GBNs has been evaluated in important target organs such as the immune system and cardiovascular system.

5.6 GRAPHENE-BASED SENSOR FOR HEALTH MONITORING

Vital physiological parameters monitoring of a human being on a real-time basis as required is collectively called physiological and health monitoring. Physiological status monitoring (PSM) provides a way to collect, record, and report a user's vital signs. It is a relatively non-intrusive measure that can occur in real time for extended periods of time. Physiological signal monitoring helps in various domains, with the majority in (1) critical care units, (2) emergency departments, (3) intermediate care units and general medical/surgical floors, (4) operating rooms, and (5) post-anesthetic care units. Physiological signal monitoring is utilized in both critical and non-critical applications. The majority of uses are related to clinicians and doctors so that they may be informed of changes in a patient's condition. Physiological systems consist of many different components, including base stations, invasive or noninvasive sensors, and connected telemetry transmitters with receivers that do not limit a person's ability to function daily. Clearly, more attention needs to be paid to the design of body protectors, which should not impede movement but be strong enough to protect against environmental or other harms. A simple system includes a variety of sensors, transmitters, receivers, and application software to display measurements needed for diagnosis. Automatic comparison of these measurements with previously recorded data detects warning signs of overwork and forgetfulness, allowing staff to make health adjustments rather than implementing changes. Sensors can be combined with electrodes, wristbands, clothing, and personal protective equipment.[4]

Continuous and uninterrupted monitoring of vital signs of heart rate, blood pressure, respiratory rate, and body temperature via GPS transponder-based positioning is the key to physical monitoring. Physical therapy can also be used in the training environment to prevent overtraining, reduce injury, and assist with self-assessment. Information gained from the audit can lead to improvements in education. Graphene's unique electrical, and optical properties enable a revolutionary bioengineering solution to create the best cardiac biosensors and actuators. Graphene has lower energy, lower intrinsic electron mobility, higher current, and greater mechanical properties; This makes it the thinnest, lightest, and strongest electron-hole pair excited by original light, making it a more useful heat generator. These products are essential for designing and manufacturing cardiovascular biosensors and actuators with new modes of action and technological resources (see Figure 5.2).

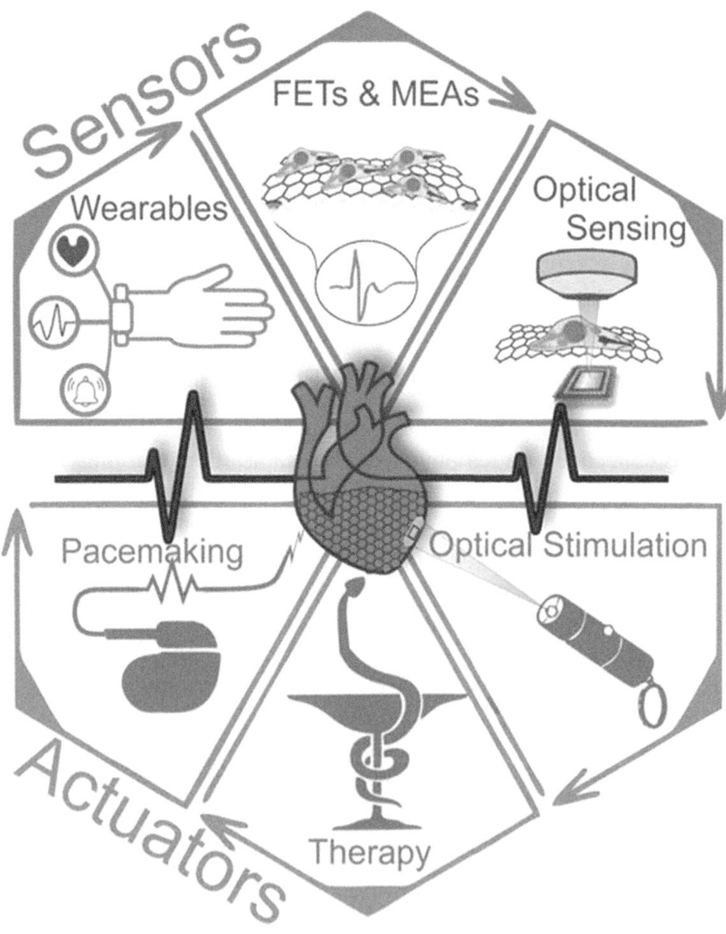

FIGURE 5.2 Overview of the graphene-based sensors and actuators for cardiac application.

A blood pressure-monitoring solution by laser-inducing graphene on a flexible polyimide substrate is showing excellent results. Graphenes have higher piezoresistive behavior that can be utilized to measure the human heartbeat by registering median cubital vein motion during blood pumping.

These new blood pressure-monitoring devices made from graphene can help people keep better track of their cardiovascular health. The array of stick-on graphene sensors, dubbed graphene electronic tattoos, can measure blood pressure continuously over the course of hours. Data from this device will give doctors a better overall picture of blood pressure than using measurements from a blood pressure cuff taken a few times a day.

A regular-wear blood pressure-monitoring platform based on bioelectrical impedance, uses atomically thin, self-adhesive, lightweight, and inconspicuous graphene electric tattoos as the human body bioelectronic interface. The graphene electric tattoo was used to monitor blood pressure for over 300 minutes, with an accuracy of 0.2±4.5 mm±Hg for diastolic pressure and 0.2±5.8 mm±Hg for diastolic pressure.[2,4]

Breath analysis is a way to understand the relationship between breathing and the physical and mental health of the human body. In addition, these sensors are portable and noninvasive, which makes them especially useful for respiratory monitoring. Many types of sensors can be configured according to four outputs: impedance, light, frequency, and voltage.

The need to monitor human breathing has become increasingly important during the COVID-19 era. Smart face masks (SFM) that can wirelessly monitor breathing are one solution. An SFM is reinforced by incorporating a graphene-based layer into a commercial surgical mask. Here, graphene nanoplatelets (GNPs) are embedded in a polycaprolactone (PCL) polymer matrix. These SFMs are well characterized, and their electrical, electromechanical, and sensory properties have been well evaluated. The SFM application has the best performance (> 1000 cycles) and fast response time (± 42 ms), and provides both normal and abnormal breathing with different parameters. Finally, a mobile application has been developed that wirelessly monitors the breathing patterns of mask users and provides alerts without compromising ease of use and comfort.

Temperature is one of the most important parameters of the human body. Human activities are based on metabolism. So real-time monitoring of body surface temperature is one of the simplest, and most effective ways to understand human health conditions. As an example, during COVID-19, the most prominent clinical indication was fever. So continuous real-time monitoring of human body temperature can not only effectively screen out infected patients, but also restrict the spread of the disease to a certain extent.

Generally, temperature is measured by a contact-type thermometer. So, it is required to measure at regular intervals. Commercially available contact temperature sensors are often rigid, have a poor fit with human skin, and are difficult to achieve accurate measurement during human movement. As a traditional non-contact temperature sensor, the infrared thermal imager can monitor the temperature of the human body in motion, but line of sight is very important for this type of sensor. So noninvasive wearable type temperature sensors with flexibility, stretchability, and biocompatibility are very effective. However, some flexible temperature sensors like skin-type thermocouples and thermistors are also used for continuous monitoring. But they are generally large in size and require complex circuitry. These are not suitable for prolonged use on the body.[7,8]

Graphene-based temperature sensors are one of the most promising noninvasive sensors available today. Graphene has been shown to be very promising in the development of flexible sensors due to its good tensile strength, flexibility, transparency, and biocompatibility. It has a large surface area and a larger thermal range than other materials. Graphene has ultra-high thermal conductivity. Therefore, it has a very fast specific temperature response. Its conductivity mainly depends on the internal carrier and carrier concentration. As the temperature increases, changes in carrier concentration become an important factor affecting conductivity. As the carrier concentration increases, the conductivity increases, and the semiconductor properties are expanded.[8]

Physical preparation of graphene involves mechanical exfoliation to obtain one or more layers of graphene. This method can prepare micron-sized graphene, which cannot be produced commercially. Chemical preparations include chemical vapor deposition (CVD), Hummer's reduction of graphene oxide, and synthesis of graphene for temperature sensors. Graphene layers can be further processed by depositing graphene flakes onto the target substrate by screen printing or 3D printing. At the same time, the properties of graphene-based temperature sensors can be precisely determined by measuring the current change with temperature change.

The temperature of graphene is related to the scattering effect of electronic phonons and the scattering effect of other charged particles at high temperatures. When high temperatures are applied to graphene, the direction of electricity changes due to the scattering of electrons and phonons as well as the coupling of electrons and other electronic components (see Figure 5.3). This leads to a change in shielding. Temperature measurement is done by changing the resistance. The reliability of the sensor is very important for the accurate measurement of human body temperature. This reliability depends not only on the sensor material (e.g., graphene) but also on the flexibility. In addition to improved data sensitivity, these sensors should also be equipped with flexible substrates characterized by uniform deformation, high elasticity, and flexibility. There is availability in a variety of substrate materials, including flexible thermoplastic polymers (polycarbonate, polyimide, polyurethane, etc.), thermoset polymers (polydimethylsiloxane), textiles, and paper. Different types of soft substrates can be selected depending on the specific application. Continuous body temperature monitoring requires access to sensor data. Cloud computing technology and data analysis provide a great opportunity for further analysis of sensor data and storage in the Internet of Things (IoT) cloud. This allows users to easily receive data from any desired Internet device. Wireless communication can be achieved using different technologies such as Bluetooth low energy, radio frequency identification (RFID), near field communication (NFC), ZigBee, and remote control in the local area.[7,8]

A wireless sensor network (WSN) consists of many sensors connected to each other by wireless communication. Unlike traditional field connections, wireless sensor networks have stable communication and ensure the integrity and confidentiality of data during transmission. A wireless sensor network consists of many sections, such as a data receiving room, data transmission room, data processing room, and power room . It collects and processes data from sensors and sends it wirelessly, while

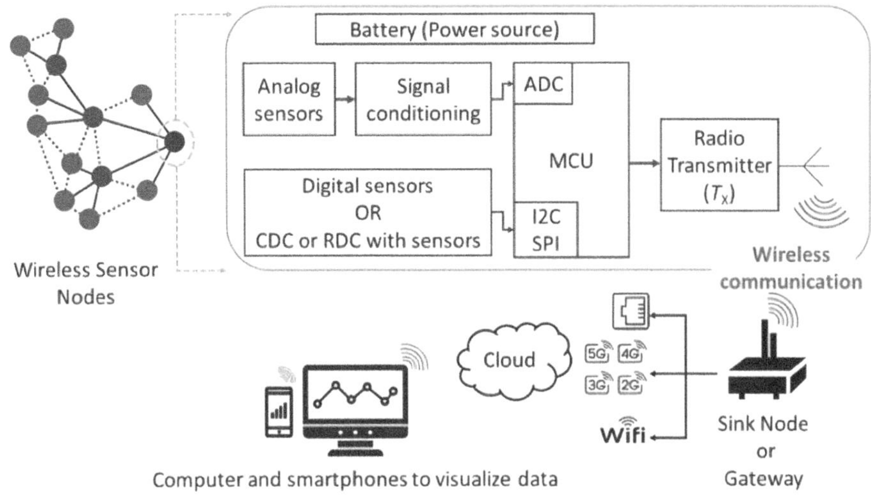

FIGURE 5.3 Block diagram of a typical wireless sensor network: represents a schematic diagram illustrating the structure and connectivity of a wireless sensor network. Graphene-based non-invasive wearable temperature sensors can be combined with wireless sensor networks to realize the remote monitoring of human health conditions.

Figure courtesy: www.sciencedirect.com/topics/engineering/wireless-sensor-network

also ensuring that energy flows to sensor nodes. Wireless sensors and transmission systems are essential for the IoT.

Graphene-based temperature sensors are widely used due to their low power consumption, usually tens to hundreds of microwatts. These integrated sensors have the advantage of instantaneous temperature measurement in different situations, whether it is personal health monitoring during physical activity or maintenance of remote areas in hazardous areas. The data collected by these sensors can be transmitted wirelessly for further analysis and decision-making.

Graphene-based noninvasive wearable temperature sensors have many advantages over traditional sensors and have great potential in physical monitoring and disease prediction. However, graphene-based noninvasive sensors are still difficult to use. These challenges include.[7,8]

The comfort of the wearable temperature sensor is also important for long-term use. In order to prepare a preferred sensor, the substrate material must be selected appropriately. Some substrates are less breathable. To solve this problem, it is important to look for other substrate materials that are lightweight, flexible, and have good skin compatibility. This will provide users with longer-term use.

The complex and expensive process involved in using graphene-based temperature sensors limits their wide application. To overcome this limitation, it is necessary to create lower-cost and simpler arrangements. We should focus on finding alternative techniques that are cost effective.

Another challenge faced by graphene-based temperature sensors is interference from different stimulus signals, so it is difficult to capture data from a specific sensor.

To overcome these problems, we need to develop signal processing methods and techniques to filter out unnecessary noise and improve the signal-to-noise ratio.

5.7 GRAPHENE-BASED PRESSURE SENSORS

Nowadays, cardiovascular diseases (CVDs) are the leading cause of death globally. CVD is group of disorders of the heart and blood vessels, and it includes coronary heart disease, cerebrovascular disease, rheumatic heart disease, and other conditions. Heart disease has a strong relation with blood flow to the heart. In cases of poor or irregular flow to heart, this can lead to cardiovascular disease. This flow depends on heartbeats or pulse rate. Pulse signals are generated by the periodic contraction and expansion of the heart muscle. So, it is deeply related with cardiovascular status. Continuous monitoring of pulse can significantly reduce accidents due to cardiovascular problems.

In the past few years, many sensors have been developed to monitor pulse signals. From the perspective of operation, pulse sensors can be classified into photoelectric pulse sensors, ultrasonic pulse sensors, and pressure pulse sensors. All of these sensors have some limitations. Photoelectric pulse sensors and ultrasonic pulse sensors cannot capture important pulse information, such as blood vessel elasticity, which directly affects the transmural pressure of blood vessels. The most widely used technology for pulse measurement in smart watches today is photoplethysmography (PPG),[9] but this technology is sensitive to changes in light and body movement. Therefore, it is difficult to measure pulse signals. Therefore, creating a low-cost, wearable pulse sensor is important for precision medicine, health engineering, chronic health care, and so on. Since pulses are weak signals, pressure sensors need a uniform location and sensitivity.[10] From this perspective, high-sensitivity, skin-compatible, flexible, and stretchable pressure sensors are required for accurate pulse measurement.[11,12]

In the past few years, significant progress has been made in noninvasive health-care devices such as continuous monitoring of intravascular blood pressure, intraocular pressure, and intra-abdominal pressure. Nanotechnology is an important application area in many fields, including the sensor industry for biomedical applications. Pressure sensors are one of the devices that have benefited from advances in nanotechnology. Ultra-miniaturization of sensors and peripherals is required to have ultra-low power and minimal processing. Graphene is the hardest, thinnest, and strongest material, and is considered one of the best electrical/electronic materials, making it an ideal candidate for noninvasive health products. There are many types of graphene pressure sensors used for health monitoring.

Researchers have developed a passive and sensitive wearable pressure sensor for health monitoring. It is made of a flexible film and graphene nanomaterials as flexible sensing elements. The electrical conductivity of the graphene content changes with the use of various pressures. The sensor device uses a flat square membrane design. The flexible film is fabricated with dimensions of 5 mm (Lg) × 5 mm (Wg) and a thickness of 130 μm (t). The length of the inter-digitated electrode (IDE) was optimized with dimensions of 4000 μm (length), 600 μm (width), and 200 μm (difference). The flexible pressure sensor and its IDE are fabricated on a flexible polyimide substrate with copper electrodes. The graphene on the sensor is fabricated by chemical vapor deposition (CVD). A 15 μm thin layer of polymethylmethacrylate (PMMA) polymer is applied on this graphene layer. The graphene on the copper foil with PMMA was immersed in ferric chloride solution for at least 6 h to remove the copper (Cu) catalyst layer. After the completion of the Cu catalyst layer, the graphene with PMMA will float on the solution. This will be called a pressure switch sensor. After the PMMA layer was completely removed, the sample was washed with isopropanol and dried with nitrogen. This graphene-based pressure sensor was used to measure blood pressure. The sensor has been used to measure changes in blood pressure up to 3 psi with an accuracy of 0.042 psi and a non-linearity of less than 1%. The device will be suitable for continuous monitoring of blood pressure measurement.[11,12]

The new pressure sensor is made of printed porous graphene (PG). The porous graphene is produced with printing ink and has the potential for mass production. The PG is protected by encapsulated plastic to absorb force and is sealed with silicone rubber to ensure conformity to the user's skin. The sensor shows unusual pressure (53.99/MPa) and high resolution (< 0.3 kPa) even at the lowest pressure. The sensor is flexible, strong, and durable, and can be attached to different parts of the body to measure heart disease, heart rate, and blood pressure. The sensor can be easily mounted on human skin and plays an important role in cardiovascular and other medical applications. When external pressure is applied, the overlapping area of the compressed portion increases. As a result, the PG resistance decreases as per the equation of contact resistance: $R_{contact} = k * d/S$, where k is the contact resistivity, d represents the interlayer spacing, and S indicates the overlapping area volume.

The fabrication process of this sensor is mentioned here. The electrodes were deposited on a polyimide substrate by electroless nickel plating and immersion gold (IG). GO ink was poured into the electrode by ink printing, and 30% (w/w) N2H4 aqueous solution was sprayed onto the dry GO film to form high-sensitivity porous graphene (PG). The oxygen groups of GO were further removed with hydrogen iodide and acetic acid (1:1)[11,12] and dried at 95°C for 2 hours. PET/EVA laminate film was used to cover the PG. Finally, the entire device is covered with silicone rubber for skin compatibility.

For testing purposes, these sensors are placed on the base of the forehead and the wrist to measure heart rate and pulse. The two sensors are connected to the skin with a polyurethane (PU) film, allowing them to capture all pulse signals. These sensors produce an output of sinusoidal pulse signals that may be affected by external noise or electromagnetic fields. Therefore, a low-pass filter is used to remove high-frequency components of noise and increase the signal.

One of the most commonly used fabrics in textiles is cotton fiber. Cotton fibers are more flexible and more durable than inorganic materials such as foam, sponge, and paper. The pressure sensor is made with the help of conductive cotton.[11,12]

Graphene-modified cotton fiber is used to measure the pressure. The fabrication of rGO cotton-based pressure sensors is very simple. The cotton was first cut into small pieces and then soaked in 2 mg/mL graphene oxide (GO) aqueous solution for 12 h. Then, the wet cotton was placed in a CVD furnace at 250°C for 5 h to convert GO into rGO in a nitrogen environment. The rGO cotton was cooled to room temperature in the furnace and then removed. The cotton rGO is then sandwiched between copper strips and fixed with silver paste. Here the copper strips act as electrodes. The dimensions of the upper electrode are 1 cm x 1 cm. Finally, the cotton-based pressure sensor is encapsulated with polyimide tape.

This cotton pressure sensor is useful for monitoring human physiological signals. The cotton pressure sensor is worn on the wrist to detect heart rate and on the throat to detect muscle contractions while talking. The rGO cotton-based pressure sensor has a sensitivity of 0.21 kPa at pressures up to 500 kPa. It also has good repeatability and stability. It can be used for pulse detection. This rGO-modified cotton-based pressure sensor has a wide range of applications in smart clothing.

Graphene is the hardest, thinnest, and strongest material, which makes it ideal for interference-free noninvasive pressure measurements. However, the process of preparing graphene films is very complex and requires expensive equipment. Chemical processing for large-scale graphene products is difficult to achieve because the large-scale production of the graphene film process often leads to defects in graphene. The process of turning nanoscale-thick graphene films into soft elastomers is another major challenge. Beyond these challenges, graphene-based noninvasive sensors have a promising future in medicine.[9]

5.8 GRAPHENE-BASED NANOMATERIALS IN BIOMEDICAL APPLICATIONS FOR THERAPEUTICS

Graphene-based nanomaterials (GBNs) are revolutionizing the field of biomedicine due to their unique properties. Their large surface area, excellent conductivity, and biocompatibility make them ideal candidates for various therapeutic applications. Some of the promising areas of graphene-based nanomaterials in therapeutics are discussed below.

GBNs can act as carriers for targeted drug delivery. They can be loaded with therapeutic drugs and then designed to release them in a specific location within the body. This can improve the efficacy of drugs while reducing side effects. GBNs can be used to deliver genes into cells for gene therapy applications. This approach has the potential to treat genetic diseases by correcting faulty genes. GBNs exhibit excellent antimicrobial properties. They can be used to kill bacteria, viruses, and fungi. This antimicrobial property makes GBNs ideal for developing new antibacterial and antiviral therapies. GBNs can be used to create scaffolds for tissue engineering. These scaffolds can provide support for cell growth and regeneration, promoting tissue repair. GBNs can be used to develop biosensors for the detection of diseases. These biosensors can detect biomarkers associated with diseases such as cancer at early stages, enabling early diagnosis and treatment. GBNs can be used as contrast agents for bioimaging techniques like magnetic resonance imaging (MRI) and photothermal therapy. This allows for better visualization of diseased tissues during diagnosis and treatment. Despite the promising potential of GBNs in therapeutics, there are still some challenges that need to be addressed. These include the following: While some GBNs have shown good biocompatibility, the long-term effects of these materials on the body are still unknown. Efficient and specific delivery of GBNs to diseased tissues remains a challenge. Large-scale production of GBNs with consistent properties is necessary for clinical applications.

5.9 EFFECT OF GRAPHENE-BASED MATERIALS ON THE HUMAN BODY

Graphene's sharp edges and high aspect ratio can cause physical damage to skin cells upon contact. Studies have shown that graphene nanoparticles can penetrate the skin's outermost layer (stratum corneum), leading to potential irritation and inflammation. Inhalation of graphene nanoparticles is a major concern due to their small size and ability to travel deep into the lungs. These nanoparticles can accumulate in the lungs, causing inflammation, oxidative stress, and potentially fibrosis (scarring). The effects of graphene on the cardiovascular system are still under investigation. Some studies suggest that graphene nanoparticles may interact with blood components, potentially causing platelet aggregation (clumping) and thrombosis (blood clot formation). Ingestion of graphene nanoparticles is another possible exposure route. Studies suggest that these nanoparticles may pass through the gastrointestinal tract without significant absorption. However, potential damage to the gut lining and associated inflammation cannot be ruled out. The possible effects of graphene on reproduction are not fully understood. Some studies have shown that graphene nanoparticles can accumulate in the testes and ovaries of laboratory animals, raising concerns about potential impacts on fertility. The possibility for graphene nanoparticles to reach the brain is a major concern. Due to their small size, they may be able to cross the blood-brain barrier, a protective layer that separates the bloodstream from the brain. The effects of graphene on brain function and development are still being investigated. However, the effect of graphene on the biological system is depending on following points.

The size, shape, and surface properties of graphene nanoparticles significantly influence their biological effects. Smaller nanoparticles with sharper edges tend to be more potent in causing harm. The severity of the effects depends on the amount of graphene exposure and the duration. Higher doses and longer exposure times are likely to cause more significant effects. Graphene can be modified with various functional groups to alter its properties. Functionalization may influence its interaction with biological systems and potentially reduce toxicity.

5.10 CONCLUSION

Graphene is a promising material with vast potential applications. However, it is crucial to understand its potential risks to human health. Further research is needed to ensure the safe development and use of graphene technologies. One should wear appropriate personal protective equipment (PPE) when handling graphene nanoparticles. Engineering controls should be implemented to minimize airborne exposure in workplaces. Safety assessments should be conducted before using graphene in consumer products. By following these guidelines, one can harness the potential of graphene while minimizing potential risks. It is possible to find more information and relevant pictures through a web search using the terms mentioned throughout this chapter. In conclusion, graphene-based nanomaterials hold tremendous potential for revolutionizing various therapeutic applications. Their unique properties offer new avenues for targeted drug delivery, gene therapy, tissue engineering, and disease diagnosis. As research continues to address the existing challenges, GBNs are poised to make a significant impact on the future of medicine. Research on the biological effects of graphene is ongoing. Scientists are working to understand the mechanisms by which graphene interacts with cells and tissues and to develop strategies to mitigate potential risks. Here are some key areas of future research:

- Developing safer forms of graphene with reduced toxicity.
- Understanding the long-term effects of graphene exposure.
- Establishing safety standards for graphene production and use.

ACKNOWLEDGMENTS

I sincerely acknowledge the support of my colleague and guide to channelizing the research in a meaningful and effective manner.

REFERENCES

1. Singh R., Jansen M., Ganguly D., Kulkarni D., Ramaprabhu S., Choudhary S.K., Pramanik C. (2022). Shellac Derived Graphene Films on Solid, Flexible, and Porous Substrates for High Performance Bipolar Plates and Supercapacitor Electrodes and Tata Steel Training Document. *Renewable Energy*, 181, 1008–1022. https://doi.org/10.1016/j.renene.2021.09.091.
2. Kulakova I., Lisichkin G. (2022). Biosensors Based on Graphene Nanomaterials. *Moscow University Chemistry Bulletin*, 77(6), 307–321. https://doi.org /10.3103/S0027131422060049.
3. Bai Y., Xu T., Zhang X. (2020). Graphene-Based Biosensors for Detection of Biomarkers. *Micromachines*, 11(1), 60. https://doi.org/10.3390/mi11010060.
4. Huang H., Shi S., Nan W., Hao W., Shu W., Hengchang B., Litao S. (2019). Graphene-Based Sensors for Human Health Monitoring. *Frontiers in Chemistry*, 7, 399. https://doi.org/10.3389/fchem.2019.00399.
5. Bahamonde P., Nguyen J., Fanourakis H. (2018). Recent Advances in Graphene-Based Biosensor Technology with Applications in Life Sciences. *Journal of Nanobiotechnology*, 16(1), 75. https://doi.org/10.1186/s12951-018-0400-z.
6. Mao P., Li H., Yu Z. (2023). A Review of Skin-Wearable Sensors for Non-Invasive Health Monitoring Applications. *Sensors*. 23(7), 3673. https://doi.org/10.3390/s23073673
7. Liu J., Wang Y., Li X., Wang J., Zhao Y. (2023). Graphene-Based Wearable Temperature Sensors. *Nanomaterials*, 13(16), 2339–2358 https://doi.org/10.3390/nano13162339.
8. Nag A., Roy B., Simorangkir b., Gawade D., Nuthalapati S., Buckley J., Brendan F., Mehmet A., Mukhopadhyay S. (2022). Graphene-Based Wearable Temperature Sensors. *Materials & Design*, 221, 110971. https://doi.org/10.1016/j.matdes.2022.110971.
9. Jiang H., Peng X., Wei L., Jiangwei S., Zhang L., Liang Y., Caofeng P., Shiao K.,Chen, T. (2019). A Universal High Accuracy Wearable Pulse Monitoring System via High Sensitivity and Large Linearity Graphene Pressure Sensor. *Nano Energy*, 59. https://doi.org/10.1016/j.nanoen.2019.02.036.
10. Shazni M., Lee M., & Lee H. (2017). Highly-Sensitive Graphene-based Flexible Pressure Sensor Platform. *Sains Malaysiana,* 46. https://doi.org/10.17576/jsm-2017-4607-19.

11. Peng Y., Zhou J.. Song X.. Pang K., Samy A., Hao Z., Wang J. (2021). A Flexible Pressure Sensor with Ink Printed Porous Graphene for Continuous Cardiovascular Status Monitoring. *Sensors*, 21(2), 485. https://doi.org/10.3390/s21020485.
12. Li P., Zhao L., Jiang Z. (2019). A Wearable and Sensitive graphene-cotton Based Pressure Sensor for Human Physiological Signals Monitoring. *Scientific Report*, 9. https://doi.org/10.1038/s41598-019-50997-1.

6 Engineered Nanoparticles for Drug Delivery System

Soumik Podder, Subhrajit Chowdhury, Bishal Sarkar, Priyanka Daw, Aneek Nath, and Anurup Nag

6.1 INTRODUCTION

Nanomaterials belong to a specific branch of materials whose at least one-dimensional measurement is between 1 and 100nm. Natural materials can be converted into nanomaterials either by a top-down or a bottom-up approach as per the user's requirements. Nanomaterials generally possess unique physicochemical properties such as optical, magnetic, catalytic, electrical characteristics and so on. The structural miniaturization in nanomaterials produces significant elevation in surface energy due to exposure of more dangling or unsatisfied chemical bonds in the enhanced surface area. The surface energy essentially brings fascination in the aforementioned properties. Nanomaterials, based on their dimensional miniaturization, are classified into three major categories: 1) Three-dimensional nanoparticles (3DNPs), 2) Two-dimensional nanoparticles (2DNPs) and 3) One-dimensional nanoparticles (1DNPs). Zero-dimensional quantum dots (0DQDs) have exhibited breakthrough optical, magnetic and catalytic properties due to a strong electron confinement effect.[1]

In practice, spherical NPs are available most as they are easy to synthesise through a simpler wet chemical process. The smaller-sized NPs are preferred in biomedical applications, particularly in the drug delivery system, due to higher drafting density, but for specifically targeting the affected cell, rod-shaped and star-shaped NPs are preferred.[2,3] Rod-like morphology exhibited enhanced luminescence properties that are helpful in cellular imaging and tracking of the drug.[4] As a whole, NPs offer excellent retention possibility, high chemical endurance, proven low cytotoxicity and biocompatibility.[5] Nowadays the term 'engineered' is often used, especially in nanotechnology. Engineered nanotechnology is conceptualized as tailoring the surface energy or surface area of bare NPs for enhancement in their activity. Tailoring in surface energy is performed by the use of surfactant or a capping agent. The use of a capping agent reduces the size of the bare NPs so as to inhibit agglomeration. Therefore, the employment of a capping agent automatically enhances the surface energy of the NPs and consequently facilitates the penetration of the NP-coated drugs to the target cell. This capping-inspired engineering in NPs is primarily observed in metal and metal-oxide NPs. Polymers are generally used as a capping agent in NPs. Other than

 DOI: 10.1201/9781003453635-6

metal and metal-oxide NPs, non-metallic NPs are drawing much attention in the drug delivery system as the inorganic NPs promote the "green drug delivery" concept.[5] Inorganic NPs such as lipids and polymer NPs are less toxic than metal and metal-oxide NPs, and the synthesis process involves green technology rather than a cluster of hazardous chemicals and related by-products.[6] In this chapter we have lluminated organic NPs in the drug delivery system. Another aspect in the significant usage of organic NPs in the drug delivery system is less catalytic activity and resultant reactive oxygen species (ROS) generation, which may adversely affect the microenvironment of targeted cells and normal cells. Another important concern in the drug delivery system is effectiveness in overcoming biological barriers in the cellular environment. A NP-induced drug delivery system has significant authority to overcome different biological barriers and quickly target the affected cell. Another perspective is hybrid NPs developed from making a heterostructure of two different NPs, hierarchical NPs and polymer-based composites and so forth, that are aggressively exploited in the therapeutic system due to cultivation of diverse properties using a single material. Thus, this chapter demonstrates the exploitation of the aforementioned engineered NPs to rejuvenate the drug delivery architecture.

6.2 BIOMEDICAL PROSPECTS OF NANOMATERIALS

The biomedical prospects of nanomaterials are a major exciting area of research nowadays as some diseases are now more drug resistant. The biomedical activities of nanomaterials can be categorized into six groups:

i. Antimicrobial therapy
ii. Cancer therapy
iii. Endocrine therapy (Anti-diabetic)
iv. Neurodegenerative disease therapy
v. Anti-inflammatory activities
vi. Anti-oxidant therapy[5]

The root causes of complex diseases are primarily classified as microbes and free radicals. Antimicrobial activity is regarded as one of the prime biomedical activities in today's health scenario. If we look into skin disease, coughs, colds, allergies, endodontic issues, dysentery, respiratory diseases, urinal tract infections, pneumonia and so forth, all are caused by gram-positive and gram-negative bacteria, fungi and viruses. The established medication system encourages uptake of antibiotics (micrometre sized) that gradually become less effective to microbes, as microbes are continuously changing their genetic structure in order to overcome the antibiotic effect, and astonishingly this game strategy makes the microbes more drug resistant. In this scenario, nanodimensional materials have proven themselves as the most significant candidate to combat the new variants of microbes. Among nanomaterials, metal-oxide NPs such as ZnO (n-type), TiO_2(n-type), CuO(p-type), Ag_3PO_4(n-type), NiO (p-type), Ni_2O_3(p-type), and V_2O_5 (n-type) significantly exhibit antimicrobial activity in an elevated manner.[6,7] On the other hand, metal NPs such as Ag NPs, Au NPs, Cu

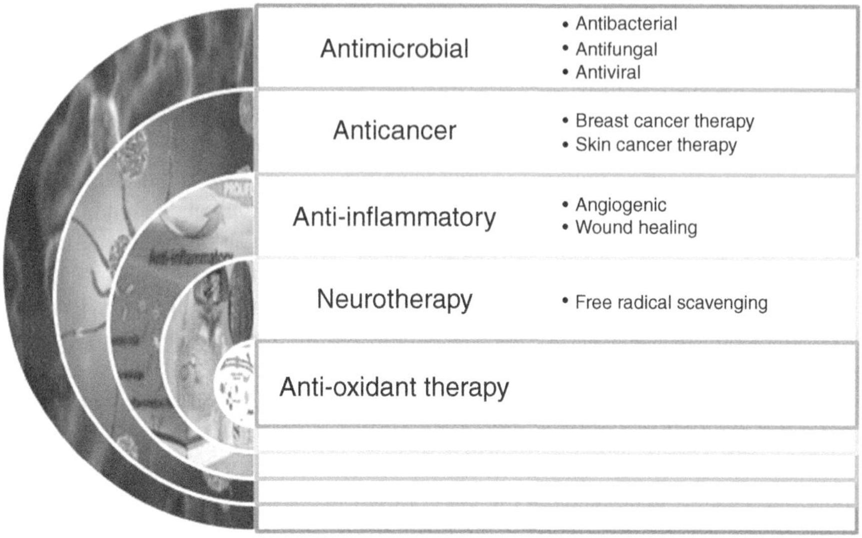

FIGURE 6.1 Biomedical prospects of nanomaterials.

NPs, Si NPs, Ni NPs and so forth also combat microbial infections in a very competitive manner.[8] The plausible mechanism for these diseases is classified into two approach: 1) Non-Reactive Oxygen Species (ROS)-mediated approach and 2) ROS-mediated approach.[9]

The non-ROS-mediated approach is generally confined to adsorption of nanomaterials, cell wall destruction, leakage of proteins, nucleic acid and finally cell death.[9]

The ROS-mediated approach is generally subjected to the production of oxygen-centric free radicals like superoxide anion (O_2^-), hydroxyl radical ($^{\circ}OH$), hydrogen peroxide (H_2O_2), singlet oxygen ($1O_2$) by aerial oxygen, and moisture reaction. Experimentally and theoretically it was proved that the foundation of such free radical production is a serious consequence of a cross-relation between NPs' band edge energy and the redox potential of free radicals.[5]

NPs' specific electronic structures are responsible for the production of ROS. Comprehensibly, redox potential (E_H) of $O_2/O_2^{\cdot-}$ is -0.2 eV and `being conduction electron mediated process, the band edge potential of conduction band, i.e. E_{CB} will be less than -0.2eV for detectable $O_2^{\cdot-}$. Similarly the redox couple (E_H) of H_2O/OH is 2.2 eV, and being valence band mediated generation the band edge potential of the valence band (E_{VB}) will be authentically greater than 2.2eV. $O_2/^1O_2$has E_H=1.88eV and successful generation of $^1O_2$ requires $E_{VB} > 1.88$eV.[8]

NPs such as Si, Ag, Au and Ni are are chosen due to their frequent usage in industrial, commercial biomedical fields, as well as to study their antibacterial properties, which are essential for environmental purposes. Si NPs have a band gap of 2.0 eV with E_{CB} and E_{VB} of 0.1 and 2.1 eV respectively, so they cannot produce .OH as well as $O_2^{\cdot-}$, as explained previously. Concomitantly, $^1O_2$ generation is possible, which we

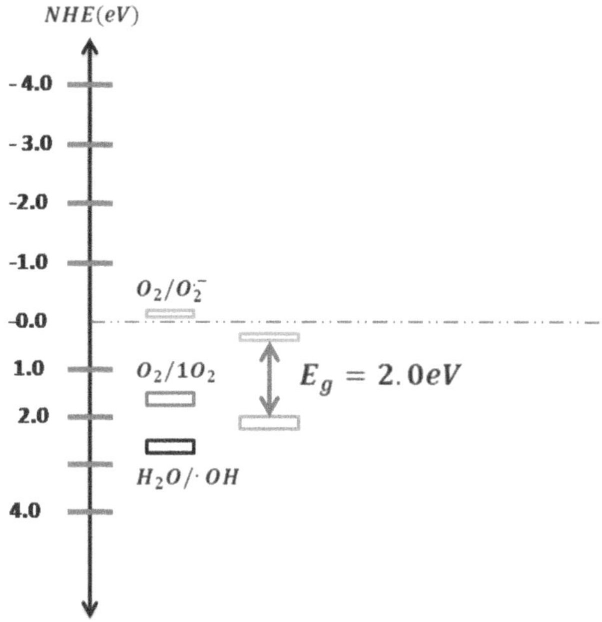

FIGURE 6.2 Selective generation of ROS depending on band edge positions of Si NP in aqueous suspension (pH = 5.6). All measurement was carried out with respect to normal hydrogen electrode (NHE).

have noted in the previous section.[8] The themes of different types of ROS generation are shown in Figure 6.2.

From metal oxide NPs, ROS are evolved after UV light (λ=365 nm) irradiation as $h\nu \geq E_g$, where E_g = band gap of metal oxide. As a result of this, electron-hole pairs are generated from metal-oxide NPs, out of which the photogenerated electrons in the conduction band (CB) and holes in the valence band (VB) react with molecular oxygen and surface adsorbed water/moisture to generate ROS (shown in Figure 6.3). Primarily three types of primary ROS are produced from metal-oxide NP, viz. superoxide anion ($O_2^{\cdot-}$), hydroxyl radical ($\cdot OH$), and singlet oxygen ($^1O_2$).[5]

There are other ROS produced in metal-oxide NPs known as secondary ROS, such as hydroperoxyl radical ($\cdot HO_2$) and hydrogen peroxide (H_2O_2), which are evolved from $O_2^{\cdot-}$ by reductive process. .OH is also produced from $O_2^{\cdot-}$ after successful reduction of H_2O_2.[9,10] The pathway of ROS generation in metal-oxide NPs is as follows:

Step:1 UV light absorption by metal oxide (MO) and electron-hole pair generation at CB and VB

$$MO + h\nu \longrightarrow e^-_{CB} + h^+_{VB} \tag{6.1}$$

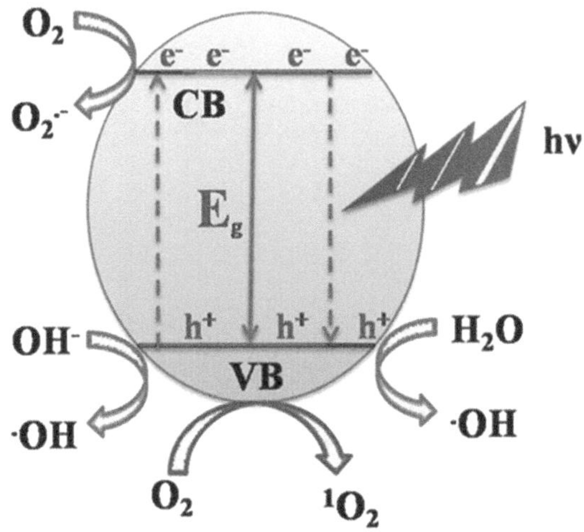

FIGURE 6.3 ROS generation in metal oxide nanoparticle.

Step:2 Oxygen ionosorption from molecular oxygen

$$(O_2)_{ads} + e^-_{CB} \longrightarrow O_2^{\cdot-} \tag{6.2}$$

Step:3 $\cdot HO_2$ generation after neutralization of $O_2^{\cdot-}$ by proton

$$O_2^{\cdot-} + H_2O \longrightarrow \cdot HO_2 + OH^- \tag{6.3}$$

Step:4 H_2O_2 generation by dismutation reaction

$$.HO_2 + \cdot HO_2 \longrightarrow H_2O_2 + O_2 \tag{6.4}$$

Step:5 Second reduction of oxygen to form $\cdot OH$

$$H_2O_2 + O_2^{\cdot-} \longrightarrow O_2 + \cdot OH + OH^- \tag{6.5}$$

Step:6 Oxidation of adsorbed H_2O and neutralization of OH^- by photoholes

$$h^+ + (H_2O)_{ads} \longrightarrow h^+ + (H^+ + OH^-)_{ads} \longrightarrow \cdot OH + H^+ \tag{6.6}$$

Step:7 Oxidation of oxygen ion by photoholes

$$h^+ + O_2^{\cdot-} \longrightarrow {}^1O_2 \tag{6.7}$$

The generation of ROS in metal-oxide NPs develops oxidative stress, which is attributed to cellular toxicity.

The values of E_{VB} and E_{CB} are derived as follows:

$$E_{VB} = \chi - 4.5 + \frac{1}{2} \times E_g \quad (6.8)$$

$$E_{CB} = E_{VB} - E_g \quad (6.9)$$

where χ is Mulliken's electronegativity, the geometric mean of absolute electronegativity of constituent elements, and E_g is the band gap of metal oxide NP (derived from Wood-Tauc's plot).

6.3 ANTIBACTERIAL ACTIVITIES OF METAL-OXIDE NPS

Among metal oxide functioning as antibacterial agents, ZnO is well versed with worldwide researchers. As a n-type semiconducting metal oxide, it has a rich level of oxygen vacancies which essentially act as electron donor sites and facilitate the generation of $O_2^{\cdot-}$ and related ROS.[11] Numerous research studies have been conducted on the antibacterial activity of ZnO and its hybrid structures. Even the Zn^{2+} ion is employed in wound healing, especially from burns, due to its anti-inflammatory, drying, antiseptic and astringent properties. Most of this research has been focused on the study of the effects of variation in morphology, shape, size, synthesis parameters and so forth on antibacterial properties, aside from the mechanism of antibacterial activity.[12,13] Even the existing mechanism of antibacterial activity of ZnO nanostructures is based on Zn^{2+} ion dissolution and internalization of nanostructures, which eventually depicts the mortality of bacteria due to consumption of excessive Zn^{2+} ions (treated as an oligoelements) and to physical damage to the cell membranes of bacteria.[14] With the advent of scientific research on the antibacterial activity of ZnO, Singh *et. al.* documented some advancements in proposing the inner mechanism of morphology-controlled antibacterial activity. The researchers designed a ZnO spherical, nanorod, cauliflower and mushroom-like architecture and investigated antibacterial activity against *S. aureus* under UV light and in the dark. The antibacterial activity was profound for the cauliflower-like structure among other assemblies and also higher under UV light than in a dark state. This extraordinary feature occurred due to the presence of large number of oxygen vacancies, as evidenced from photoluminescence (PL) spectroscopy, and eventually these native point defects produced enhanced ROS in UV light.[12] Even for ZnO-based composite materials, viz. Ag-ZnO/Chitosan complex, Polyaniline/$Cu_{0.05}Zn_{0.95}O$ complex, and graphene oxide (GO), ZnO illustrated potential antibacterial and antifungal activity with the predicted mechanism, viz. cell membrane permeability change due to strong electrostatic interaction (estimated from scanning electron microscopy (SEM), absorption spectroscopy at 260 nm to evaluate membrane integrity respectively); enhanced H_2O_2 generation from Polyaniline/$Cu_{0.05}Zn_{0.95}O$ surface due to charge separation; enhanced Zn^{2+} ion release from ZnO due to presence of GO as storage sites; and so forth but lacked experimental evidence of ROS generation.[14] Apart from ZnO nanostructures, other metal oxides like TiO_2, Fe_2O_3,Fe_3O_4, CuO, MgO and so on showed potential antibacterial activity along with their hybrid structures. Hypotheses have been proposed to explain the underlying mechanism of antibacterial activity, viz. ROS production in other metal oxides under various environmental conditions, viz. UV illumination, dark, visible light irradiation and so forth.[5] Raut *et. al.* reported such underlying mechanisms of antibacterial activity of nitrogen-doped TiO_2 (N-TiO_2) relative to native point defects (oxygen vacancies measured by X-ray photoelectron spectroscopy) and ROS production under UV (380 nm), sunlight and room light. The excellent bactericidal property of N-TiO_2 under room light and sunlight is due to increased charge carrier separation owing to oxygen vacancies which facilitate 1) the disruption of cell membranes by ROS, and 2) the disordering of inner cytoplasmic membranes and decomposition of bacterial cells; however, no direct evidence of ROS production was documented.[15] Similar charge carrier separation-induced high bacterial (*S. aureus*) inactivation by carbon-doped anatase-brookite TiO_2 nanoheterojunction was reported by Etacheri *et. al.*[16] The charge carrier separation, monitored by PL spectra, would enable enhancement in ROS generation, which eventually empowered antibacterial activity.[16] This visible light-induced antibacterial activity could be extended to dark after successful conjugation of silver halide/TiO_2 with hydroxyapatite (Hp). The phosphate (PO_4^{3-}) group under UV illumination acts as an electron trapping centre due to the formation of oxygen vacancies which facilitate the generation of $O_2^{\cdot-}$, and in the dark there are plenty of covalent bonds formed between bacterial surfaces and PO_4^{3-} groups that prevent bacteria from taking nutrition from the environment; thus, the inhibition of bacterial growth instead of killing occurs.[17]

Similar dark- and visible light-induced antibacterial activity of the CuO/$Cu(OH)_2$ hierarchical structure was documented by Akhvan *et. al.*, where visible light-driven antibacterial activity is attributed to·OH radical generation from adsorbed .OH species on CuO surfaces due to reaction with photogenerated holes, whereas dark-induced antibacterial properties were developed due to an increased Cu^{2+} ion release from enlarged surface density and physical damage to cell membranes.[18]

The NP drug carriers should have antibacterial activity apart from drug targeting to protect wounded cells.

6.4 ADVANTAGES OF NPS IN DRUG DELIVERY

NPs present a promising avenue for overcoming limitations associated with traditional drug delivery methods. These limitations include poor solubility, rapid degradation, non-specific targeting and unintended side effects. NPs offer unique physical and chemical properties that make them attractive for drug delivery applications. By encapsulating or attaching drugs to NPs, researchers can design innovative drug delivery systems with enhanced therapeutic efficacy and improved safety profiles.

In the context of drug delivery, NPs are submicron colloidal particles, typically measuring less than 1 micron (< 1μm). This category encompasses nanospheres, where the drug is adsorbed, dissolved or dispersed throughout the matrix, and nano capsules, where the drug is confined to an aqueous or oily core surrounded by a shell-like wall. Additionally, drugs can be covalently attached to the surface or within the matrix of NPs.

Various biocompatible and biodegradable materials are employed to create NPs, including natural polymers such as gelatine and albumin, synthetic polymers like polylactides and polyalkylcyanoacrylates, as well as solid lipids. These materials ensure that the NPs are compatible with the body's biological systems.

In the body, the release of drugs from NPs is typically facilitated by processes such as diffusion, swelling, erosion or degradation of the NP matrix. This controlled release mechanism enhances the therapeutic benefits of drug delivery, allowing for sustained and targeted delivery of medications while minimising adverse effects. Overall, the use of NPs in drug delivery systems represents a promising strategy to address the shortcomings of traditional drug administration methods.[6]

6.5 ORAL ADMINISTRATION OF NANOPARTICLE-BASED TB DRUGS

The stability of NPs makes them suitable for oral administration, and their fate in the gastrointestinal tract has been extensively studied. NP uptake typically occurs through various mechanisms, including transcytosis via M cells; intracellular uptake and transport by epithelial cells lining the intestinal mucosa; and uptake via Peyer's patches.[19]

NPs have potential for sustained drug release and enhanced efficacy after oral administration. The researchers co-encapsulated three frontline anti-TB drugs (rifampin, isoniazid, and pyrazinamide) in poly(lactide-co-glycolide) (PLG) NPs. After a single oral administration in mice, therapeutic drug concentrations were maintained in circulation for several days, ranging from 4 days for rifampin to 9 days for isoniazid and pyrazinamide. In contrast, free drugs were cleared from the plasma within 12 to 24 hours. The nanoparticle-bound drugs, administered every tenth day, resulted in complete bacterial clearance from the organs. Free drugs required daily administration of 46 doses to achieve similar efficacy. Similar results were observed in guinea pigs.[20]

However, the incorporation of drugs into microparticles was less effective, with lower drug-loading capacity and shorter plasma half-life. The bio-adhesive properties of polymeric NPs, influenced by lectins, play a crucial role in their behaviour in the gastrointestinal tract. Lectins enhance mucoadhesion by recognising glycosylated structures in the intestine, thereby improving drug absorption and bioavailability.

To further enhance the efficacy of PLG-based formulations of anti-TB drugs, wheat germ agglutinin was covalently attached. This modification extended the serum half-life of the drugs, with detectable levels observed for 6 to 7 days for rifampin and 13 to 14 days for isoniazid and pyrazinamide. The lectin-modified formulations, administered every 14 days, achieved bacterial clearance in organs after only 3 doses, compared to 45 daily doses of free drugs. The prolonged circulation of drugs in wheat germ agglutinin-grafted NPs was attributed to enhanced adhesion, allowing increased absorption time and localized concentration gradients in the intestine.[20]

6.6 POTENTIAL FOR THE INHALATION FORM APPLICATION

Direct delivery of TB drugs to the lungs through pulmonary administration offers several potential advantages, including the possibility of reducing systemic

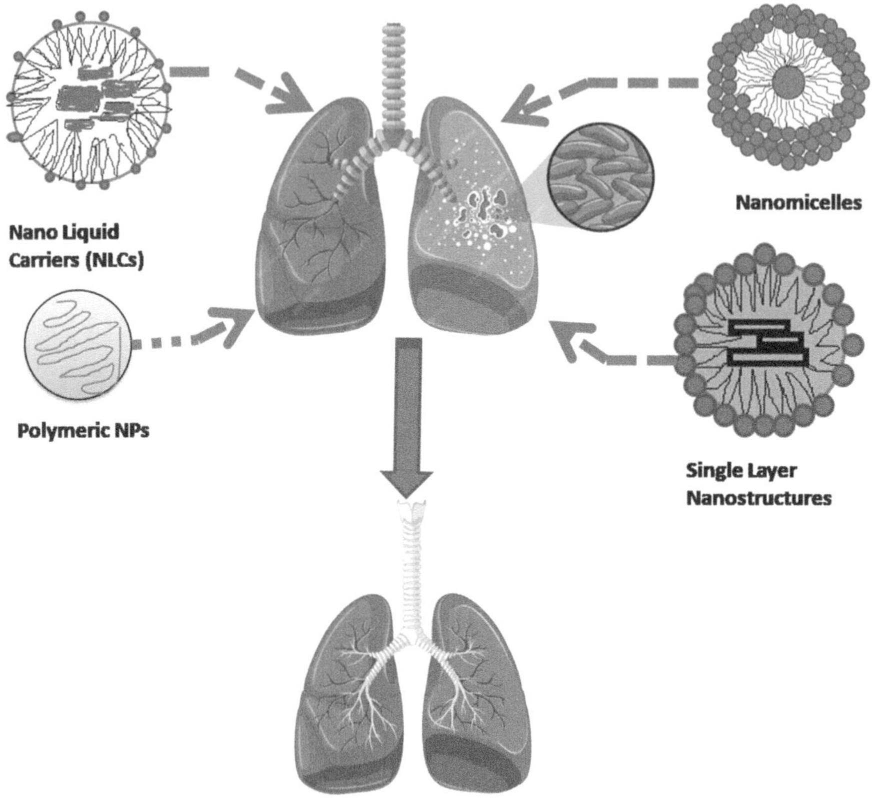

FIGURE 6.4 Oral administration of NPs.

toxicity and achieving higher drug concentrations at the primary site of infection. Additionally, inhaled drugs avoid first-pass metabolism, a process encountered with oral administration.

Despite potential challenges related to the mass median aerodynamic diameter of nanocarriers, studies have demonstrated the effectiveness of pulmonary drug delivery using NPs. In guinea pigs, the pharmacokinetics and antibacterial effects of anti-TB drugs encapsulated in poly(lactide-co-glycolide) (PLG) NPs were investigated. A single nebulization of rifampin (RMP), isoniazid (INH), and pyrazinamide (PZA) co-encapsulated in PLG NPs resulted in sustained therapeutic drug levels in the plasma for 6 to 8 days and in the lungs for up to 11 days. This was comparable to the effects observed after oral administration of the nanoparticulate formulation of the same drugs. Nebulization of NPs to M. tuberculosis-infected guinea pigs at every 10th day achieved a significant therapeutic benefit, with no detectable tubercle bacilli in the lungs after only 5 doses of treatment, compared to 46 daily doses required for orally administered drugs.[20]

The effectiveness of pulmonary drug delivery was further enhanced by using wheat germ agglutinin-functionalized PLG NPs. Three doses administered fortnightly for

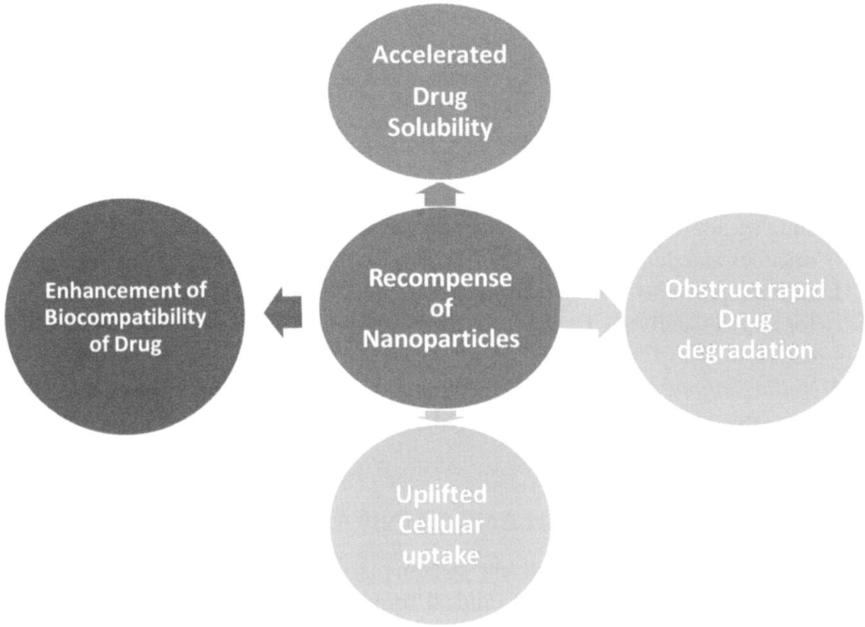

FIGURE 6.5 Flow chart of advantages of nanoparticles.

45 days produced a sterilizing effect in the lungs and spleen of infected guinea pigs. Solid lipid NPs loaded with anti-TB drugs also achieved a sterilizing effect, with no detectable tubercle bacilli in the lungs and spleen after seven doses of treatment. Solid lipid NPs offer advantages such as a composition of physiological compounds and the feasibility of large-scale production without the use of organic solvents.[20]

- **Enhanced Solubility**: Many drugs are poorly soluble in water, limiting their absorption and bioavailability. NPs can increase drug solubility by providing a larger surface area for interaction with water molecules. This allows for greater drug loading and ultimately higher drug concentration at the target site.
- **Increased Bioavailability**: Bioavailability refers to the fraction of an administered drug that reaches the bloodstream and becomes available to exert its therapeutic effect. NPs can protect drugs from degradation in the bloodstream and facilitate their passage through biological barriers, leading to improved bioavailability and ultimately lower effective doses.
- **Targeted Delivery**: NPs can be functionalized with ligands (targeting molecules) that specifically bind to receptors on diseased cells or tissues. This allows for targeted delivery of the drug, minimizing exposure to healthy tissues and reducing side effects.
- **Controlled Release**: NPs can be designed to release their drug cargo slowly and steadily over a prolonged period, reducing the need for frequent dosing and improving patient compliance. This controlled release profile can also optimize drug concentration at the target site, enhancing therapeutic efficacy.

- **Reduced Side Effects**: By delivering drugs directly to the target site, NPs can minimize exposure to healthy tissues and organs, thereby reducing systemic side effects associated with conventional drug delivery methods.

6.7 TYPES OF NANOPARTICLES IN DRUG DELIVERY

Polymeric nanoparticles: These are biodegradable and biocompatible, making them suitable for various drug delivery applications. Examples include liposomes, micelles and dendrimers.

Lipid nanoparticles: These offer high drug-loading capacity and can be modified for targeted delivery. Examples include solid lipid NPs and nanostructured lipid carriers.

Metallic nanoparticles: These have unique optical and magnetic properties, allowing for imaging and controlled drug release. Examples include gold NPs and iron oxide NPs.

6.8 CURRENT RESEARCH AND CLINICAL APPLICATIONS

Research on NP-based drug delivery systems is rapidly advancing, with numerous promising candidates currently in clinical trials. These systems are being explored for the treatment of various diseases, including cancer, infectious diseases, neurodegenerative diseases and diabetes.

Mesoporous silica NPs for drug delivery

Mesoporous silica materials (MSMs) were initially discovered in the 1990s by researchers in Japan and the United States. These materials are created by utilising self-assembled surfactant molecules as templates for silica precursors, resulting in a structure filled with interconnected cavities upon template removal. MSMs are known for their ordered pore distribution, uniform pore sizes ranging from 2 to 20 nm, high pore volume (approximately 1 cm^3 g^{-1}), significant surface area (around 1000 m^2 g^{-1}), and a dense presence of silanol groups on their surface, facilitating subsequent functionalization processes. Due to these characteristics, MSMs have found applications in various fields, including drug delivery systems, as suggested by the Vallet-Regí group in 2001.

MSMs also exhibit bioactive properties due to the silanol groups on their surface and a chemical composition similar to bioactive glasses. Consequently, they have been utilized as starting materials for manufacturing 3D scaffolds for bone tissue engineering (Figure 6.1). Inspired by the remarkable textural properties and biomedical potential of bulk MSMs, researchers ventured into the realm of nanoscale materials, leading to the development of mesoporous silica nanoparticles (MSNs). MSNs have garnered significant attention as drug delivery systems, with applications spanning cancer treatment, infectious diseases and various bone ailments. Moreover, MSNs serve as effective imaging agents, given their ability to encapsulate diverse dyes and contrast agents within their pores.[21]

The exceptional properties of MSNs have spurred the creation of advanced multifunctional materials for diverse biotechnological applications. Recent breakthroughs in biomedical research utilizing MSNs offer promising avenues for personalized treatments and diagnostic techniques with exceptional selectivity. Advances in nanotechnology, encompassing synthesis and characterization techniques, enable the development of NPs capable of intricate interactions within the biological milieu. This progressive attribute will delve into recent advancements in MSNs and their applications in drug delivery technologies.[21]

Preparation of mesoporous silica nanoparticles

Silica NPs are formed of colloidal amorphous metal oxide, which can be synthesised via the sol-gel technique, exhibiting diameters ranging from 50 to 300 nm. Various methods exist for producing MSNs, typically involving three key processes: the sol-gel method for silica production, the utilization of surfactants as structure-directing agents for generating mesoporous materials, and a modified Stöber method under diluted conditions to produce spherical NPs.

Traditionally, silica NPs are synthesized through the hydrolytic sol-gel route, where silicon alkoxide precursors undergo hydrolysis and condensation under acidic or basic catalysts. Polycondensation occurs around the surfactant templates, resulting in an oxide network that forms a colloidal sol, gradually progressing to gel or discrete particles based on the reaction conditions. Employing extremely diluted conditions facilitates the production of monodisperse spherical silica particles.

The synthetic process involves initially dissolving the surfactant in water at a basic pH, followed by the dropwise addition of silica precursors at low concentrations. Finally, the surfactant template is removed via solvent extraction, yielding mesoporous silica NPs.[22]

Properties of mesoporous silica NPs

Mesoporous silica nanoparticles (MSNs) possess several key properties that make them highly advantageous for various applications, particularly in biomedicine and nanotechnology. Some of the important properties of MSNs include:

i. **Tuneable Physicochemical Characteristics**: MSNs offer tuneable properties such as particle size, pore size, pore volume, surface area, pore structure and surface functionality. These characteristics can be adjusted to meet specific application requirements, making MSNs versatile platforms for various purposes.[21]
ii. **Porous Structure**: The mesoporous structure of MSNs provides a network of interconnected pores with well-defined sizes and shapes. This porous architecture allows for high surface area and large pore volume, enabling efficient loading and controlled release of guest molecules.
iii. **Drug Delivery Capabilities**: MSNs are widely used as drug delivery vehicles due to their ability to encapsulate and deliver therapeutic agents in a controlled manner. The porous nature of MSNs allows for high drug-loading capacity, while their tuneable properties enable customization of drug release kinetics.

iv. **Biocompatibility**: MSNs are generally biocompatible and exhibit low toxicity, making them suitable for biomedical applications such as drug delivery, imaging and tissue engineering. Surface modifications can further enhance their biocompatibility and targeting specificity.
v. **Versatility in Synthesis**: MSNs can be synthesized using various methods, including sol-gel processes, template-assisted techniques and surface functionalization strategies. This versatility in synthesis allows for the production of MSNs with tailored properties and functionalities.

Mesoporous silica NPs as smart drug delivery systems

The success of MSNs as drug delivery systems is attributed to their distinctive physical and chemical properties. These properties offer a promising arsenal for tackling the shortcomings of conventional medicine. They address issues like drug non-specificity, limited efficacy windows, low solubility or stability of drugs, unfavourable pharmacokinetic profiles, and potential side effects. In general, any NP serving as a drug delivery system must meet certain essential criteria. These criteria include efficiently loading cargo molecules, releasing cargo on-demand to prevent premature release, targeting diseased tissue to deliver cargo precisely where needed, and, particularly in cancer treatment, penetrating deeply into tumours.[21]

Loading and protecting therapeutic cargo

MSNs serve as versatile carriers for loading and protecting therapeutic cargo, embodying the essence of smart delivery systems in biomedical applications. Their intricate mesoporous structure, characterized by high surface area and tuneable pore size, facilitates efficient encapsulation of diverse therapeutic agents ranging from small molecules to biomacromolecules. This loading capability, coupled with precise control over drug release kinetics, makes MSNs an attractive platform for tailored drug delivery strategies.

One of the key advantages of MSNs lies in their ability to shield encapsulated cargo from premature degradation and clearance within the body. The porous architecture of MSNs acts as a physical barrier, safeguarding therapeutic molecules from enzymatic and chemical degradation in physiological environments. Furthermore, surface functionalization of MSNs with biocompatible polymers or biomolecules enhances cargo protection by mitigating nonspecific interactions and immune recognition, thereby prolonging circulation time and improving bioavailability.

Controlled release of therapeutic cargo represents another hallmark feature of MSNs as smart delivery systems. By engineering stimuli-responsive MSNs, researchers can orchestrate precise spatio-temporal release profiles in response to endogenous or exogenous triggers such as pH, temperature, light or enzymes. This on-demand release mechanism enables targeted drug delivery to specific tissues or cells, minimizing off-target effects and maximizing therapeutic efficacy.

In biomedical therapeutics, MSNs find diverse applications across various domains including cancer therapy, gene delivery, regenerative medicine and infectious disease treatment. In cancer therapy, MSNs enable targeted delivery of chemotherapeutic drugs to tumour sites, mitigating systemic toxicity while enhancing therapeutic outcomes. Similarly, in gene delivery, MSNs offer a stable and protective environment for nucleic acid payloads, facilitating efficient transfection and gene modulation. Moreover, in regenerative medicine, MSNs serve as carriers for growth factors and biomolecules, promoting tissue regeneration and repair. In infectious disease treatment, MSNs deliver antimicrobial agents or vaccines to combat pathogens, offering new avenues for combating drug resistance.[21]

Transporting the Therapeutic Cargo to the Right Place

Transporting therapeutic cargo to the precise location within the body represents a critical challenge in drug delivery, and MSNs emerge as intelligent solutions in this endeavour. MSNs offer a sophisticated platform for targeted drug delivery, ensuring the precise delivery of therapeutic cargo to the intended site of action while minimizing off-target effects. Leveraging their unique properties, MSNs enable the efficient transport of therapeutic molecules through biological barriers to reach specific tissues or cells, making them invaluable tools in biomedical applications.

One of the key features of MSNs is their ability to navigate through physiological barriers to deliver therapeutic cargo to the right place. The mesoporous structure of MSNs allows for the encapsulation of therapeutic agents within their pores, providing protection against degradation and facilitating controlled release. This loading capability, combined with surface functionalization strategies, enables MSNs to evade immune surveillance and traverse biological barriers such as the blood-brain barrier or the mucosal epithelium, reaching target sites with precision.

Furthermore, MSNs can be engineered to respond to specific cues in the microenvironment, facilitating targeted drug delivery to diseased tissues or cells. By incorporating stimuli-responsive moieties into MSNs, researchers can design smart delivery systems that release therapeutic cargo in response to changes in pH, temperature or enzyme activity characteristic of diseased tissues. This spatio-temporal control over drug release ensures that therapeutic molecules are released only at the desired location, maximizing efficacy while minimizing systemic toxicity.

In biomedical therapeutics, the application of MSNs as smart delivery systems holds immense promise across a wide range of conditions. In cancer therapy, MSNs enable targeted delivery of chemotherapeutic drugs to tumour sites, enhancing treatment outcomes while minimizing adverse effects on healthy tissues. Similarly, in neurological disorders, MSNs offer a means to deliver therapeutic agents across the blood-brain barrier, opening new avenues for the treatment of neurodegenerative diseases. Moreover, in infectious diseases, MSNs can deliver antimicrobial agents to specific sites of infection, overcoming drug resistance and improving therapeutic efficacy.[21]

Tumour Penetration

Tumour penetration represents a critical challenge in cancer therapy, where effective delivery of therapeutic agents to deep-seated tumour regions is often hindered by physiological barriers such as dense extracellular matrix and elevated interstitial fluid pressure. MSNs emerge as promising candidates for overcoming these obstacles and enhancing tumour penetration as smart delivery systems in cancer therapy.

The unique physicochemical properties of MSNs make them well suited for facilitating tumour penetration of therapeutic cargo. Their nanoscale size and tuneable pore structure enable efficient extravasation from leaky tumour vasculature, allowing MSNs to accumulate in the tumour micro-environment. Furthermore, the mesoporous architecture of MSNs provides a reservoir for encapsulating therapeutic agents, protecting them from premature degradation and facilitating sustained release within the tumour.

Surface functionalization of MSNs with targeting ligands further enhances tumour penetration by promoting specific binding to tumour-associated receptors or antigens. By selectively targeting tumour cells or tumour vasculature, functionalised MSNs can navigate through the complex tumour matrix and penetrate deep into the tumour tissue. Additionally, stimuli-responsive MSNs can be designed to release therapeutic cargo in response to tumour-specific cues, further optimising drug distribution and efficacy within the tumour.

In cancer therapy, enhanced tumour penetration facilitated by MSNs offers several advantages. It allows for more uniform distribution of therapeutic agents throughout the tumour mass, overcoming spatial heterogeneity and reducing the likelihood of resistance development. Moreover, improved penetration enables better access to cancer stem cells and metastatic niches, addressing key drivers of tumour recurrence and metastasis. By enhancing the depth and extent of drug penetration, MSNs as smart delivery systems hold promise for improving the overall efficacy of cancer treatment and enhancing patient outcomes.[23]

Polymeric Nanoparticle-Based Drug Delivery

Increasing the complications of some diseases, the inherent toxicity of certain drugs results in the increasing need for the development and optimization of drug delivery systems. Here, polymeric NPs acts a tool to improve drug bioavailability. The adaptability of polymers makes them potentially ideal for meeting the requirements of each particular drug delivery system. There are different anatomic routes of administration for drug delivery currently available. The potential toxicity inherent to active ingredients or the high dosage required to achieve pharmacological effect, is a disadvantage shown by the systemic administration routes. The oral pathway limits the use of pH-resistant or highly hydrophilic drugs to ensure the required absorption by the intestinal epithelium cells. That is why nanotechnology is gaining high relevance as it could solve some problems. NPs can be optimized to improve drug bioavailability by increasing their absorption through enhanced solubility. Polymers are macromolecules formed by the covalent union of one or different sorts of units, called monomers. Polymeric materials are gaining great relevance in nanotechnology in general and are being used as NP precursors for drug delivery systems (DDSs). For the preparation of polymeric NPs, the use of surfactants may be a requirement.[24]

Polymeric Nanocarriers for Ocular Drug Delivery:

Nanotechnological systems come in handy for ocular therapy, especially those derived from biodegradable polymers. These systems are mainly designed to achieve an effective dose in the action site, by improved formulation solubility properties, enhanced bioavailability, targeted delivery or sustained release.

Micelle Nanocarriers for Ocular Delivery – The most commonly used polymer for the synthesis of micelles as DDSs is poly(lactic-co-glycolic acid) (PLGA), due to its high biocompatibility and biodegradability. After various researches it has its wild application in treatment of retinal and choroidal neovascularization. Chitosan (CH) polymer, which is also used for the preparation of micelles, has excellent mucoadhesion and penetration properties, making it ideal for drug delivery in mucosa and ophthalmic areas.[24]

Dendrimeric Nanocarriers for Ocular Delivery – The archetypical dendritic polymer for the preparation of nanosystems for drug delivery to the posterior segment of the eye is based on polyamidoamines. Various research studies have shown it can be applied for the treatment of ocular diseases by regulating the expression of target proteins and genes in cells.[24]

Other Types of Polymeric Nanocarriers for Ocular Delivery – (CDs) and polymeric vesicles (PVs) are another type of polymeric nanocarriers which are less utilized. The best option for DDS of latanoprost was the propylamino-β-CD, which entails an enhanced ocular tolerance and hence, better patient compliance. It had better drug stability and low eye irritation.[24]

Polymeric Nanoparticles in Cancer Diagnosis and Imaging – According to the World Health Organization, cancer is the second leading cause of death worldwide, with an estimated 9.6 million deaths in 2018. Early detection and treatment has become a great objective for NPs as DDSs.[24]

Gold-Based Polymeric Nanoparticles Used in Cancer Diagnosis – Gold metallic NPs (AuNPs) and their derivates are the most important compounds used, being extremely versatile, and can be used in multiple imaging methods. These are non-toxic and present up to threefold more efficiency in X-ray absorption.[21]

Gadolinium Polymeric Nanoparticles (GdNPs) Used in Cancer Diagnosis – Gadolinium-based materials are the most widely used and mainly are those formed by chelated metal. As they are easily eliminated due to low molecular weight, a GdNPs-based contrast agent (Anti-VEGF PLA–PEG–PLL–GdNP) was synthesized(here obtained NPs were 70–80 nm-sized). During experimentation, it showed excellent biocompatibility and photostability, good water solubility, low toxicity, strong PA signal intensity and a good performance as a contrast agent.[21]

Perfluorocarbons Polymeric Nanoparticles (PFCNPs) Used in Cancer Diagnosis – Perfluorocarbons (PFCs) are molecules whose structure is similar to common organic compounds (e.g. alkanes). The results of various research and development of new systems have led to the numerous designs of NP systems having enhanced imaging effects and enhanced NMR effect.[21]

Polymeric Nanoparticles in Oncologic Treatment – The main objective of nanomedicine in oncology is to transport a drug only to cancer cells selectively in order to improve its efficacy and reduce its toxicity.[23,24]

Advantages of Nanotechnological Drug-Delivery Systems – Nanotechnological drug delivery in oncology excels through using targeted precision, improving drug solubility, extending drug circulation, enabling multi-drug delivery, overcoming biological barriers, fostering personalized medicine, integrating imaging capabilities, reducing systemic toxicity, facilitating sustained release, and synergizing combination therapies—collectively enhancing treatment efficacy and minimizing side effects.[23,24]

Challenges Associated with Nanoparticulate Drug Delivery Systems – Despite various advantages, some inconveniences related to DDSs limited shape, chemistry, wide size distribution, agglomeration state and electromagnetic properties can lead to poor oral bioavailability, instability in circulation and inadequate tissue distribution.[23,24]

The Enhanced Permeability and Retention (EPR) Effect – NPs play a pivotal role in harnessing the enhanced permeability and retention (EPR) effect for targeted drug delivery in cancer treatment. Tailored to exploit tumour-specific characteristics, these nanocarriers capitalize on leaky vasculature and impaired lymphatic drainage within tumours. The unique size and surface properties of NPs enable passive accumulation at tumour sites, enhancing drug concentration while sparing healthy tissues. This strategic use of NPs leverages the EPR effect to optimize the efficiency and precision of cancer therapies.[23,24]

Active Targeting – This method, based on molecular recognition processes, consists of modifying the NPs' surface with one or more required moieties to achieve their functionalization and, in consequence, raising the drug concentration in tumour tissues. The main goal here is to improve the incorporation of the nanocarriers to the cancer cells.[23,24]

Stimuli-Responsive and Triggered Release Systems – The objective of these systems is the controlled release of antineoplastic drugs provoked by stimuli that develop a change in the nanocarrier. The use of these polymer NPs can provide improvements in cancer therapy, gene therapy, brain tumour, breast cancer and so forth.[23,24]

Polymeric Nanoparticles as Nutraceutical Agents – Polymeric NPs serve as efficient carriers for nutraceutical agents, enhancing their stability and bioavailability. This innovative approach optimizes nutrient delivery, promoting improved absorption and overall efficacy in nutraceutical applications.

Bioavailability and Nanoparticles – Polymeric NPs significantly enhance the bioavailability of nutraceutical agents by protecting them from degradation, improving solubility and facilitating controlled release. The nanoscale size and surface modifications optimize absorption, ensuring efficient delivery and maximizing the therapeutic benefits of nutraceuticals.

Toxicity – Polymeric nanoparticles used in nutraceutical delivery systems generally exhibit low toxicity. However, comprehensive safety assessments are crucial due to potential variations in polymer composition and NP characteristics. Rigorous studies have ensured the safe integration of polymeric NPs as carriers for nutraceutical agents.[21]

6.9 CANCER DRUG DELIVERY

Cancer is a group of diseases characterized by the uncontrolled growth and spread of abnormal cells in the body. These cells can invade nearby tissues and organs, disrupting their normal functions. Cancer can arise from almost any tissue or organ in the body and can metastasize, spreading to other parts of the body through the bloodstream or lymphatic system.

There are more than 100 different types of cancers, each with its own set of causes, risk factors and treatment options. Common types of cancer include breast cancer, lung cancer, prostate cancer and skin cancer. Treatment for cancer typically involves a combination of approaches, including surgery, chemotherapy, radiation therapy, targeted therapy and so on. Early detection and advances in treatment have improved survival rates for many types of cancer, but it remains a significant public health challenge worldwide.

Chemotherapy is a type of cancer treatment that uses drugs to destroy cancer cells. It works by stopping or slowing the growth of cancer cells, which can shrink tumours and help control the spread of cancer throughout the body. It often comes with side effects such as nausea, hair loss, fatigue and so forth. Multimodal cancer treatment is a frequent approach for oncologic act against specific properties of the cancer cells, such as their accelerated metabolism, proteins on the membrane, associated immune cells and so forth. Involvement of 3D tumour cell models in the in vitro biological testing of novel nanotechnology-based strategies for cancer management can provide in-depth information on the real behaviour of tumour cells in complex biomimetic architectures. Here, we used polyethylene glycol-encapsulated iron oxide NPs for the controlled delivery of a doxorubicin chemotherapeutic substance (IONPDOX), and to enhance cytotoxicity of photon radiation therapy. A drug delivery system refers to a technology or method used to deliver therapeutic agents, such as drugs or biologics, to a specific site in the body for their desired effects. These systems can range from simple oral tablets to complex NPs.

The goal of drug delivery systems is to enhance the efficacy and safety of drugs by improving their pharmacokinetics, targeting specific tissues or cells, and reducing side effects. They play a very crucial role in medical science. NPs are promising for drug delivery in cancer treatment due to their ability to target specific cells, enhance drug efficacy and minimize side effects. They can be engineered to carry various types of drugs, such as chemotherapy agents or targeted therapies, directly to cancerous cells, improving treatment outcomes while reducing systemic toxicity. NP-based therapies can be applied in the management of certain cancers in order to directly deliver small quantities of drugs at the tumour site, which concentrates the dose of the active substance at the desired location, while sparing the healthy tissues met in systemic circulation. Here, iron oxide NPs are used, due to their magnetic transport ability, as well as biocompatibility for normal tissues, proven through their clinical use. We tested the biological effect of iron oxide-based drug delivery systems of adoxorubicin chemotherapeutic substance in 3D tumour cell models in vitro, as well as a clear cytotoxic differentiated effect between normal and cancer cells. Nanoscale capsules can deliver DOX only inside cancer cells using oncomarker signatures. Nanoscale capsules consist of a DNA-origami shell covered by immune factor with molecular binding sites on its surface. When nanocapsules conjugate with several markers, then the DOX is released.[23,24]

Lipid-based NPs include various subset structures but are most typically spherical platforms comprising at least one lipid bilayer surrounding and at least one internal aqueous compartment. As a delivery system, lipid-based NPs offer many advantages, including formulation simplicity, self-assembly, biocompatibility, high bioavailability,

ability to carry large payloads and a range of physicochemical properties that can be controlled to modulate their biological characteristics. For these reasons, lipid-based NPs are the most common class of FDA-approved nanomedicines. For liposomes—one of the subsets of lipid-based NPs that has the most members—the NPs are typically composed of phospholipids, which can form unilamellar and multilamellar vesicular structures. This allows the liposome to carry and deliver hydrophilic, hydrophobic and lipophilic drugs, and they can even entrap hydrophilic and lipophilic compounds in the same system, thereby expanding their use. Because they can be rapidly taken up by the reticuloendothelial system, liposomes often include surface modifications to extend their circulation and enhance delivery, which has enabled their clinical use. Another notable subset of lipid-based NPs is commonly referred to as lipid nanoparticles (LNPs)—liposome-like structures widely used for the delivery of nucleic acids. They differ from traditional liposomes primarily because they form micellar structures within the particle core, a morphology that can be altered based on formulation and synthesis parameters. LNPs are typically composed of four major components: cationic or ionizable lipids that complex with negatively charged genetic material and aid endosomal escape; phospholipids for particle structure; cholesterol for stability and membrane fusion; and PEGylated lipids to improve stability and circulation. The efficacy of their nucleic acid delivery along with their simple synthesis, small size and serum stability have made LNPs particularly important in personalized genetic therapy applications. Ionizable LNPs are an ideal platform for the delivery of these nucleic acid therapies as they have a near-neutral charge at physiological pH but become charged in acidic endosomal compartments, promoting endosomal escape for intracellular delivery. However, despite these advantages, LNP systems can still be limited by low drug loading and bio-distribution that results in high uptake to the liver and spleen. NPs have been extensively investigated for drug delivery for decades. Lipid-based nanoparticles such as liposomes, solid lipid nanoparticles (SLNs), and nanostructured lipid carriers (NLCs) have demonstrated tremendous clinical success in delivering both hydrophobic and hydrophilic therapeutics. The first FDA-approved nanodrug, Doxil, is a doxorubicin (DOX)-loaded PEGylated liposome for treating breast cancer, ovarian cancer and other solid tumours. LNPs containing cationic lipids or pH-responsive lipids have been employed for nucleic acids encapsulation and delivery since the 1980s. However, cationic lipids cause undesirable toxicity. In contrast, ionizable cationic lipids, having positive charges at lower pH ($pH < 6.0$) but neutral at physiological pH, are favourable for formulating LNP systems.[25]

6.10 DIFFERENT CLASSES OF LIPID-BASED NANOPARTICLES

Among various classes of lipid-based nanoparticles for drug delivery, four typical classes include liposomes, SLNs, NLCs and hybrid lipid-polymeric NPs.

Liposomes

Liposomes, discovered in 1965, are spherical structures typically consisting of an amphipathic phospholipid bilayer and an internal aqueous core. The core–shell nanostructure of liposomes makes them suitable for loading both hydrophobic and hydrophilic molecules. Normally, hydrophobic drugs are encapsulated in the lipophilic bilayers of the shell, and hydrophilic drugs are to be entrapped in the aqueous phase of the core.[25]

Solid Lipid Nanoparticles (SLN)

Solid lipid nanoparticles (SLNs) are nanoscale drug delivery systems composed of lipids that are in a solid state at both room and body temperature. They offer advantages such as improved drug stability, controlled release and enhanced bioavailability. SLNs have gained attention in pharmaceutical research for their potential in targeted drug delivery and for minimizing side effects. SLNs suffer from two major problems, that is, poor long-term drug retention and low drug-loading capacity. During the storage, the lipid matrix undergoes polymorphic transition from high energy state to low energy state, leading to the formation of a more organized crystalline lattice and the gradual expulsion of the encapsulated drugs.[25]

Nanostructured Lipid Carriers (NLCs)

Nanostructured lipid carriers (NLCs) are advanced drug delivery systems designed to improve the bioavailability and stability of pharmaceutical compounds. These carriers consist of a blend of solid and liquid lipids, forming a nano-sized structure. NLCs offer advantages such as enhanced drug solubility, controlled release and improved targeting, making them promising for delivering various therapeutic agents.[25]

Hybrid LNPs

Hybrid LNPs were introduced as an integrated system composed of at least two types of materials to achieve multifunctions or to address the limitations of single-component nanomaterials, combining the advantages of the two individual components. Recently, a novel drug delivery system called hybrid lipid-polymeric NPs has been reported with promising applications. Basically, it consists of a therapeutic-containing polymeric core enveloped by an inner lipid layer and a PEGylated lipid outer layer. Owing to the characteristics of both lipids and polymers, hybrid lipid-polymeric NPs present great stability, sustained release and high biocompatibility.[25]

Application

LNPs have been widely used for drug delivery applications not only in preclinical studies but also in clinical contexts. Many LNPs have been approved for clinical uses, demonstrating their unique advantages compared to other drug delivery systems.

Cancer Therapy

In cancer therapy, LNPs play a crucial role in drug delivery. These NPs consist of lipid bilayers and are designed to encapsulate therapeutic agents, such as chemotherapy drugs. The lipophilic nature of LNPs allows them to efficiently encapsulate hydrophobic drugs, enhancing their solubility and stability. LNPs are advantageous for cancer treatment due to their ability to selectively accumulate in tumour tissues. This is often attributed to the EPR effect, where LNPs exploit the leaky vasculature of tumours for targeted drug delivery. This targeted approach minimizes drug exposure to healthy tissues, reducing systemic side effects.[25]

Vaccine Delivery

LNPs are extensively used in vaccine delivery, particularly for mRNA vaccines. LNPs serve as effective carriers to protect and deliver fragile mRNA molecules into cells. In this context, LNPs encapsulate the mRNA, forming LNPs that facilitate cellular uptake. LNPs provide several advantages in vaccine delivery, including enhanced stability of mRNA, protection against enzymatic degradation and improved cellular delivery. They help the mRNA entering into the cells, where it instructs the production of viral proteins, triggering an immune response without using live viruses. The success of mRNA Covid-19 vaccines, such as Pfizer-BioNTech and Moderna, highlights the efficacy of LNPs in vaccine delivery. LNPs contribute to the rapid development and deployment of vaccines, offering a versatile platform for delivering genetic material to elicit protective immune responses. [25]

Infectious Diseases

LNPs find application in infectious diseases beyond vaccines, particularly in the delivery of antiviral drugs. LNPs can encapsulate various therapeutic agents, such as small-molecule antivirals or nucleic acid-based drugs, facilitating targeted delivery to infected cells. In antiviral drug delivery, LNPs enhance the solubility and stability of drugs, ensuring their effective delivery to the site of infection. This targeted approach helps minimize off-target effects and systemic toxicity while improving the therapeutic efficacy against viral pathogens. Moreover, LNPs can be engineered to exploit the EPR effect, allowing them to accumulate selectively in infected tissues. This targeted delivery enhances the concentration of antiviral agents at the site of infection, improving the overall effectiveness of the treatment. [25]

Cardiovascular Diseases

LNPs are being explored in the context of cardiovascular diseases for targeted drug delivery. One application is in delivering drugs aimed at managing conditions like arteriosclerosis. LNPs can encapsulate therapeutic agents, such as anti-inflammatory drugs or lipid-lowering medications, to specifically target affected areas within blood vessels. In arteriosclerosis, for example, LNPs can be designed to exploit the inflammatory processes associated with plaque formation. These NPs may accumulate in the inflamed regions, delivering drugs to mitigate inflammation or reduce cholesterol levels. LNPs can also play a role in delivering genetic material or nucleic acid-based therapies to modulate gene expression related to cardiovascular health. This targeted approach aims to address specific molecular pathways involved in cardiovascular diseases. [25]

These applications highlight the versatility and potential of LNP-based drug delivery in improving therapeutic outcomes across various medical fields.

Magnetic Nanoparticles in Cancer Treatment

Magnetic hyperthermia therapy is very useful in cancer treatment. In this technique we can burn the cancer cells. Magnetic NPs can generate heat through the oscillation of their magnetic moments. Magnetite (Fe_3O_4) can be used as a magnetic nanoparticle, while cobalt and nickel have shown promise in magnetic hyperthermia. Additionally, nickel serves as a contrast agent for magnetic resonance imaging (MRI) in cancer diagnosis. We need to gather more information about their safety before widespread clinical application.[26]

Gold nanoparticles

Gold NPs are used in cancer treatment because of their unique properties, including their absorption ability, scatter ability and so on. These NPs can be designed to specifically target cancer cells, either by attaching antibodies, or other targeting molecules that recognize cancer cells due to their size and shape.[27]

Gold nanoparticles can be used in various ways for treatment

Photothermal therapy—Gold NPs can absorb light energy and convert it into heat, which can selectively kill cancer cells when nanoparticles are irradiated with laser light. This is known as photothermal therapy, which can be used to destroy tumours with minimal damage to healthy tissues.

Drug delivery – Gold NPs can be loaded with chemotherapy drugs and delivered directly to cancer cells. This targeted drug delivery approach minimizes systemic toxicity.

Overall, the unique properties of gold NPs make promising candidates for cancer treatment, offering targeted therapy and minimal side effects.[27]

6.11 CONCLUSIONS AND FUTURE DIRECTIONS

In conclusion, nanomaterials are proven candidates for new-generation drug delivery for their unique characteristics and easy attachment to drug carriers. The surface potential of nanomaterials (NPs) are useful for attachment of NPs with drugs. Apart from solo drug delivery, NPs offer antibacterial activities that may be a synergistic action to the drug delivery system, and also the antioxidant activities of NPs shape them as protectors of normal cells from affected cells. Among NPs, mesoporous silica nanoparticles (MSNs) have emerged as highly promising platforms for drug delivery due to their unique physicochemical properties and versatile functionality. MSNs offer a combination of high surface area, tuneable pore size, excellent biocompatibility and facile surface modification, making them well suited for encapsulating and delivering a wide range of therapeutic agents. The mesoporous structure of MSNs provides ample space for drug loading, allowing for high drug-loading capacities and controlled release profiles. Furthermore, their biocompatibility and low toxicity make them suitable candidates for in vivo applications. Surface functionalization of MSNs enables targeted drug delivery, enhancing specificity and reducing off-target effects. Stimuli-responsive MSNs allow for triggered drug release in response to specific physiological cues, offering spatio-temporal control over drug delivery. Moreover, MSNs have shown promise in overcoming biological barriers, such as the blood-brain barrier and tumour microenvironment, facilitating drug delivery to otherwise inaccessible sites. Their ability to protect encapsulated cargo from degradation and prolong circulation time further enhances their efficacy as drug delivery vehicles. Overall, MSNs hold tremendous potential for revolutionizing drug delivery in various therapeutic areas, including cancer therapy, gene delivery, regenerative medicine and infectious disease treatment. Continued research and development efforts are essential to optimize MSN-based drug delivery systems, address remaining challenges and translate these innovations into clinical applications for improved patient outcomes. As such, MSNs represent a promising avenue for advancing precision medicine and personalized healthcare in the future. The unique properties of Gold NPs make them promising candidates for cancer treatment, offering targeted therapy and minimal side effects. LNPs can be designed to exploit the inflammatory processes associated with plaque formation. These NPs may accumulate in the inflamed regions, delivering drugs to mitigate inflammation or reduce cholesterol levels. LNPs can also play a role in delivering genetic material or nucleic acid-based therapies to modulate gene expression related to cardiovascular diseases. Nanostructured lipid carriers (NLCs) are also a successful candidate in drug delivery due to their unique attributes.

Future challenges in drug delivery systems encompass optimizing nanocarrier biocompatibility, addressing potential immunogenic responses, enhancing targeted delivery to specific cell types, and ensuring regulatory approval for advanced technologies. Overcoming these hurdles will advance personalized medicine, improve treatment outcomes and usher in a new era of precision in drug delivery for diverse therapeutic applications.

NP-based drug delivery systems hold immense potential to revolutionize the pharmaceutical industry. Further research and development are needed to address remaining challenges, including cost-effectiveness, scalability and regulatory approval. However, the potential benefits of this technology are significant, offering opportunities for personalized medicine, improved treatment efficacy and reduced side effects for patients.

REFERENCES

1. Chattopadhyay K. K. & Banerjee A. N. (2009) Introduction to nanoscience and nanotechnology, *PHI Learning*, ISBN: 9788120336087, 8120336089. Introduction to Nanoscience and Nanotechnology – ... (phindia.com).
2. Adnan N. N. M., Cheng Y. Y., Ong N. M. N., Kamaruddin T. T., Rozlan E., Schmidt T. W., Duonga H. T. T. & Boyer C. (2016) Functional gold nanoparticles with different shapes for photothermal therapy and drug delivery, *Polym. Chem.*, 7, 2888–2903. https://doi.org/10.1039/C6PY00465B
3. Zhao Y., Wang Y., Ran F., Cui Y., Liu C., Zhao Q., Gao Y., Wang D. & Wang S. (2017) A comparison between sphere and rod nanoparticles regarding their in vivo biological behavior and pharmacokinetics, *Sci. Rep.*, 7, 4131. https://doi.org/10.1038/s41598-017-03834-2
4. Zhang J., Xu B., Tian W. & Xie Z. (2018) Tailoring the morphology of AIE gen fluorescent nanoparticles for optimal cellular uptake and imaging efficacy, *Chem. Sci.*, 9, 2620. https://doi.org/10.1039/C7SC05130A
5. Podder S., Ghosh C. K., Das A. & Hardy J. G. (2022) Light-responsive nanomaterials with pro-oxidant and anti-oxidant activity, *Emerg. Mater.*, 5(2), 455–475. https://doi.org/10.1007/s42247-022-00361-3
6. Mitchell M. J., Billingsley M. M., Haley R. M., Wechsler M. E., Peppas N. A. & Langer R. (2021) Engineering precision nanoparticles for drug delivery, *Nat. Rev. Drug Discov.*, 20, 101. https://doi.org/10.1038/s41573-020-0090-8.
7. Dey S., Podder S., Roychowdhury A, Das D. & Ghosh C. K. (2018) Facile synthesis of hierarchical nickel (III) oxide nanostructue: A synergistic remediating action towards

water contaminants, *J. Environ. Manage.*, 211, 356–366. https://doi.org/10.1016/j.jenvman.2018.01.009

8. Zhang W., Li Y., Niu J. & Chen Y. (2013) Photogeneration of reactive oxygen species on uncoated silver, gold, nickel, and silicon nanoparticles and their antibacterial effects, *Langmuir*, 29(15), 4647–4465. https://doi.org/10.1021/la400500t
9. Prasanna V. L. &Vijayaraghavan R. (2015) Insight into the mechanism of antibacterial activity of ZnO: Surface: Defects mediated reactive oxygen species even in the dark. *Langmuir*, 31(33), 9155–9162. https://doi.org/10.1021/acs.langmuir.5b02266.
10. Kiseleva V. M., Kislyakov I. M. & Burchinova A. N. (2016) Generation of Singlet Oxygen on the surface of Metal Oxides, *Opt. Spectrosc.*, 120, 520–528. https://doi.org/10.1134/S0030400X16040123.
11. Yang L., Zhang W., Niu J. & Chen Y. (2012) Mechanism of photogenerated reactive oxygen species and correlation with the antibacterial properties of engineered Metal-Oxide Nanoparticles. *ACS Nano*, 6(6), 5164–5173. https://doi.org/10.1021/nn300934k.
12. Singha S., Barick K. C. & Bahadur D. (2013) Shape-controlled hierarchical ZnO architectures: Photocatalytic and antibacterial activities, *CrystEngComm*, 15, 4631–4639. https://doi.org/10.1039/C3CE27084J.
13. Podder S., Chanda D., Mukhopadhyay A. K., De A., Das B., Samanta A., Hardy J. G. & Ghosh C. K. (2018) Effect of morphology and concentration on crossover between antioxidant and pro-oxidant activity of MgO nanostructures, *Inorg. Chem.*, 57(20),12727–12739. https://doi.org/10.1021/acs.inorgchem.8b01938.
14. Podder S., Halder S., Roychowdhury A., Das D & Ghosh C. K.(2016) Superb hydroxyl radical-mediated biocidal effect induced antibacterial activity of tuned ZnO/chitosan type II heterostructure under dark. *J. Nanopart. Res.*, 18, 294. https://doi.org/10.1007/s11051-016-3605-9.
15. Raut N. C., Mathews T., Ajikumar P. K, George R. P., Dash S. & Tyagi A. K. (2012) Sunlight active antibacterial nanostructured N-doped TiO2 thin films synthesized by an ultrasonic spray pyrolysis technique, *RSC Adv.*, 2, 10639–10647. https://doi.org/10.1039/C2RA21024J.
16. Etacheri V., Michlits G., Seery M. K., Hinder S. J. & Pillai S. C. (2013) A highly efficient $TiO_{2-x}C_x$ Nano-heterojunction photocatalyst for visible light induced antibacterial applications, *ACS Appl. Mater. Interfaces,* 5(5), 1663–1672. https://doi.org/10.1021/am302676a
17. Podder S., Paul S., Basak P., Xie B., Fullwood N. J., Baldock S. J., Yang Y., Hardy J. G. & Ghosh C. K. (2020) Bioactive silver phosphate/polyindole nanocomposites, *RSC Adv.*, 10, 11060–11073. https://doi.org/10.1039/D0RA01129K.
18. Akhavan O., Azimirad R., Safad S. & Hasani E. (2011) CuO/$Cu(OH)_2$ hierarchical nanostructures as bactericidal photocatalysts. *J. Mater. Chem.*, 21, 9634–9640. https://doi.org/10.1039/C0JM04364H.
19. Sadeghi S., Lee W. K., Kong S. N., Shetty A. & Drum C. L.(2020) Oral administration of protein nanoparticles: An emerging route to disease treatment, *Pharmacol. Res.*, 158, 104685. https://doi.org/10.1016/j.phrs.2020.104685.
20. Pandey R., Zahoor A., Sharma S. & Khuller G. K. (2003) Nanoparticle encapsulated antitubercular drugs as a potential oral drug delivery system against murine tuberculosis. *Tuberculosis (Edinb),* 83(6), 373–378. https://doi.org/10.1016/j.tube.2003.07.001.
21. Manzano M. & Vallet-Regí M. (2019) Mesoporous Silica nanoparticles for drug delivery,*Adv. Funct. Mater.*, 30(2), 1902634. https://doi.org/10.1002/adfm.201902634.
22. Bharti C., Nagaich U., Pal A. K. & Gulati N. (2015) Mesoporous silica nanoparticles in target drug delivery system: A review. *Int J Pharm Investig.*, 5(3), 124–133. https://dx.doi.org/10.4103/2230-973X.160844.

23. R. C.Popescu, V.Kopatz, E.Andronescu, D. I.Savu, & W.Doerr, Nanoparticle-mediated drug delivery of Doxorubicin induces a differentiated clonogenic inactivation in 3D tumor spheroids In Vitro, *Int. J. Mol. Sci.*, 24, 2198–2023. https://doi.org/10.3390/ijms24032198.
24. Begines B., Ortiz T., Pérez-Aranda M., Martínez G., Merinero M., Argüelles-Arias F. & Alcudia A. (2020) Polymeric nanoparticles for drug delivery: recent developments and future prospects, *Nanomaterials,* 10, 1403. https://doi.org/10.3390/nano10071403B.
25. Xu L., Wang X., Liu Y., Yang G., Falconer R. J. & Xia Zhao C. (2022) Lipid nanoparticles for drug delivery, *Adv. Nano Biomed Res.*, 2(2), 2100109. https://doi.org/10.1002/anbr.202100109.
26. Tian X., Ruan L., Zhou S., Wu L., Cao J., Qi X., Zhang X.& Shen S.(2022) Appropriate size of Fe_3O_4 nanoparticles for cancer therapy by Ferroptosis, *ACS Appl. Bio Mater.,* 5(4), 1692–1699. https://doi.org/10.1021/acsabm.2c00068.
27. Sztandera K., Gorzkiewicz M. & Klajnert-Maculewicz B. (2019) Gold nanoparticles in cancer treatment, *Pharmaceutics*, 16(1), 1–23. https://doi.org/10.1021/acs.molpharmaceut.8b00810.

7 Two-Dimensional Material-Based Sensors for Healthcare Applications

Rejaul Ali and Arnab Hazra

7.1 INTRODUCTION

In our daily lives, sensors are becoming an integral part. They find their application in human–machine interaction, machine-to-machine interaction and healthcare applications. A sensor is a single or a series of devices using chemical and physical principles such as electrochemistry, fluorescence, thermal, surface plasmon resonance, piezo, reflectometry and bioluminescence. Two things are mainly involved in the sensing mechanism: first is the sensor, and second is the stimulus that will be sensed, which can be light, heat, and matter adsorbed on the sensing layer. Stimuli (material, light, heat, etc.) should come into contact with a sensor or sensors to initiate sensing. We can consider sensors as the transducers that help to generate signals in response to certain stimuli like heat, light and gas molecule adsorption. Later, when these produced signals are processed, it gives us valuable information about the surroundings. After sensing the surroundings of the sensor, crucial decisions can be made according to our needs.

There are sensors available in the market specifically designed to record specific vitals related to the human body, such as temperature, blood pressure, pH, blood glucose level, respiratory rate, heart rate and so forth. Most current healthcare systems are bulky and designed for regular monitoring of the vitals of the human body locally. To ensure good health, one must regularly visit healthcare facilities, which is time-consuming and inconvenient. So, there is a need to develop sensors that are small, light and flexible enough to attach to skin and clothes. This is possible by incorporating 2D materials as a sensing layer because of their unique properties.

2D Materials consist of such materials having a few nanometre-thick layered structured structures like graphene, black phosphorus, transition metal dichalcogenide, MXene and so on.[1,2] These materials have specific properties that can help miniaturize existing sensor sizes. These nanometre-size layer structures in 2D material have a very high surface-to-volume ratio because of being single-layered. That ultimately helps the sensitivity of the sensors. The sensor's sensitivity is the parameter that determines how much the sensor will respond when in contact with certain stimuli in terms of specific parameters like resistance, current, fluorescence and so forth. For

DOI: 10.1201/9781003453635-7

most healthcare applications, sensors should be flexible, small and lightweight, and 2D materials are suitable for such applications because they can be exfoliated to very thin layers and withstand a certain amount of strain and stress without permanent distortion. Besides these properties, it is observed that the bandgap is a function of layers. As the number of layers in the bulk increases, their bandgap decreases. That can help to tune their conductivity.[3] The conductivity of 2D materials can also be tuned by controlling the structural defect and functionalization with foreign atoms.[4–6]

After discovering graphene in 2004, 2D materials started attracting researchers to explore the vast number of applications because of their unique properties.[7] But before graphene, more than 100 years ago, black phosphorus was synthesized from white phosphorus by applying high temperature and pressure.[8] However, because of its high instability and the unavailability of large-scale synthesis techniques, it did not get much attention then. There is a subclass of 2D materials named transition metal dichalcogenides (TMDCs). These include MoS_2, WS_2, $MoSe_2$, WSe_2, etc. Unlike graphene, TMDCs have tuneable bandgap and astonishing optical properties. Along with graphene, they are well-studied materials. MoS_2, like graphene, is a well-explored member of TMDCs.[9] MXene is a comparatively newly added subclass of 2D materials derived from the MAX phases. These are metal carbides and metal nitride-based materials. Their structure is discussed in detail in section 2. There is a vast list of 2D materials besides those mentioned above, for example, hexagonal boron nitride, silicene, antimonide, inorganic perovskites, organic-inorganic hybrid perovskites, metal-organic frameworks and graphitic carbon nitride. However, we will discuss graphene, TMDCs and MXene sensors for healthcare applications.

In this chapter, we will examine three types of sensors: (i) flexible sensors, (ii) chemical sensors and (ii) biosensors. Each section is divided into three to four subsections to discuss specific material-based sensors used for healthcare applications.

7.2 FLEXIBLE SENSORS

A flexible sensor is made of flexible stretchable material equipped with specific electronic components for sensing stimuli from the surroundings. For healthcare applications, flexible sensors have a wide range of applications and have the potential to revolutionize the healthcare industry. The existing healthcare instruments are bulky, and continuous human body monitoring is very inconvenient. Flexible sensors that can be easily attached to skin and clothing can significantly reduce the headache of regularly reaching out to a bulky healthcare system. The basic structure of a flexible sensor consists of active material, a flexible substrate and electrical contacts. A 2D material is an excellent choice for the active layer because of its high surface-to-volume ratio, which helps to increase its sensitivity, and it can undergo temporary deformation on a flexible substrate. When these sensors are stretched or compressed, their electrical properties like capacitance and resistance change, leading to variation in output electrical signals through the sensor. We can make conclusive health decisions for a person using appropriate data processing techniques. The following section explains how 2D materials like graphene and transition metal chalcogenide are being utilized in flexible sensors.

7.2.1 Graphene-Based Flexible Sensors

Graphene is a 2D material made of sp^2-bonded carbon atoms. The thickness of each graphene sheet is 1 to 2 nm and has an interlayer spacing of 1.41 Å.[10] Though graphene is made of the non-metallic element carbon because of its structure, it shows some astonishing chemical, electrical and mechanical properties, for example, tensile strength of 130 GPa, Young's modulus of (1.02 TPa), and a large surface area of 2630 m^2/g (about the area of a large mansion/g). There are a few processes for synthesizing graphene: chemical vapor deposition (CVD), chemical or plasma exfoliation from natural graphite and mechanical cleavage. CVD is the method that can be utilized to grow large-scale graphene. If graphene is deposited on a flexible substrate like PDSM (Polydimethylsiloxane) or PI (polyimide), because of its inherent properties, graphene can make a strain and stress sensor. However, repeated cycles of stress and release can produce cracks that can increase resistance.

MoS_2 being a 2D transition metal dichalcogenide (TMDC), has a sheet-like structure like graphene, and it shows metallic to semiconductor properties depending on the crystalline phase. MoS_2-decorated graphene shows improvement in Vanderwaal forces between neighbouring layers, and mechanical properties in terms of reusability of sensors.[11] For making a sensor, graphene is synthesized with the help of a laser-scribing method, which uses PI, wood and CO_2 as carbon sources, as shown in Figure 7.1(a). This method causes sp^3 hybridized carbon atom conversion to sp^2 hybridized carbon atom, making the graphene layer porous. This strain-type sensor will recognize specific movements on human skin because of pulses in blood arteries,

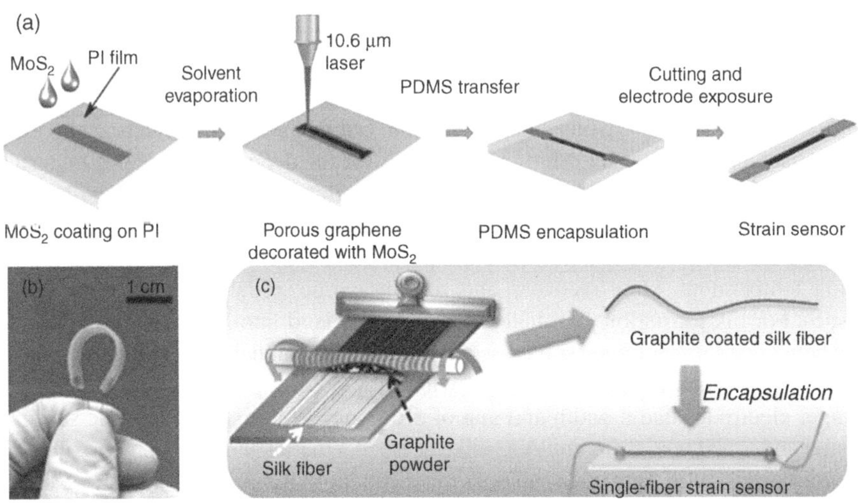

FIGURE 7.1 Structure and fabrication steps for graphene-based flexible sensors, (a) Flow chart for fabrication of the MDS-LIG strain sensor (adapted with permission from[11]. Copyright 2019 American Chemical Society); (b) Silk-graphene fibre-based sensor; (c) Fabrication of sheath-core structured graphite/silk strain sensors through a dry-Meyer-rod-coating process (adapted with permission from[12]. Copyright 2016 American Chemical Society).

eye blinking and fingers closing and opening. This sensor shows reliable durability because MoS_2 decoration on a porous graphene layer provides sensor electromechanical stability. It has a gauge factor (GF) of 1242, a minimal hysteresis of 2.75%, a minimum detectable limit of 0.025%, and an ability > 12000 cycles. This sensor is believed to have a robust application in e-skin, real-time human body movement.

Sensors integrated with clothing are another domain of flexible sensors. They can help to detect certain body movements without caring about the cytotoxicity of the material when it is in intimate contact with the skin. Examples of this kind of sensor are silk fibre-based graphene sensors, which are strain-type sensors. Graphene flakes were used to fabricate this sensor with the help of the dry-Meyer-rod-coating process, and the fabrication process is shown in Figure 7.1(c).

The obtained GSF fibers were covered with ecoflex, and each fibre end was connected with copper wire, as in the device structure shown in Figure 7.1(c). The sensor was attached with limbs and fingers for sufficient stretching force, and the substrate and corresponding electrical signal were recorded. This sensor has a GF of 14.5 and can withstand more than 3,000 bending cycles, making it a durable and reliable choice for soft robotics and prosthetic limbs.[12]

A graphene/alginate composite is a resistive-type flexible temperature sensor with a sensitivity of TCR = $-1.5\ ^{o}C^{-1}$. Along with excellent sensitivity, it has a response time of 26.3 seconds and a recovery time of 95 seconds, and is stable and reliable. For fabrication, graphene dispersion was made in sodium alginate, and later, this dispersion was used for deep coating on calcium alginate, which was created using calcium chloride and sodium alginate. The non-woven fabric sensor was used to observe the temperature of the palm's back and front, which showed temperatures of 30.6 and 32.4°C. The temperature of the human body is one of the critical parameters for deciding one's health condition. That is why this type of flexible temperature sensor may play a vital role in measuring the temperature of the human body and thus in flexible healthcare electronics.[13]

Besides CVD, drop casting and spin coating, printing techniques can be categorized as contact and non-contact printing techniques for the deposition of active materials. For example, there is an inkjet-printed CMOS-integrated graphene-metal oxide-based sensor. This graphene-based sensor is developed to overcome the disadvantages of metal oxide-based sensors, such as high-power consumption due to very high operating temperature, slow recovery and response time, reliability, and repeatability issues. For ink formation, exfoliated graphene is mixed into ZnO nanoparticles with the help of certain inorganic solvents and binders. The graphene ZnO ink is transferred to the Si_3N_4 Substrate of CMOS μHP, where the gold finger electrode is deposited. Finger electrodes had a width and gap of 5 μm and 250 μm diameter. The sensor is tested under NH_3 and different VOCs like acetone, ethanol and carbon monoxide. The device successfully showed the response to different VOCs, but it produced the distinguishable better response in case NH_3 is a biomarker for kidney and liver problems.[14]

Graphene aerogel is a very light sponge-type thermally stable material obtained from graphene oxide. This material can retain its shape even after applying 50% strain. Their light weight and hydrophobicity make them float on the surface of water. GO dispersion solution is used to make GO hydrogel sphere in a coagulation bath with the help of wet spinning. GO hydrogel is reduced by sodium ascorbic acid (VcNa) for 4 hours at 95°C. Reduced GO is frozen at -18°C for 1 hour and dried in an oven at 60°C after washing it in DI water. In the end, thermal treatment of reduced graphene aerogel is done in argon ambient at 2500°C. Reduced graphene oxide aerogel can be fabricated in various shapes, like spheres and cylinders.[15] Twenty random 3-D graphene aerogel spheres (GASs) were picked and used to make temperature and pressure sensors. Temperature is made on PDMS substrate. GAS was placed inside the PDMS hole, and copper contact was made from the open sides of the hole. The temperature heat ranges from 30°C to 90°C on the sensor provided through the programmable hotplate. On application, the heat resistance of GAS starts increasing. This sensor showed a temperature coefficient of resistance (TCR) of 2.2% per degree. GAS-based temperatures withstand extreme conditions like temperature ranges of -196°C to 2000°C and vacuum. The TCR of the GAS-based sensor can be modulated while synthesizing the GAS with the help of doping materials like ZnO. Moreover, the annealing temperature is also responsible for influencing the TCR. The pressure sensor based on GASs was tested using a force gauge. A force gauge was applied to the pressure while electrical measurements were recorded. On increasing pressure, the GAS gets squished, leading to a decrease in the resistance because of an increase in the conducting path. A cylindrical PDMS chamber was constructed to fabricate pressure. Inside this cylinder, the GAS was kept, and pressure was applied from the top, which was the roof made of PDMS over the GAS. The range of force sensing can be modulated by varying the thickness of the cylindrical wall of the PDMS. This sensor can sense the force from 4 mN to 2 N. The GAS's softness can be applied in tissue elasticity sensing and finger gesture sensing.[16]

7.2.2 Chalcogenides-Based Flexible Sensors

Transition metal dichalcogenides have a general formula of MX_2, where M is an element from transition metals, and X is from the chalcogenide group (S, Se, Te etc.). TMDCs, like graphene, also have 2D nanosheets, and their bandgap is the function of the thickness. MoS_2 is an extensively explored TMDC and has a bandgap range of 2.219 eV to 1.441 eV depending on its bandgap, contrary to graphene, which has zero bandgaps.[17] MoS_2 sheets are separated by a distance of 0.615 nm. Researchers are exploring this material well for chemical sensing, pressure sensing and photodetection. A tactile sensor is designed to sense the pressure while the sensor is being touched so that mimicking the basic functionality of the skin like touch sensitivity can be achieved, as shown in Figure 7.2 (b) and (d). For making a MoS_2-based active-matrix tactile sensor (MoS_2 AMTS), MoS_2 was synthesized using the MOCVD (metal-organic chemical vapor deposition) technique. Molybdenum hexacarbonyl (source of Mo) and dimethyl sulphide (source of sulphur) are used in MOCVD. PMMA/SiO_2/Si substrate is covered with SU-8, then Al_2O_3 is deposited. With the help of photolithography, the source (Cr) lines of 3nm and drain (Au) of 30 nm lines patterned on the Al_2O_3 and MoS_2 layers are transferred to active sites. After the deposition of Al_2O_3 on the MoS_2 layer, this array of TFT was covered with a 500 μm-thick layer of SU-8. The sensing layer peels off from the SiO_2 layer by removing the PMMA sacrificial layer. The final structure will like the one shown in Figure 7.2(c). One unit of the sensing layer consists of 1 transistor and one pressure gauge; this circuit is replicated 8 × 8

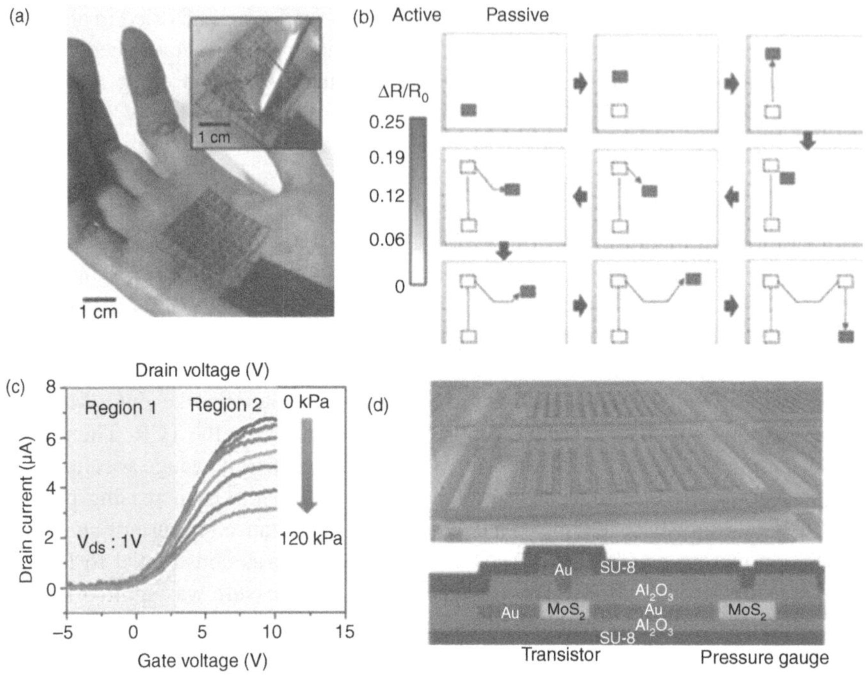

FIGURE 7.2 Device structure and responses of AM-MoS_2 tactile sensor, (a) AM MoS_2 tactile sensor on a human palm, (b) Real-time tracking of stylus position on tactile sensor, (c) Transfer characteristics of single unit of tactile sensor at different applied pressures (from 0 to 120 kPa) at a drain voltage of 1 V (adapted with permission from[18]. Copyright 2019 American Chemical Society).

times all over the sensing matrix unit (see Figure 7.2(a)). The pressure gauge in the circuit has a much more significant aspect ratio (W/L: 300/4) than the TFT channel (W/L: 300/4). This is done to increase the effective pressure sensing area. The performance of this AMTS remained intact after 1,000 cycles of applying pressure and reduced crosstalk during single processing. This AMTS can identify the object's shape in contact, making it useful in e-skin and robotic applications.[18]

Similar to MoS_2, $SnSe_2$ belongs to the 2D TMDC family. This is an n-type TMDC and has mobility of 85 cm^2/Vs. $SnSe_2$ shows sensitivity towards light, pressure and gases. The sensor mentioned above is paper-based. Bulk $SnSe_2$ is sonicated for 8 hours in NMP(N-Methyl-2-pyrrolidone) solvent. The procedure for the fabrication of this sensor is shown in Figure 7.3(a). This process is done to exfoliate atomically thin sheets of $SnSe_2$. The prepared solution is then used to coat the Whatman filter paper. The papers were completely dried and stacked together to make a piezoelectric sensor. Contact was taken from copper plates on the bottom and top of stacked $SnSe_2$-coated sheets. When pressure is applied to the $SnSe_2$ stacked papers, the current increases as the contact area between sheets increases (see Figure 7.3(c)). It was observed that sensitivity is directly related to number of sheets in a stack. More sheets in the stack

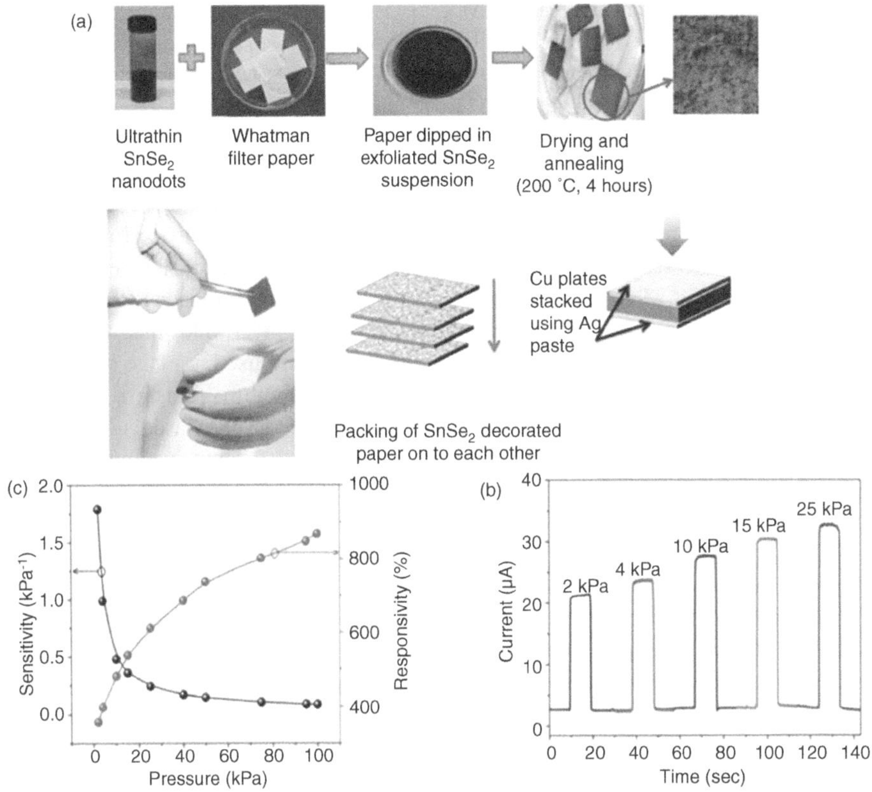

FIGURE 7.3 $SnSe_2$ functionalized paper-based sensor structure and its response, (a) Fabrication of device, (b) Sensitivity and Responsivity vs. Pressure for 2–100 kPa, (c) current response when momentary pressure applied (2 to 25 kPa) (adapted with permission from [19]. Copyright 2020 American Chemical Society).

resulted in an increase in sensitivity to pressure. The paper-based graphene sensor showed a sensitivity of 0.1 kPa^{-1} for a pressure range of 2–20 kPa. However, the paper-based $SnSe_2$ nanosheet pressure sensor has a sensitivity of 1.79 kPa^{-1} for a pressure range of 2–100 kPa (Figure 7.3(b)). It can withstand 5,000 loading cycles of 25 kPa. A 2D-$SnSe_2$ nanosheet functionalized piezo-resistive flexible sensor can respond to finger tapping and even human breath. Researchers claim that this can be helpful in the online diagnosis of patients.[19]

7.2.3 2D Metal Carbides/Nitrides (or MXene)-Based Flexible Sensors

MXene is a class of materials obtained from selectively etching metal layers from the MAX. M stands for transition metal, A belongs to an element from group 13 and group 14, and X is carbon or nitrogen. MXene is symbolized as $M_{n+1}X_nT_x$, where T_x is the terminal group, such as hydroxyl, oxygen or fluorine.[20] Let us consider $Ti_3C_2O_2$ for pressure-sensing application. Pristine MXene (P MXene) shows a change in resistance under applied pressure that can be utilized for pressure sensors. Moreover, it is seen that when P MXene is oxidized, it shows more sensitivity towards pressure. Oxidized MXene is further classified into O1 MXene (lower degree of oxidized MXene) and O2 MXene (higher degree of oxidized MXene). When MXene is oxide, the space between interlayers is increased because of the formation of TiO_2, which acts as the isolation between MXene 2D nanolayers. This structure causes more deformation while pressure is applied, resulting in more sensitivity for the sensor. However, oxidation reduces the sensitivity because of the excess formation of TiO_2 in the monolayer. It is seen that when O_2 MXene is subjected to pressure, it leads to a disconnect between the conducting path and the MXene nano 2D layer, and thus, O2 MXene shows lesser sensitivity. The dipping-drying methods were used to make paper-based MXene sensors. Once the paper is coated with O1 MXene, it is transferred to a flexible substrate containing interdigital electrodes. The O_1 MXene-based sensor shows sensitivity as high as 28.83 kPa^{-1}, a response time of 98.3ms, and sustained repeated applied pressure up to 5,000 cycles. These make reliable and durable choices for body movement trackers and remote health monitoring.[21]

PDMS (polydimethylsiloxane), which is a flexible material, can be used instead of paper for making piezoresistive sensors. The substrate was engraved with a honeycomb-like structure with the help of a silicon mould. For a mould, silicon was etched using the Ti: sapphire-based laser system, which was incident at 15 fs, and silicon lines were separated by a distance of 100 μm. After peeling off the PDMS from this mould, MXene was deposited on the PDMS and transferred to the PET (Polyethylene terephthalate), where silver interdigitated electrodes were deposited. The device showed the response to the applied pressure and stress because of the increase in the conducting path due to the honeycomb-like structure of the PDMS. This MXene-based sensor can reach a sensitivity as high as 0.61 KPa^{-1}, have a response time of 160 ms, and sustain repeated loading cycles. Researchers claim it can be significant in future healthcare systems.[22]

7.2.4 Black Phosphorus-Based Flexible Sensors

Phosphorus has many allotropes, such as black phosphorus, red phosphorus, white phosphorus and violet phosphorus. These allotropes are known for their instability and vigorous reaction when exposed to atmosphere and heat. Among all of these, black phosphorus (BP) is comparatively the most stable. However, it is not enough to make a durable sensor when the sensing layer is exposed to the atmosphere. It readily oxidizes in the presence of oxygen and moisture. Black phosphorus has attracted the attention of researchers because of its 2D honeycomb puckered-like structure, which goes in a zigzag fashion on a horizontal plane, a high mobility of ~ 1000 cm^2/Vs, and a direct bandgap. Like a TMDC, the bandgap of black phosphorus is also the function of the number of 2D nanosheets. The bandgap decreases as the number of layers increases, the bandgap of the monolayer is near 1.5ev, and the bulk has a bandgap near 0.3 eV. There are methods like mechanical exfoliation, chemical methods and chemical vapor deposition for synthesizing black phosphorus. The main challenges for fabricating black phosphorus-based devices are finding the method for large-scale synthesis and stabilizing the BP in ambient conditions. BP was synthesized from red phosphorus using the solvothermal method to fabricate black phosphorus strain sensor. First, red phosphorus was dispersed in DI water and sonicated for 10 minutes. The sonication process is repeated ten times. The sediment of black phosphorus was collected at the bottom of the beaker.

These sediments were transferred to an autoclave containing N, N-dimethylformamide (DMF), and a cellulose paper. The solvothermal process was set for 6 hours at 140°C. In this way, black phosphorus was grown on cellulose paper. Cellulose paper is encapsulated with the PDMS, and contact is taken from copper wire. Under the strain, the electrical properties like resistance changed. This sensor can detect various body movements successfully, and electrical characteristics were intact for the 2000 cycle. Because black phosphorus was encapsulated, its electrical properties did not degrade even after 90 days, which acknowledges its durability and reliability.[23]

Instead of a flexible polymer material like PDMS, a melamine sponge with black phosphorus can be considered for making strain-type flexible sensors. This sponge has to be soaked in an exfoliated BP solution. For that, black phosphorus crystal is exfoliated in boiled DI water with the help of sonication. Boiled DI water was taken to remove the dissolved oxygen from DI water. Chloroauric acid ($HAuCl_4$) is used as the source of gold nanoparticles. A black-coloured solution of exfoliated black phosphorus is mixed with chloroauric acid. The solution turns reddish purple, symbolizing the formation of black phosphorus/gold nanocomposites. The sponge is soaked in this solution and dried at 70°C. A dried sponge containing black phosphorus/nanocomposites is placed on paper with a double-sided copper tap. Electrical measurements were done through that copper tap. As the pressure increases, the resistance of the sponge rises sharply, and then it gets saturated. This is because, after specific pressure formation, the conducting path in the sponge gets saturated. This showed different sensitivity depending on the pressure range. The sensor showed more sensitivity for the initial increasing pressure of 2.047 kPa^{-1} and later at near zero. This sensor is subject to different applications like finger tapping, body movement, robotic hand movement, and even breath rate monitoring. Response times of 10ms and 20ms were recorded for pressing and releasing the pressure, respectively.[24]

7.3 CHEMICAL SENSORS

This section will discuss how 2D materials are implemented to recognize biomarkers of different diseases. In human body, several chemical substances are produced in metabolic reactions. Sometimes, the amount of certain chemicals or molecules crosses its maximum permissible limit, and sometimes, unexpected chemical compounds may be made inside the body.[6,25] This may be because of some abnormal health conditions or underlying diseases. The critical idea of chemical sensors is that detecting these biomarkers in a human breath will help us identify diseases from which a person is suffering. The basic mechanism is when an ion or molecule is adsorbed on the surface of an active material (in our case, it is a certain 2D material). Since 2D materials have a high surface-to-volume ratio, this ensures the maximum adsorption of molecules on the material. After the adsorption of molecules, a charge transfer (electron or hole transfer) happens between the molecule and the surface of the active layer. that causes the change in carrier concentration locally on the surface of the active layer. which eventually leads to a change in resistance. In the following section, we mainly focus on the molecules present in breath and how the targeted molecule is expected to be a biomarker for a particular disease.

7.3.1 Graphene-Based Chemical Sensors

Pure graphene can be used for gas sensing-related applications. Electron-accepting gas like ammonia NH_3 can accept electrons from graphene and generate a hole on the top layer. Hence, it reduces the resistance of the graphene layer.[26] The sensitivity of graphene towards different gases can be increased by doping or generating defects in it. Instead of pure graphene, reduced graphene oxide (rGO) is used for better sensitivity. For the synthesis of rGO, first graphite is oxidized and then partially reduced to regain the properties of graphene.[27,28] Graphene oxide is covered by oxygen-containing groups such as hydroxyls, epoxies, carboxyls and lactones—these oxygen groups on rGO act as the active sites for different gas adsorption.

An MoS_2-based sensor shows sensitivity towards NO gas but cannot be used for repeated cycles because its electrical properties change over time.[29,30] The CVD-fabricated graphene sensor can detect ammonia up to 17 ppm, and the recovery time for this sensor was 20 min.[31] For ammonia sensing, a MoS_2/graphene hybrid was used; this sensor can detect 50 ppb of ammonia. However, it had a slow recovery time.[32] The above discussion shows that using TMDCs with graphene can enhance the sensitivity and selectivity of the sensors. The composite of graphene oxide with MoS_2 has several surface defects that make it suitable for sensing different types of gases. Since it can operate at low temperatures, a polyimide substrate provides flexibility to this sensor. The PI substrate was used to provide the flexibility to device. The device can show reliable sensitivity when the substrate is deformed at 30, 45 and 90° angles. The sensor can withstand 1000 bending. This sensor can detect the NH_3 over the range of 50–1500 ppm under the applied temperature of 141°C since ammonia is a chronic kidney disease biomarker in patients.[33]

In the previous example, MoS_2 functionalized reduced graphene oxide showed sensitivity towards NH_3. However, it is evident that with the help of certain functionalization materials, we can tune the sensitivity of graphene towards particular molecule or molecules. Similarly, ZnO functionalized reduced graphene oxide showed sensitivity towards various biomarkers in the breath, like NH_3, CO and NO. Ammonia, as mentioned in the previous section, is the biomarker of chronic kidney diseases. CO and NO in breath can be the cause of smoking and inflammatory diseases of the lungs and other organs.[34] ZnO is itself able to detect NH_3, CO and NO_2. However, a sensor made of only this material is not operable at room temperature. rGO also has excellent potential for the detection of the biomarkers mentioned above. However, ZnO-rGO oxide improved performance compared to the rGO-based sensor in terms of responsitivity and operating temperature. ZnO-based sensors require high operating temperatures, while ZnO-functionalized rGO operates at room temperature. The rGO sensor showed a response for 5 ppm NH_3, while the ZnO-functionalized rGO responded for 1 ppm NH_3 with more sensitivity. GO oxide was synthesized by following the Hummers method. The dispersed solution of GO in ethanol was mixed with zinc acetate. Later, the solution is sonicated for a few minutes and left for a day to ultimately reduce GO. Here, ZnO is doing the sensing part of the sensor, and graphene acts as a conducting mesh for charge transport. The material was deposited on the ITO glass. In response to gas exposure to the sensor, I–V measurements were taken.[35]

Graphene on paper can detect NO_2 gas. Growing graphene on paper directly is not an easy task. That is why CVD would be suitable for graphene growth before transferring it onto paper. CVD-grown graphene was deposited on paper, and silver contacts were taken on film to observe the relative change in the conductance. The sensor obtained on the paper showed a p-type nature under ambient conditions. When NO_2 comes in contact with the sensor, it extracts electrons and creates holes in the graphene paper, thereby increasing its conductance. This sensor showed a response time of 65% in the 1400s when NO_2 molecules were adsorbed on the surface. However, desorption of NO_2 from the surface took a longer time. A 15% recovery was observed in the 1500s. Recovery time can be reduced to 30s when the sensor is exposed to UV. A strip of G-paper is exposed to 2.5 ppm of NO_2. After keeping the sensor inside the deep UV setup, it gets back to the baseline in 30 seconds. This sensor can detect concentrations as low as 0.5 ppm of NO_2 with a responsivity of 325% ppm^{-1}. When this paper-based sensor was stretched, a decrease in the resistance was observed. This is expected as the cracks produced because of the strain cause an increase in the sensing surface area.[36]

The human breath contains a specific composition of certain volatile organic compounds (VOCs). When a person is suffering from a particular disease, the concentration of particular VOCs may change. Chemical sensors are very useful in detecting these changes without using any invasive techniques. Chronic obstructive pulmonary disease (COPD) is a lung disease that causes improper flow in or out of air. COPD has multiple biomarkers, for example, acetaldehyde, benzaldehyde, benzene, butanal, isoprene, isopropanol, limonene, nonanal and many other VOCs.[37] Here, we discuss the sensor based on rGO-M for the detection of acetaldehyde. M symbolizes elements from the transition metals (M = Mn, Fe, Co, Ni). A GO dispersion solution was mixed with different precursors of transition metals, for example, $Mn(CH_3COOH)_2.4H_2O$, $FeCl_3.6H_2O$, $CoCl_2.6H_2O$, $NiCl_2.6H_2O$. The mixture is stirred at 25°C for 24 hours and then drop cast on the ITO glass. Material on the ITO glass was heated at a temperature of 80°C. The high temperature applied on the substrate ensures the thermal reduction of the GO-M composite to rGO-M hybrid.

In this way, three layers of material were deposited on the ITO glass. Oxide and hydroxide functional groups have a negative charge, and transition metal ions have a positive charge. Thus, a stable electrostatic bond is formed between the metal and the oxygenated site. In the presence of the metal ion, this oxygenated site becomes immune to reduction. Finally, rGO-M on ITO substrate was annealed at 70°C for 24 hours in hydrazine gas, followed by vacuum drying to bring stability to the sensor. Gold contacts were taken 4 mm apart to observe the change in resistance in the presence of applied gas. As oxygen molecules come in contact with the surface where the transition metal is present, they get electrons because of the interaction of the d orbital of the transition metal with oxygen's molecular orbital. This leads to the formation of O^{1-} anion. Acetaldehydes react with this oxygen anion, producing acetic acid and a free electron. Free electrons reduce the conductivity because of the recombination of the electron-hole pair. This mechanism also suggests that synthesized material shows a p-type nature. Out of different rGO-M-based sensors, rGO-Ni has the highest conductivity and sensitivity towards acetaldehyde. This happened because during the synthesis of rGO-Fe, rGO-Mn and rGO-Co, some of the oxygen functional groups did not reduce on GO. All rGO-M-based sensors showed improved performance in terms of sensitivity compared with rGO-based sensors when subjected to the same ambient.[38]

The presence of acetone in the human breath is the biomarker of type 1 diabetes. When acetone levels in the breath exceed 1.8 ppm, this confirms that the person has type 1 diabetes. Fe_2O_3 exhibits the sensing towards various volatile compounds. However, its operating temperature is very high and has poor conducting properties. Graphene provides a conducting mesh to charges for moving across the sensor, and the presence of noble acts as the catalyst in dissociating gas molecules into more reactive species. Synthesis of Ag/Fe_3O_4/rGO is shown step by step in Figure 7.4 (a). GO was dissolved in ethanol, and a dispersion solution is created with the help of sonication. Another NaF solution was mixed with Fe_2O_3. The final solution was mixed drop by drop in GO dispersion. After heating, the residue was washed with distilled water and freeze-dried to get β-FeOOH/GO. This residue was mixed with Tris HCl, and dopamine hydrochloric acid was added to the solution. The mixture was stirred for 12 hours, and the precipitate was filtered out to get PDA/β-FeOOH/GO. Dispersion of PDA/β-FeOOH/GO was created in ethanol with sonication's help. Later, $AgNO_3$ was slowly poured while constantly stirring dispersion. At the end, after annealing and vacuum drying, Ag0.1/Fe_3O_4/rGO was obtained. The prepared sample was ground and mixed with distilled water to create a paste. The paste of Ag/Fe_3O4/rGO was deposited on the ceramic tube with the help of spin coating. Gold electrodes were deposited at the end of the ceramic tube, and platinum wire was connected to gold electrodes as a current collector (Figure 7.4 b). A nickel-chromium alloy heater was inserted inside the tube to provide the temperature. The final sensor structure was left for 10 days before starting the sensing. This sensor was exposed to different concentrations of acetones ranging from 5 ppm to 800 ppm at various operating temperatures ranging from 220 °C to 320 °C. Response for the sensor was R_a/R_g, where Ra is the resistance of the sensor in air, and R_g was the resistance of the sensor in the target gas. In the air, oxygen molecules caption the electrons from Fe_2O_3 and get ionized into O_2^{1-}, O^- and O^{2-}). Since this is an n-type MOS structure, it causes the formation of the depletion layer and increases the resistance. Though acetone is a reducing gas, it will take electrons from ionic oxygen species to acetone molecules. It oxidizes ionic oxygen to H_2O and CO_2, which again decreases the resistance. The responses for Ag0.05/Fe_3O_4/rGO, Ag0.1/Fe_3O_4/rGO, and Ag0.1/Fe_3O_4/rGO were 14.89, 35.81 and 23.03, respectively. The response of Fe_3O_4/rGO for acetone was 2.5 lower compared with Ag0.1/Fe_3O_4/rGO. Notably, the selectivity of the Fe3O4/rGO for ethanol and acetone was not as good as Ag0.1/Fe_3O_4/rGO (see Figure 7.4 (c)). Ag0.1/Fe_3O_4/rGO was also studied under a humid environment. It showed the response of 35.8, 26.7, 22.0, 7.8 and 3.0 when subjected to the relative humidity of 10%, 30%, 50%, 70% and 90%, respectively. When the sensor is placed under the

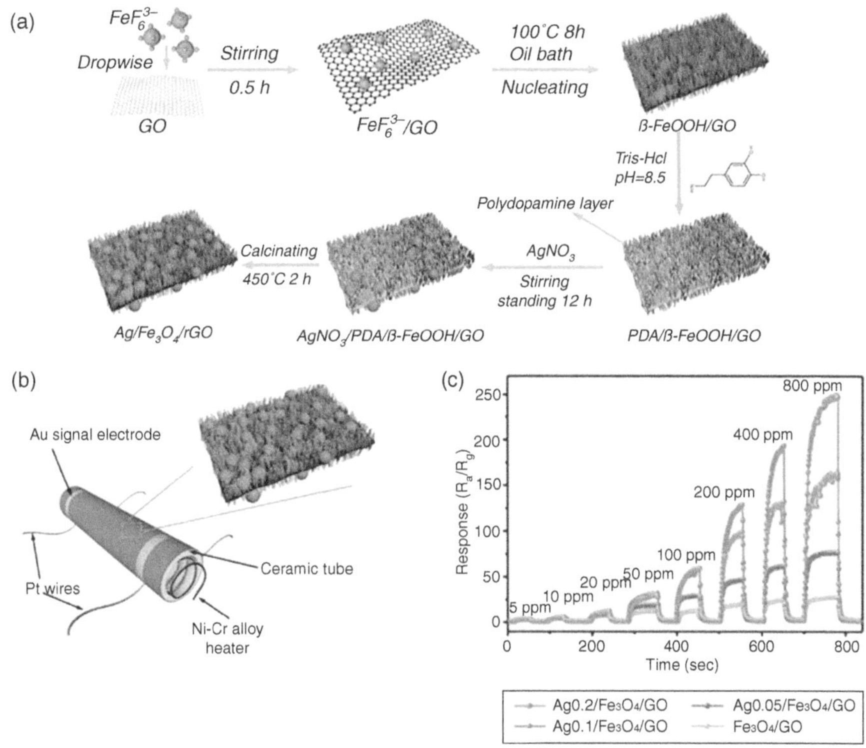

FIGURE 7.4 (a) $Ag/Fe_3O_4/rGO$ synthesis procedure flow chart, (b) Fabricated device structure of $Ag/Fe_3O_4/rGO$-based sensor, (c) Dynamic response and recovery cycles during acetone exposure ranging from 5 to 800 ppm at 220°C (Reproduced with permission from Materials Chemistry and Physics[38]).

humid atmosphere, the water molecule occupied a few areas where the oxygen was already adsorbed. This n-type MOS sensor, in the presence of reducing gas, accepts electrons from acetone and water molecules and causes a reduction in the sensor response.[39]

7.3.2 2D Chalcogenide-Based Chemical Sensor

H_2S is a very toxic gas for humans, having a rotten egg-like smell. Exposure to H_2S can badly damage the nervous system and cause the loss of consciousness. According to the US Scientific Advisory Board, the permissible level of H_2S to human exposure is 100 ppb. Besides toxicity, it is one of the biomarkers of different diseases like asthma, halitosis, irritable bowel and so on.

$MoSe_2$ nanoflakes showed the potential for ppb-level detection of H_2S. $MoSe_2$ was exfoliated in acetone, isopropanol and DI water for sensor fabrication. The dispersed solution of the $MoSe_2$ was drop cast on a Si/SiO_2 wafer where interdigitated electrodes were already sputtered. The interdigitated electrodes had a 5μm width and gap. Inherently pure $MoSe_2$ is an n-type semiconductor, but because of the adsorption of O_2 and H_2O, it got p-type. Since H_2S is a reducing gas, the presence of H_2S on the surface of $MoSe_2$ causes an increase in the resistance. That leads to a decrease in the sensor's current response. The operation of this sensor was also checked under different operating temperatures. It was observed that the sensor's response linearly increased as the temperature increased to 200°C, and reaching this temperature, the sensor's responsivity began to decrease. The fabricated sensor was exposed to different VOCs, including ammonia. It showed minor or no response to these VOCs and ammonia compared to H_2S. A response of 18.57% was recorded for 50 ppb of H_2S gas. After testing the sensor's repeatability, stability and long-term reliability, researchers claimed that this can detect diseases like halitosis and replace expensive gas chromatography techniques.[40]

Isoprene is a potential biomarker for chronic liver disease. While cholesterol biosynthesis happens in the liver, isoprene is produced as a by-product. Advanced fibrosis in the liver caused the reduction in isoprene levels in the breath. It was observed that the patients without advanced fibrosis had isoprene in their breath at 40.4 ppb, and the patients with advanced fibrosis had isoprene in their breath at 13.5 ppb. Based on obtained data from the pilot study, 25 ppb is the threshold limit, and the above persons were less likely to have advanced fibrosis in their liver.[41] In_2O_3/MoS_2 is a promising material that detects ppb levels of isoprene in the human breath. The required composite is synthesized using the hydrothermal method. A solution of ammonium molybdate, thiourea and ammonium fluoride was transferred to the PTFE-lined reactor. The reactor was kept at 200°C for 24 hours to synthesize MoS_2. The precipitate of MoS_2 was filtered out and cleaned with ethanol and DI water. Again, a solution of $InCl_3$ and sodium hydroxide was created and transferred to the PTFE line reactor. The reactor was heated at 160°C for 18 hours to synthesize $InCl_3$. The obtained MoS_2 and $InCl_3$ were mixed and heated at 190°C to make the required composite. Suspension of the composite was created in the ethanol and drop cast on the ceramic sheet containing the electrodes. The backside of the ceramic sheet also contains the heating coil. Four terminals of the ceramic sheet were tin-shouldered with the TO package. The adsorption of electrons on surface nanocomposite produces ionized oxygen species. When the sensor comes in contact with isoprene, it dissociates and produces free electrons. These free electrons help form the electronic depletion region and oxygen faster. Extra electron formation causes a faster recombination rate, increasing the sensor's response. The sensor was more selective to isoprene compared to other gases. It had a response time of 38 s at 100 ppb level. The sensor showed the measurable response for the range of isoprene concentration of 1.8 ppb to 100 ppb. The optimum operation temperature for the sensor was 200°C. Long-term stability was also conducted, and the sensors' response decreased after a few days.[42]

The presence of formaldehyde is generally related to the sick building syndrome (SBS). SnS_2 showed the potential to detect ppb levels of formaldehyde. For the sensor, multilayer SnS_2 is synthesized with the help of chemical vapor deposition; for CVD, SnO_2 and elemental sulphur crystal were used. SnS_2 was grown on the Si/SiO_2 substrate; the device structure is shown in Figure 7.5(a). A sonication substrate in IPA created SnS_2 suspension in isopropyl alcohol (IPA). This suspension was transferred to SiO_2, and later, Ti/Au patterns were developed on the surface with the help of photolithography and lift-off techniques. Contact was taken to make the back gate FET

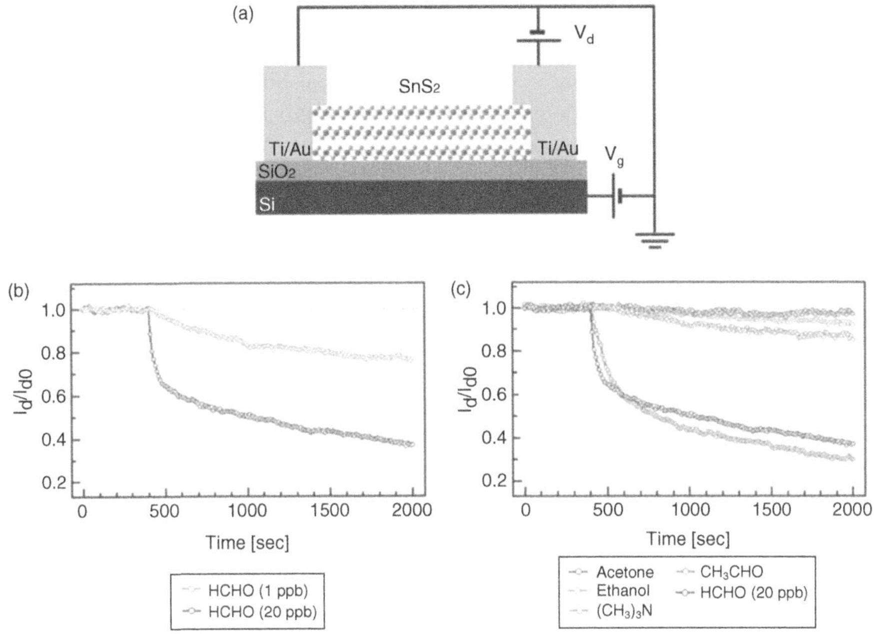

FIGURE 7.5 (a) Device structure of SnS_2 sensor, (b) Normalized drain current of SnS_2 at different concentrations of HCHO ranging from 1 to 20ppb, (c) Normalized drain current of the CVD-grown SnS_2 sensor in response to various test gases. The gas concentrations were 200 ppb, except for HCHO (20 ppb) (adapted with permission from [43]. Copyright 2020 American Chemical Society).

base sensor at the back gate. Gas sensing was performed in the nitrogen and oxygen ambient. The sensor showed the n-type response with excellent selectivity towards the formaldehyde and the NO_2. Sulphur vacancies on the surface of SnS_2 caused the upward shift of 0.5 eV in the fermi level, making it n-type. When SnS_2 interacts with HCHO, sulphur vacancies get filled by oxygen because HCHO dissociates into oxygen and C_2H_4, and the fermi level moves downward. Thus, it increases the resistance and reduces the current. However, it is an irreversible process, so the sensor did not recover from exposure to HCHO or oxygen. The exposure to 20 ppb of HCHO caused a 70% change in the initial current within 1 minute. Even one ppb sensor showed a 20% change in the current (Figure 7.5 (b) and (c)). All these sensing tests were performed in the nitrogen ambient. When sensing tests were conducted in the oxygen ambient, researchers observed the performance degradation.[43]

7.3.3 2D Metal Carbides/Nitrides (or MXene)-Based Chemical Sensor

$Ti_3C_2T_x$ can sense acetone at room temperature. However, the sensor's response is unstable, and heating improves the performance, but a vast fluctuation is still visible in the response. Heating the sensor to 300°C does not help and sometimes leads to no response. $W_{18}O_{49}$, a nonstoichiometric compound, contains many oxygen vacancies, improving its acetone-sensing properties. Despite its suitable sensing, it cannot detect whether a person has diabetes because of its low detection limit, and the presence of acetone at a level of 2ppm in breath is a potential biomarker for diabetes. The composite of $W_{18}O_{49}/Ti_3C_2Tx$ can overcome the disadvantages of both materials. A solvothermal process is used for the synthesis of $W_{18}O_{49}$. WCl_6 was wholly dissolved in ethanol by mechanical stirring. Later, acetylacetone was also added, and the final solution was transferred to an autoclave and heated for 25 hours at 150°C. The precipitate of $W_{18}O_{49}$ was collected and washed with DI water. Similarly, $W_{18}O_{49}/Ti_3C_2T_x$ was synthesized with the help of a solvothermal process.

One difference is that $W_{18}O_{49}/Ti_3C_2T_x$ was added to the solvothermal reaction solution. For the material deposition on the sensor, a dispersion solution of $W_{18}O_{49}$ nano rode, $W_{18}O_{49}/Ti_3C_2T_x$, and $Ti_3C_2T_x$ was created in the ethanol and deposited on the ceramic plates containing gold electrodes. A nickel-chromium heater was placed inside the ceramic plate to operate the sensor at different temperatures. The $W_{18}O_{49}/Ti_3C_2T_x$ interface creates a Schottky barrier because electrons from $W_{18}O_{49}$ were transferred to $Ti_3C_2T_x$, since $Ti_3C_2T_x$ has a higher work function than $W_{18}O_{49}$. In air below 150 °C, O_2 accepts electrons from oxide semiconductors to form O_2^- ions. When temperatures exceed 150 °C, oxygen ions create O^- and O^{2-} species. Since the operating temperature of the $W_{18}O_{49}/Ti_3C_2T_x$ sensor was around 200 °C, which caused an increase in the concentration of oxygen anions, it led to an increase in the depletion width of the Schottky barrier. On exposure of acetone to the sensor, these anions of oxygen react with acetone molecules and release electrons, causing a decrease in the depletion width. Ultimately, resistance decreases in response to acetone. Based on the above discussion, the fabricated sensor showed n-type behaviour. This sensor showed an excellent response for 11.2 to 20 ppm acetone with a detection limit of 170 ppb acetone. It had recovery and response times of 6 s and 5.6 s, respectively.[44]

$Ti_3C_2T_x$ is not the only MXene used for chemical detection; there are other MXenes, such as $V_4C_3T_x$. Its sensor activity is studied in a more realistic environment, such as in the presence of water vapours. $V_4C_3T_x$ MXene is obtained from V_4AlC_3 MAX. V_4AlC_3 is synthesized by heating the mixture of V, Al, Co, and C in a molar ratio of 4:28:6:3 at 1500 °C over 10 h under flowing argon and cooling to 900 °C for 30 hours. Finally, V_4AlC_3 is obtained by cooling the mixture to room temperature within 20 hours. $V_4C_3T_x$ was synthesized by selectively etching Al from $V_4C_3T_x$ with HF. To exfoliate $V_4C_3T_x$ MXene sheets sonication was followed by centrifugation. To increase the yield, the delamination agent tetrabutylammonium hydroxide was also used in the mixture. After vacuum filtration, the film was freeze-dried, stripped and cut into pieces. Silver contacts were taken, and copper wire was contacted to them (see Figure 7.6 (a)), for taking electrical measurements. This sensor was studied in the presence of acetone and water vapours at room temperature. Figure 7.6 (b) shows a response to different acetone concentrations. It can sense the 1 ppm of the level of acetone by showing the 5 % change in the sensor's resistance. When the sensor is subjected to a humid atmosphere and acetone vapours, it shows ten times more response to acetone vapours than water vapours for the same concentration of acetone and water. The response time for acetone was in the 40 s, which is much faster than the response for water vapours. The above discussion shows the potential of the $V_4C_3T_x$-based acetone sensors in diabetes detection.[45]

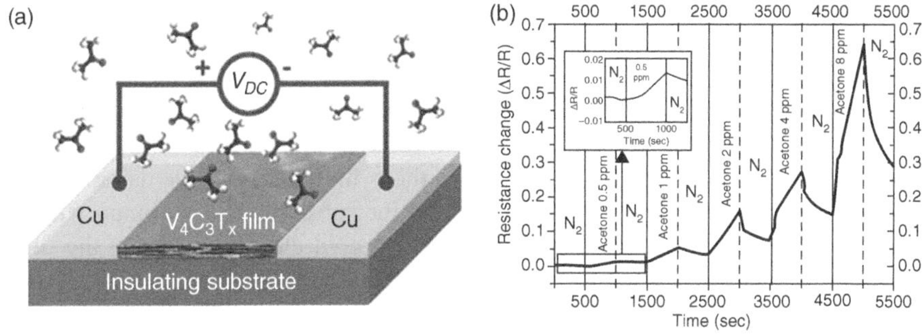

FIGURE 7.6 (a) Device structure of $V_4C_3T_x$ film-based sensor, (b) V4C3Tx film-based chemical sensor response to different concentrations of acetone ranging from 0.5 ppm to 8 ppm at room temperature; the response to 0.5 ppm is shown in inset (Reproduced with permission from Royal society of chemistry [45]).

Ammonia is the potential biomarker for the peptic ulcer. The presence of 0.250–1500 ppb ammonia level in the breath signifies that a person suffers from a peptic ulcer. Since ammonia is highly toxic, the exposure limit of ammonia is 22 ppm for 8 hours. For $Ti_3C_2T_x$ MXene/SnO_2 nanocomposites, MXene was synthesized using the delamination method. LiF and HCl were used for this. SnO_2 nanoparticles were synthesized using the hydrothermal method. Stannous chloride was dissolved in the 40 ml distilled water, and after stirring, an appropriate amount of sulfuric acid was added to the solution. The final solution was transferred to the autoclave and heated at 200 °C for 24 hours. In this way, nanoparticles for SnO_2 were synthesized. MXene was added to a well-dispersed SnO_2 solution and sonicated to obtain $Ti_3C_2T_x$ MXene/SnO_2. $Ti_3C_2T_x$ MXene is inherently n-type, but because of the presence of SnO_2 on the surface of the MXene, it showed the p-type behavior, that is, its resistance increased in the presence of reducing gas like ammonia. To check the sensitivity of $Ti_3C_2T_x$ MXene/SnO_2 nanocomposites for NH_3, the sensor is exposed to different VOCs like formaldehyde, octane, Dimethyl formamide, toluene, methanol, ethanol, acetone, and ammonia. Among all of these, the sensor has a distinguishable response for ammonia. The limit of detection of this sensor, 3.14 ppb, was observed. Recovery and response for $Ti_3C_3T_x$ MXene/SnO_2 were observed at 16 s and 86 s, respectively, better than the response and recovery of 61 s and 77 s SnO_2-based gas sensors. Its recovery time was also better than that of the based gas sensor, which had a recovery time of 300 s.[46]

Isopropanol is the potential biomarker for lung cancer, which is a life-threatening disease.[47] When the breath of patients with lung cancer was subjected to sensor array analyses, isopropanol was detected in a range of 250 ppb to 1000 ppb.[48] Here, we discuss the $MoO_3/TiO_2/Ti_3C_2T_x$-based sensor having a detection limit of 50.45 ppb, which can detect isopropanol in the breath for lung cancer detection. One of the measure drawbacks of MXene-based sensors is that they oxidize for a few hours to days. Because of the oxidation, a layer of TiO_2 on the MXene is formed, increasing the sensor's resistance and ultimately affecting the sensor's performance. Since TiO_2/MXene plays a significant role in sensing when it degrades because of oxidation, the sensor's performance also gets affected. For synthesizing the nanocomposite, a stirred solution of (NH_4) 6Mo_7 O_{24}.4H_2O in DI water was mixed with $Ti_3C_2T_x$ solution in DI water. Later, HCl was also added to the mixture, and the final solution was transferred to the autoclave. The autoclave was heated at 18 °C for 24 hours. After filtering the sediment, it was vacuum-dried at 120 °C to obtain $MoO_3/TiO_2/Ti_3C_2T_x$. The nanocomposite is deposited on the cylindrical substrate using the drop-cast method. Gold electrodes were already on the substrate, and platinum wire was taken for electrical measurement through the Au gold electrodes. When the sensor comes in contact with reducing gases, its resistance decreases, showing p-type behaviour. The researcher of this sensor claims that MoO_3 is the primary material through which reducing gas interacts with the sensor. Because TiO_2 has more thickness than the Debye length of mobile charge carriers, its effect on the main conducting path is negligible. MoO_3 is the primary active layer for reducing gases because it has a lot of oxygen deficiency on the surface. This showed a response of 245% to 50 ppm isopropanol, with response and recovery time of 100 and 40 s.[49]

7.4 BIOSENSORS

Like any other sensor, biosensors generate measurable electrical signals when they encounter any biomolecules, such as enzymes, specific proteins, DNA, RNA and so on. Biosensors are revolutionary approaches for monitoring a person's health condition without undergoing expensive medical tests. The basic structure of the biosensor consists of a bioreceptor and a transducer. Bioreceptors must be sensitive enough and fast enough to detect the biomolecules or biomarkers nearby for real-time monitoring. Transducers need to produce signals based on the changes in specific physical, chemical, electrochemical and optical conditions. The produced signals from biosensors are later processed for medical treatments.[50] The biosensor's structure consists of three main parts: a bioreceptor, a transducer and a signal processing unit. Bioreceptors are very specific for analytes or pathogens, and are the first layer interacting with the biomolecule.

The bioreceptor layer is followed by the transducer, which generates the corresponding output signal based on the interaction of analytes with the bioreceptor. This signal is later processed by a signal processing unit to identify the biomolecule. The transducer should be compatible with a bioreceptor. That is why sometimes the

tethering layer is introduced between the transducer and the bioreceptor to increase the conjugation of these layers.[51] The transducer's choice is crucial since it influences the device's characteristics. The use of 2D material as a transducer layer brings a lot of benefits like high volume-to-surface ratio, thin 2D monolayers in nanometre range layered structures, good biocompatibility, and strong interfacial interaction with biomolecules, proteins, DNA, cells and microorganisms. Similar to previous sections, the following section briefly discusses different kinds of 2D material-based biosensors.[52]

7.4.1 Graphene for Biosensing Application

Tuberculosis is a disease that targets the respiratory system. Though medical treatment for TB is available, if it is not detected at early stages, it can be fatal. There are several biomarkers for tuberculosis, but (Heat shock protein) Hsp16.3 is crucial. Since TB is caused by mycobacterium tuberculosis (Mtb), and Hsp16.3 is a capsule protein produced by Mtb, detecting this protein can show the presence of Mtb. Research on this sensor developed a paper-based analytical device, PAD. This

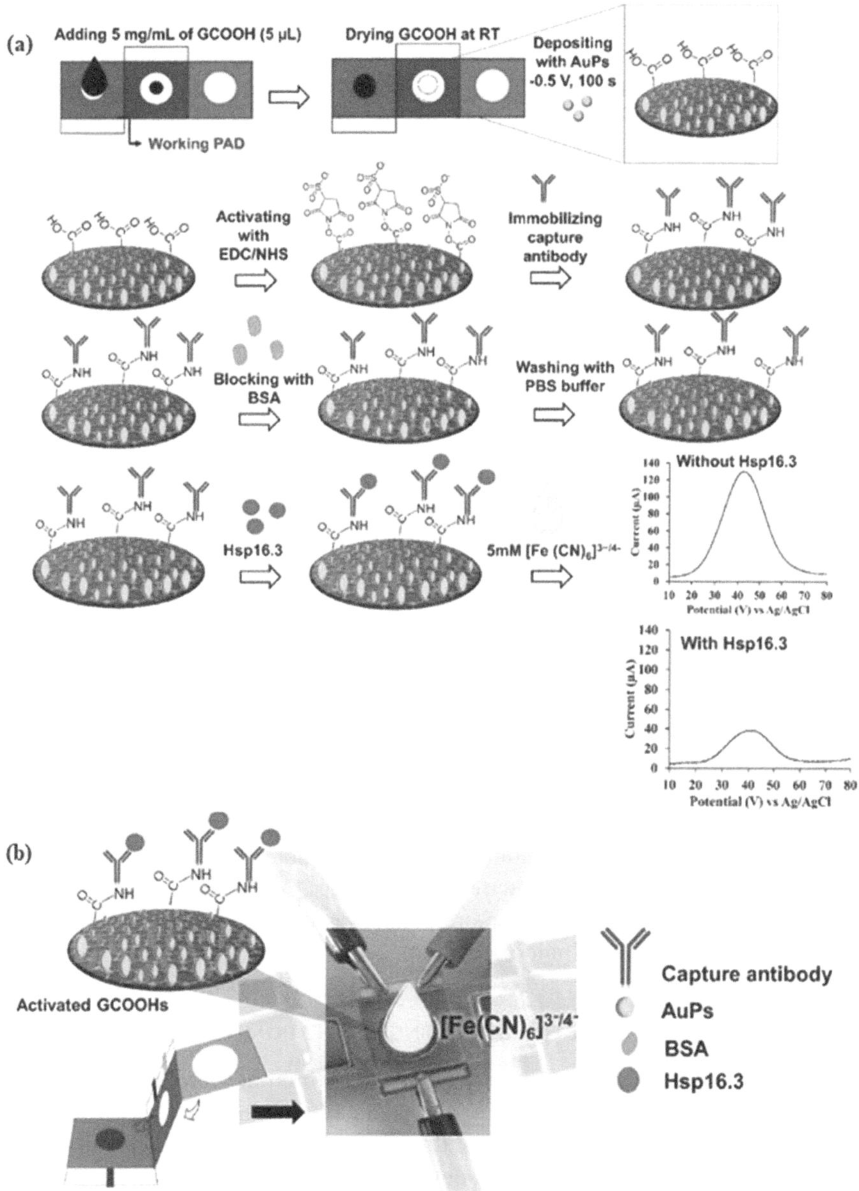

FIGURE 7.7 Developed electrochemical immune-sensing device for sensing Hsp16.3, (a) Fabrication steps, (b) Working operation, and electrochemical measurement (Reproduced with permission from Talanta publications [51]).

pad contains three parts or folds, each with a specific role towards the sensing (see Figure 7.7 (a)). The first folding, a red pad, includes a biorecognition layer and the working electrode (WE) (5 mm diameter). A counter electrode has reference and counter electrodes (CE) in the second folding blue pad. The third folding green pad was a closing electrode. The closing layer is designed in such a way that it can store the solution containing the biomolecules (see Figure 7.7 (b)). For this reason, wax printing was used to define PAD's hydrophobic and hydrophilic regions; for PAD, Whatman grade 1 chromatography paper was chosen. The centre part of the layer containing the reference electrode (RE) and the CE was removed to include the solution directly in the biorecognition area. For the biorecognition layer, the first water paper was washed multiple times with milli-Q water, and carboxyl graphene was soaked and dried at room temperature. Then, gold nanoparticles were transported on it through $AuCl_4$ solution in 0.1M HCl solution. Then, the layer was activated by deposition of 1-ethyl-3-(3-dimethylamino- propyl) carbodiimide (EDC) (20 mM)/ N-hydroxy succinimide (NHS) (40 mM) in phosphate-buffered saline (PBS) at pH 7.4 (10 μL) for 1 hour. Finally, the antibody was dropped on the surface and left for an hour for coupling reaction, and PBS washing was done. Bovine serum albumin solution was cast on the EDC/NHS surface to prevent interference through other biological molecules, followed by PBS washing. In this way, the biorecognition layer was prepared. Square wave voltammetry was used to detect the biomolecules. The solution was implemented as the redox probe for this five mM $[Fe (CN)6]^{3-/4-}$. A solution containing a specific concentration of Hsp16.3 is transferred to the PAD and incubated for 20 minutes, followed by washing with PBS. The RE/CE layer and closing layer were folded. The pad was held with top and bottom polyacrylamide lids, and the top layer had a circular opening. On the pad, 100 μL of 5 mM $[Fe(CN)_6]^{3-/4-}$ was cast through the opening of polyacrylamide. As the concentration of Hsp16.3 increases, the current response of the sensor decreases. That was expected because Hsp16.3 formed the immune complex on PAD, which is usually a poor conductor. The sensing response of PAD was also studied with different interfering substances like albumin, bilirubin, Alpha-fetoprotein, ferritin, c-reactive protein and IL-6. Concentrations of interfering substances were taken almost 20 times higher than Hsp16.3. Even at these concentrations of interfering molecules, the sensor showed a minor or negligible response to interfering substances. After preserving the sensor at 4°C, it was observed that the sensor's response was reduced by 10% after ten days. This sensor had a detection limit of 10 ng/ml.[53]

7.4.2 2D Chalcogenide for Biosensing Application

Covid-19 is caused by the severe acute respiratory syndrome coronavirus 2 (SARS-CoV-2). It emerged as a pandemic in 2020 and caused a lot of chaos in the world. This signifies how vital the detection of this virus is. However, there are outstanding solutions out there. Still, researchers want to explore more feasible and fast detection of this virus. This section discusses WSe_2-based sensors for detecting the SARS-CoV-2 spike protein. For devices, interdigitated electrodes were deposited on the triangular-shaped crystals of WSe_2 using photolithography. Si/SiO_2 was taken as the substrate. The device structure is shown in Figure 7.10. To provide the gate terminal with an ionic solution (0.01 X PBS in deionized water), we can give the SARS-CoV-2 spike protein for detection purposes. First, electrical characteristics of pristine WSe_2 were taken. I_{DS}-V_{DS} plates of this device were taken at negative gate bias. It was also observed that only negative gate biases could turn on the device and, later, functionalized WSe_2 with the chemical linker 11-mercaptoundecanoic acid (MUA). The SH end of MUA mainly interacts with the chalcogen of WSe_2 and vacancies on the surface of WSe_2. The carboxylic group on MUA was activated through n-hydroxy succinimide (NHS) and carbodiimide hydrochloride solution at room temperature. Again, electrical characteristics were measured. It was seen that functionalized WSe_2 with MUA showed a decrement in the slope of I_{DS}-V_{GS} because of the reduction in the mobile carrier. Further, on the addition of SARS-CoV-2 antibody and spikes, slope I_{DS} vs V_{GS} again increased. The possible reason for the decrease in the conductivity of the device because of the addition of WSe_2 is that the vacancies present on the WSe_2 were being compensated by S and H atoms present on MUA. That caused the reduction in the holes of p-type WSe_2. The SARS-CoV-2 antibody causes the charges to transfer to MUA, increasing conductivity. The device can detect the SARS-CoV-2 spike proteins up to 25 fg/μL in 0.01X PBS. This WSe2-based FET device was also exposed to bovine serum albumin (BSA). However, it showed negligible response to the BSA as the response was compared to the SARS-CoV-2 spike proteins. Hence, it was selective to SARS-CoV-2. This fabricated device also showed great potential for detecting the biomarkers of SARS-CoV-2, such as cytokines.[54]

7.4.3 Metal Carbides/Nitrides (or MXene)-Based Biosensors

In human sweat, many biomarkers can be analysed for specific health conditions of a person, for example pH of sweat, glucose and lactate levels. The researcher developed an MXene-based sensor with the multifunctionality of detecting multiple analytes. For sensing material, a $Ti_3C_2T_x$ and Prussian blue (PB) composite were created and intercalated with CNTs. Since oxygen is required to work the sensor correctly, porous conducting super hydrophobic carbon fibre bottom layer CFMs were used. CNTs-intercalated $Ti_3C_2T_x$/PB was transferred to CFMs. To detect glucose level, the prepared CNTs/$Ti_3C_2T_x$/PB/CFMs electrode was covered with glucose oxidase (Gox). Similarly, for the lactate sensor used, the enzyme lactate oxidase was used to cover the entire layer of the CNTs/$Ti_3C_2T_x$/PB/CFMs electrode. For the sensing of the pH, a polyaniline (PANI)-coated working electrode was used to make ion-sensitive electrodes, and a reference electrode made up of Ag/AgCl was used. The potential difference between the ISE and reference electrodes was generated because of the different pH of the medium between the electrodes. The sensing module comprises three layers: the sweat uptake layer, the sensing layer and the cover layer. The sweat uptake layer for collecting sweat was made with porous fabric and serpentine structures. Serpentine tunnels were also etched on the second layer, creating spaces for the sensors to be inserted.

The twisting tunnels between the second and first layers were filled with the toxic liquid metal alloy galinstan. Soft silicon rubber was used for the cover layer, which had holes for oxygen flow to the sensing layer. The structures at the back and front

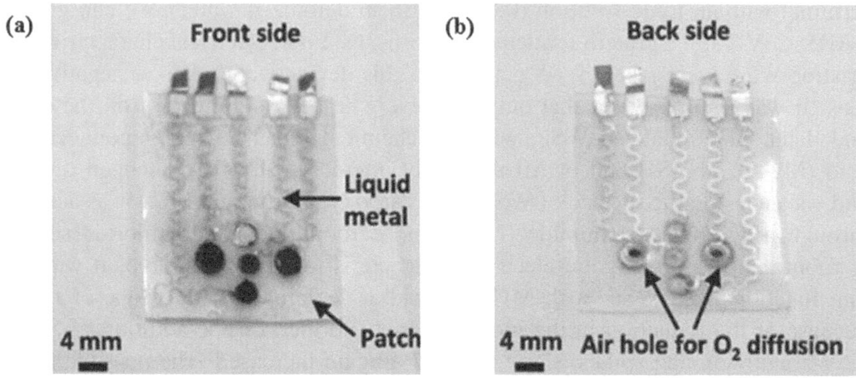

FIGURE 7.8 Wearable biosensor patch images, (a) Front side of sensor array (left and right), reference electrode (top), counter electrode (middle), and pH sensor (bottom), (b) Back side of sensor array (Reproduced with permission from Wiley publications [55]).

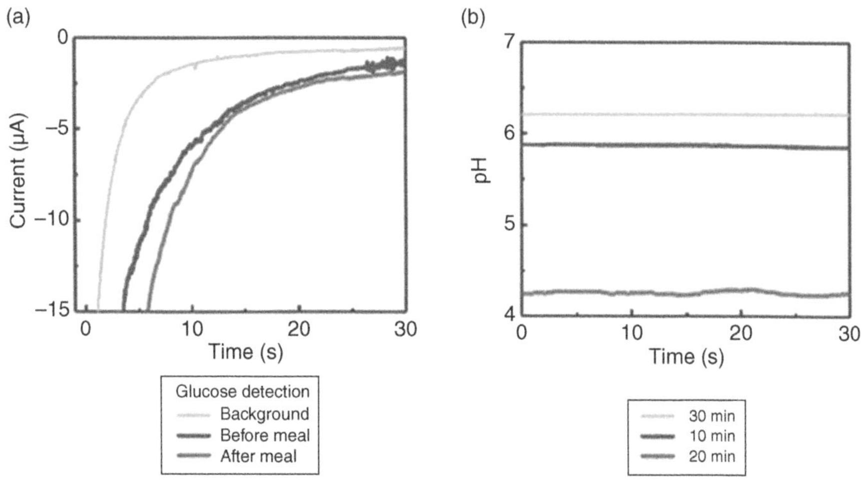

FIGURE 7.9 (a) Measured chronoamperometric responses of glucose sensors and pH changes recorded before and after meals with the help of three glucose sensors, (b) pH-level measurement readings by pH sensor while doing exercise (Reproduced with permission from Wiley publications [55]).

are shown in Figure 7.8 (a) and (b). The glucose sensor, lactate sensor and pH sensor showed a sensitivity of 35.3 $\mu AmM^{-1}\ cm^{-2}$ for the glucose concentration in the range of 10×10^{-6} M to 1.5×10^{-3} M, 11.4 $\mu A\ mM^{-1}cm^{-2}$ for the lactate concentration in the range of 10^{-3} to 22×10^{-3} M, -70 mV pH^{-1}, respectively. For testing purposes, the subject was instructed to engage in intense cycling to enable sweat analysis. After sweat was generated, the sensor patch took 2 minutes to show the response to glucose and lactate levels.[55] The pH of sweat also changes while cycling because of lactic acid in the sweat (see Figure 7.9 (a) and (b)).

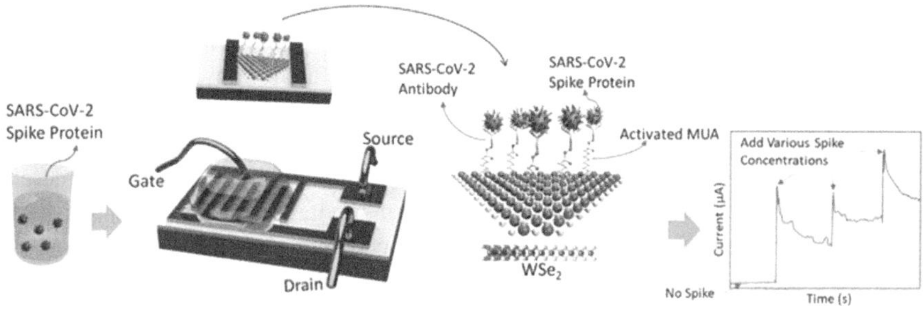

FIGURE 7.10 Device structure of the fabricated monolayer WSe_2-based FET Covid-19 sensor. Monolayer WSe_2 crystals were transferred on the fabricated interdigitated electrodes, and the SARS-CoV-2 antibody was immobilized onto the WSe_2 via 11-mercaptoundecanoic acid (MUA). MUA activation is done with n-hydroxysuccinimide (NHS) and carbodiimide hydrochloride (EDC) solution. The FET structure helps to sense the spike proteins in the form of current fluctuations (adapted with permission from American Chemical Society[54]).

7.5 CONCLUSION

In this chapter, we have discussed various 2D material-based sensors for healthcare applications. Materials like flexible polymer, paper, sponge and fabric substrates are mainly used to deposit the 2D material as an active layer for flexible sensors. We have also mentioned a sensor incorporating aerosol formation with 2D material to achieve flexibility. Since flexible substrates cannot withstand high-temperature fabrication processes, most researchers use low-temperature deposition techniques to transfer active material on their flexible substrates, like drop casting, screen printing, RF sputtering and so forth. The flexible sensors are mainly strain and stress sensors, and that is why they are meant to bend and stretch for body movement detection. Continuous efforts are being made to improve or tune the performance of sensors in terms of selectivity, sensitivity, response time and recovery time by using different device structures and functionalizing pristine 2D material with other organic or inorganic molecules. Chemical sensors can be designed to detect the presence of certain VOCs in breath since human breath contains more than 1000 VOCs in breath. The spectrum of the concentration of different VOCs, if analysed thoroughly, can be utilized to determine different health issues and diseases. We have reviewed sensors for detecting glucose, lactose and pH levels, and identifying specific biomarkers like isoprene, acetaldehyde, toluene, NO_2 and so forth since most VOCs in breath require ppb-level detection. Sometimes, that kind of high sensitivity is not achievable at room temperature. That makes using the heating coil on the back side of the sensor necessary for most chemical sensors for gas sensing. This provides an optimum heat level for activating the sensor, increasing the sensitivity. Direct VOC exposure to the sensor is enough to mimic the breath, and that is why most researchers of chemical sensors are expected to study the sensors in humid conditions. Biosensors' structure and working mechanism are more complicated than flexible and chemical sensors. It is important to note that while two-dimensional materials are utilized in biosensors, they are not capable of directly detecting biomolecules such as proteins, lipids, DNA, RNA, and glucose. They need a bioreceptor layer on top of 2D material to interact with biomolecules. Two-dimensional material is there, like a flat transducer or conducting mesh, which facilitates the transport of signals to the data processing units. Since most healthcare applications require a sensor to be in very close contact with the skin of a human body, their cytotoxicity should also be well explored before commercialization. Therefore, continuous monitoring of human vitals should not cause irritation or infection. Though these sensors are not advanced enough to replace existing bulky healthcare systems, they have the potential to revolutionize healthcare systems soon.

ACKNOWLEDGEMENTS

This work was supported in part by Core Research Grant of SERB-DST (Letter No. CRG/2021/005550), and PURSE research grant of DST (Letter No. SR/PURSE/2020/20), Govt. of India. Mr. Rejaul Ali would like to acknowledge the University Grants Commission for the Junior Research Fellowship to pursue his PhD work at BITS-Pilani.

REFERENCES

1. A. Hazra, and T. Gakhar, "GO/p-TiO2 hybrid channel based depletion-mode field-effect transistors with On/Off ratio higher than 103 at room temperature," *J. Mater. Sci. Mater. Electron.* **34**(26), 1–10 (2023).
2. T. Gakhar, Y. Rosenwaks, and A. Hazra, "Fullerene (C60) functionalized TiO2 nanotubes for conductometric sensing of formaldehyde," *Sensors Actuators B Chem.* **364**(March), 131892 (2022).
3. A. Chaves, J.G. Azadani, H. Alsalman, D.R. Costa, R. Frisenda, A.J. Chaves, S.H. Song, and Y.D. Kim, "Bandgap engineering of two-dimensional semiconductor materials," *Npj 2D Mater. Appl.* **4**, 29 (2020).
4. D. Rathore, and A. Hazra, "FlatChem Surface trapping induced negative photoconductivity in Au nanoparticles functionalized SnO 2 / SnSe 2 nanosheets under visible and NIR light," *FlatChem* **45**(January), 100650 (2024).
5. T. Gakhar, and A. Hazra, "Development of Graphene-Doped TiO2-Nanotube Array-Based MIM-Structured sensors and its application for methanol sensing at room temperature," *Chem. Proc.* **5**(1), 74 (2021).
6. P. Bindra, H. Mittal, B.R. Sarkar, and A. Hazra, "Synthesis of highly ordered TiO2 nanorods on a Titanium substrate using an optimized hydrothermal method," *J. Electron. Mater.* **51**(4), 1707–1716 (2022).
7. K.S. Novoselov, A.K. Geim, S.V. Morozov, D. Jiang, Y. Zhang, S.V. Dubonos, I.V. Grigorieva, and A.A. Firsov, "Electric field effect in atomically thin carbon films," *Science (80-).* **306**(5696), 666–669 (2004).
8. P.W. Bridgman, "Two new modifications of phosphorus," *J. Am. Chem. Soc.* **36**(7), 1344–1363 (1914).
9. W. Choi, N. Choudhary, G.H. Han, J. Park, D. Akinwande, and Y.H. Lee, "Recent development of two-dimensional transition metal dichalcogenides and their applications," *Biochem. Pharmacol.* **20**(3), 116–130 (2017).
10. A. Pareek, and S. Venkata Mohan, "Chapter 1.4 – Graphene and Its Applications in Microbial Electrochemical Technology," in *Microb. Electrochem. Technol.*, edited by S.V. Mohan, S. Varjani, and A. Pandey, (Elsevier, 2019), pp. 75–97.

11. A. Chhetry, M. Sharifuzzaman, H. Yoon, S. Sharma, X. Xuan, and J.Y. Park, "MoS2-decorated laser-induced graphene for a highly sensitive, hysteresis-free, and reliable piezoresistive strain sensor," *ACS Appl. Mater. Interfaces* **11**(25), 22531–22542 (2019).
12. M. Zhang, C. Wang, Q. Wang, M. Jian, and Y. Zhang, "Sheath–core graphite/silk fiber made by dry-meyer-rod-coating for wearable strain sensors," *ACS Appl. Mater. Interfaces* **8**(32), 20894–20899 (2016).
13. F. Wang, J. Jiang, F. Sun, L. Sun, and T. Wang, "Flexible wearable graphene / alginate composite non-woven fabric temperature sensor with high sensitivity and anti- interference," *Cellulose* **27**(4), 2369–2380 (2020).
14. T.-C. Wu, A. De Luca, Q. Zhong, X. Zhu, O. Ogbeide, D.-S. Um, G. Hu, T. Albrow-Owen, F. Udrea, and T. Hasan, "Inkjet-printed CMOS-integrated graphene-metal oxide sensors for breath analysis," *Npj 2D Mater. Appl.* **3**(1), 42 (2019).
15. X. Zhao, W. Yao, W. Gao, H. Chen, and C. Gao, "Wet-Spun superelastic graphene aerogel millispheres with group effect," *Adv. Mater.* **29**(35), 1701482 (2017).
16. R. Mao, W. Yao, A. Qadir, W. Chen, W. Gao, Y. Xu, and H. Hu, "Sensors and actuators A : Physical 3-D graphene aerogel sphere-based flexible sensors for healthcare applications," *Sensors Actuators A. Phys.* **312**, 112144 (2020).
17. Z.-Y. Zhao, and Q.-L. Liu, "Study of the layer-dependent properties of MoS 2 nanosheets with different crystal structures by DFT calculations," *Catal. Sci. Technol.* **8**(7), 1867–1879 (2018).
18. Y.J. Park, B.K. Sharma, S.M. Shinde, M.-S. Kim, B. Jang, J.-H. Kim, and J.-H. Ahn, "All MoS2-based large area, skin-attachable active-matrix tactile sensor," *ACS Nano* **13**(3), 3023–3030 (2019).
19. M. Tannarana, G.K, Solanki, S.A. Bhakhar, K.D. Patel, V.M. Pathak, and P.M. Pataniya, "2D-SnSe2 nanosheet functionalized piezo-resistive flexible sensor for pressure and human breath monitoring," *ACS Sustain. Chem. Eng.* **8**(20), 7741–7749 (2020).
20. R. Bhardwaj, and A. Hazra, "MXene-based gas sensors," *J. Mater. Chem. C* **9**(44), 15735–15754 (2021).
21. Y. Ma, Y. Cheng, J. Wang, S. Fu, M. Zhou, Y. Yang, B. Li, X. Zhang, and C. Nan, "Flexible and highly-sensitive pressure sensor based on controllably oxidized MXene," *InfoMat* **4**(9), e12328 (2022).
22. Y. Su, K. Ma, F. Yuan, J. Tang, M. Liu, and X. Zhang, "High-performance flexible piezoresistive sensor based on Ti3C2Tx MXene with a honeycomb-like structure for human activity monitoring," *Micromachines* **13**(6), 821 (2022).
23. V. Selamneni, A. BS, and P. Sahatiya, "Highly air-stabilized black phosphorus on disposable paper substrate as a tunnelling effect-based highly sensitive piezoresistive strain sensor," *Med. Devices Sensors* **3**(4), e10099 (2020).
24. H. Jiang, J. Zhang, M. Qin, J. Zhang, X. Zou, and X. Weng, "A flexible piezoresistive strain sensor based on black phosphorus / gold nanocomposites interspersed sponge for motion sensing," *Sensors Actuators A. Phys.* **356**(February), 114359 (2023).
25. R. Bhardwaj, and A. Hazra, "SrTiO3 passivated MXene (Ti3C2Tx) for efficient VOC detection in hazardous humid ambient," *Sensors Actuators B Chem.* **401**(August 2023), 134967 (2024).
26. U.N. Thakur, and A. Hazra, "Feature extractions from transfer characteristics of hybrid GO FETs for selective detection of volatile organic compounds," *Meas. J. Int. Meas. Confed.* **222**(November 2022), 113593 (2023).
27. S.S. Varghese, S. Lonkar, K.K. Singh, S. Swaminathan, and A. Abdala, "Recent advances in graphene based gas sensors," *Sensors Actuators B. Chem.* **218**, 160–183 (2015).

28. A. Hazra, "Amplified methanol sensitivity in reduced graphene oxide FET using appropriate gate electrostatic," *IEEE Trans. Electron Devices* **67**(11), 5111–5118 (2020).
29. H. Li, Z. Yin, Q. He, H. Li, X. Huang, G. Lu, D.W.H. Fam, A.I.Y. Tok, Q. Zhang, and H. Zhang, "Fabrication of single-and multilayer MoS2 film-based field-effect transistors for sensing NO at room temperature," *Small* **8**(1), 63–67 (2012).
30. U.N. Thakur, R. Bhardwaj, and A. Hazra, "Statistical analysis for selective identifications of VOCs by using surface functionalized MoS2 based sensor array," *Chem. Proc.* **5**(1), 35 (2021).
31. Y.H. Kim, S.J. Kim, Y.-J. Kim, Y.-S. Shim, S.Y. Kim, B.H. Hong, and H.W. Jang, "Self-activated transparent all-graphene gas sensor with endurance to humidity and mechanical bending," *ACS Nano* **9**(10), 10453–10460 (2015).
32. H. Long, A. Harley-Trochimczyk, T. Pham, Z. Tang, T. Shi, A. Zettl, C. Carraro, M.A. Worsley, and R. Maboudian, "High surface area MoS2/graphene hybrid aerogel for ultrasensitive NO2 detection," *Adv. Funct. Mater.* **26**(28), 5158–5165 (2016).
33. Z. Ren, Y. Shi, T. Song, T. Wang, B. Tang, H. Niu, and X. Yu, "Flexible low-temperature ammonia gas sensor based on reduced graphene oxide and molybdenum disulfide," *Chemosensors* **9**(12), 345 (2021).
34. S.W. Ryter, and A.M.K. Choi, "Carbon monoxide in exhaled breath testing and therapeutics," *J. Breath Res.* **7**(1), 17111 (2013).
35. G. Singh, A. Choudhary, D. Haranath, A.G. Joshi, N. Singh, S. Singh, and R. Pasricha, "ZnO decorated luminescent graphene as a potential gas sensor at room temperature," *Carbon N. Y.* **50**(2), 385–394 (2011).
36. S. Kumar, S. Kaushik, R. Pratap, and S. Raghavan, "Graphene on paper: A simple, low-cost chemical sensing platform," *ACS Appl. Mater. Interfaces* **7**(4), 2189–2194 (2015).
37. M. Cazzola, and G. Novelli, "Biomarkers in COPD," *Pulm. Pharmacol. Ther.* **23**(6), 493–500 (2010).
38. Y. Murashima, M.R. Karim, R. Furue, T. Matsui, H. Takehira, K. Wakata, K. Toda, R. Ohtani, M. Nakamura, and S. Hayami, "Reduced graphene oxide-transition metal hybrids as p-type semiconductors for acetaldehyde sensing," *Inorg. Chem. Front.* **3**(6), 842–848 (2016).
39. X. Jia, S. Yu, C. Cheng, J. Yang, Y. Li, S. Wang, and H. Song, "Ag nanoparticles modified Fe 3 O 4 / reduced graphene oxide and their acetone sensing properties," *Mater. Chem. Phys.* **276**(August 2021), 125378 (2022).
40. R.K. Jha, J.V. D'Costa, N. Sakhuja, and N. Bhat, "MoSe2 nanoflakes based chemiresistive sensors for ppb-level hydrogen sulfide gas detection," *Sensors Actuators, B Chem.* **297**(June), (2019).
41. N. Alkhouri, T. Singh, E. Alsabbagh, J. Guirguis, T. Chami, I. Hanouneh, D. Grove, R. Lopez, and R. Dweik, "Isoprene in the exhaled breath is a novel biomarker for advanced fibrosis in patients with chronic liver disease: a pilot study," *Clin. Transl. Gastroenterol.* **6**(9), e112 (2015).
42. C. Zhang, J. Wang, Z. Zhang, J. Gong, and H. Wang, "An ultrasensitive isoprene gas sensor based on the In2O3/MoS2 nanocomposite prepared by hydrothermal synthesis," *RSC Adv.* **13**(23), 15826–15832 (2023).
43. K. Hayashi, M. Kataoka, H. Jippo, M. Ohfuchi, and S. Sato, "Vacancy-assisted selective detection of low-ppb formaldehyde in two-dimensional layered SnS2," *ACS Appl. Mater. Interfaces* **12**(10), 12207–12214 (2020).
44. S. Sun, M. Wang, X. Chang, Y. Jiang, D. Zhang, D. Wang, Y. Zhang, and Y. Lei, "W 18 O 49 / Ti 3 C 2 T x Mxene nanocomposites for highly sensitive acetone gas sensor with low detection limit," *Sensors Actuators B. Chem.* **304**(October 2019), 127274 (2020).

45. Zhao WN, Yun N, Dai ZH, Li YF, "A high-performance trace level acetone sensor using an indispensable V 4 C 3 T x MXene," *RSC Adv.* **10**(3), 1261–70 (2020).
46. K. Govindharaj, M. Manoharan, K. Rajavel, Y. Haldorai, and R.T. Rajendrakumar, "Enhanced room temperature selective ammonia sensing based on SnO2 decorated MXene," *J. Mater. Nanosci.* **9**(1), 68–73 (2022).
47. A.G. Dent, T.G. Sutedja, and P.V. Zimmerman, "Exhaled breath analysis for lung cancer," *J. Thorac. Dis.* **5**(Suppl 5), S540 (2013).
48. R.F. Machado, D. Laskowski, O. Deffenderfer, T. Burch, S. Zheng, P.J. Mazzone, T. Mekhail, C. Jennings, J.K. Stoller, and J. Pyle, "Detection of lung cancer by sensor array analyses of exhaled breath," *Am. J. Respir. Crit. Care Med.* **171**(11), 1286–1291 (2005).
49. Y. Yao, Y. Han, M. Zhou, L. Xie, X. Zhao, Z. Wang, N. Barsan, and Z. Zhu, "MoO3/TiO2/Ti3C2Tx nanocomposite based gas sensors for highly sensitive and selective isopropanol detection at room temperature," *J. Mater. Chem. A* **10**(15), 8283–8292 (2022).
50. J.A. Goode, J.V.H. Rushworth, and P.A. Millner, "Biosensor regeneration: A review of common techniques and outcomes," *Langmuir* **31**(23), 6267–6276 (2015).
51. N. Masurkar, S. Varma, and L. Mohana Reddy Arava, "Supported and suspended 2D material-based FET biosensors," *Electrochem* **1**(3), 260–277 (2020).
52. K. Shavanova, Y. Bakakina, I. Burkova, I. Shtepliuk, R. Viter, A. Ubelis, V. Beni, N. Starodub, R. Yakimova, and V. Khranovskyy, "Application of 2D non-graphene materials and 2D oxide nanostructures for biosensing technology," *Sensors* **16**(2), 223 (2016).
53 T. Pornprom, N. Phusi, P. Thongdee, B. Pakamwong, J. Sangswan, P. Kamsri, A. Punkvang, K. Suttisintong, J. Leanpolchareanchai, P. Hongmanee, P. Lumjiaktase, S. Jampasa, O. Chailapakul, and P. Pungpo, "Toward the early diagnosis of tuberculosis: A gold particle-decorated graphene-modified paper-based electrochemical biosensor for Hsp16.3 detection," *Talanta* **267**(September 2023), 125210 (2024).
54. P. Fathi-Hafshejani, N. Azam, L. Wang, M.A. Kuroda, M.C. Hamilton, S. Hasim, and M. Mahjouri-Samani, "Two-dimensional-material-based field-effect transistor biosensor for detecting COVID-19 virus (SARS-CoV-2)," *ACS Nano* **15**(7), 11461–11469 (2021).
55. Y. Lei, W. Zhao, Y. Zhang, Q. Jiang, J. He, A.J. Baeumner, O.S. Wolfbeis, Z.L. Wang, K.N. Salama, and H.N. Alshareef, "A MXene-based wearable biosensor system for high-performance in vitro perspiration analysis," *Small* **15**(19), 1901190 (2019).

8 FET-Based Biosensors

Vikash Kumar, Aditya Kumar Singh, Dilip Kumar, and Basanta Bhowmik

8.1 INTRODUCTION

Biosensors are a rapidly growing field that is considered the interface between biology, chemistry and physics with the aim of improving medical diagnostics and controlling biotechnological processes.[1] They integrate a biological recognition element with a transduction component, enabling the swift and easy transformation of biological processes into detectable signals. The development of nanomaterials and sophisticated instruments for nanofabrication, including ion beam technology, electron beam lithography and nano-imprint lithography, has significantly advanced biosensor technology over the past two decades.[2] Researchers have been developing various electrical biosensors comprising nanomaterials and innovative strategies for ultrasensitive biosensing. Notable nanomaterials include carbon nanotubes, nanowires, nanoparticles, nano pores, nanoclusters and grapheme.[3] These nanomaterial-based electronic biosensors offer advantages over traditional techniques including high sensitivity, unique sensing mechanisms, precise spatial resolution for focused detection, smooth integration with the conventional semiconductor process, and label-free, immediate detection in a non-destructive way, which are examples of optical, biochemical and biological techniques.[4]

8.2 FET-BASED BIOSENSORS

Field effect transistors, or FETs, have attracted a lot of attention for a variety of uses, most notably biomolecule identification. FETs offer numerous benefits, such as parallel sensing capabilities, scalability, simplicity of incorporation with digital procedures, rapid response times, and they are compatible to complementary cost-effectiveness and metal-oxide semiconductor (CMOS) technologies.[2] As electrical biosensors, FETs function by utilizing a charge carrier. Mobility is measured by applying an electric field through conductive gate electrodes to a channel. Analyte collection in the channel is made more efficient by the gate voltage, which modifies the material's charge carrier mobility. Equation (1) describes the connection between the mobility of carriers and gate voltage, provided other circumstances remain stable.

 DOI: 10.1201/9781003453635-8

$$\mu = \frac{\Delta I_{ds}}{\Delta V_g} \frac{l}{w} \frac{1}{C_{TG} V_{ds}}. \quad (8.1)$$

The electrolyte concentration influences the overall gate capacitance (C_{TG}). In this context, essential parameters include the differential drain-source current (ΔI_{ds}), source-drain voltage (V_{ds}), channel width (w), channel length (L), and differential gate voltage (ΔV_g).[5]

Devices using field-effect transistors (FETs) typically feature a gate, source, and drain terminal. The current (I_D) flowing through the electric field generated by the gate voltage (V_G) controls the semiconducting channel connecting the source and drain, as well as the voltage (V_{SD}) between the power supply and drain. Such top-gating and bottom-gating (also known as back-gating) were used to apply voltage known as gate terminal. Depending on the semiconductor material used in the channel, the gate voltage modulates the electric field, controlling the accumulation or depletion of carrier charges such as electrons and holes, thus altering the channel's conductivity. The transfer characteristics of FETs can be influenced through the target molecules' or analytes' adsorption on the channel's surface, as these change the channel's charge carriers. FET biosensors are most favourable due to their ultra-sensitivity and rapid response times. Biomolecules typically carry electrostatic charges, meaning even small biological activities can alter the electrical potential.[3] These alterations can be detected by FET-based biosensing devices, which convert interactions between the FET surface and the relevant biological molecules into electrical signals. As a result, FET-based sensors, which offer incredibly sensitive detection capabilities, are seen as the biosensing industry's future.[4,6,7]

The conducting channel in traditional complementary metal oxide semiconductor (CMOS) transistors is embedded within the substrate. However, in FET biosensors, it is directly exposed to the environment, allowing for more precise control at surface charges. This configuration means that biological activity increases the sensitivity of FET-based biosensors. Near the surface of the channel can alter the channel surface potential, thereby modulating the semiconductor channel conductance.[8,9] Combined with the ease of integrating device arrays on a chip and the affordability of fabricating the device, the high surface sensitivity makes current biosensor technologies an attractive alternative.

FET biosensors enable label-free electrical detection, as well as being applied to numerous biomedical applications.[10] This approach can reduce costs by eliminating the need for biological labelling and pretreatment processes of conventional methods. Additionally, the FET technique offers high sensitivity and reliable performance for detecting bioanalytes such as proteins and DNA.[8,9] As electrical biosensors, FETs function by utilizing a charge carrier mobility that is measured by applying an electric field through a conductive gate electrodes to a channel, as shown in Figure 8.1.

BioFETs are particularly advantageous for bacterial detection because of their direct electrical signal output capabilities, which are label-free, fast and very sensitive.[11] These sensors can be used in high-throughput microarrays for various applications, including food safety, environmental surveillance, medical diagnostics,

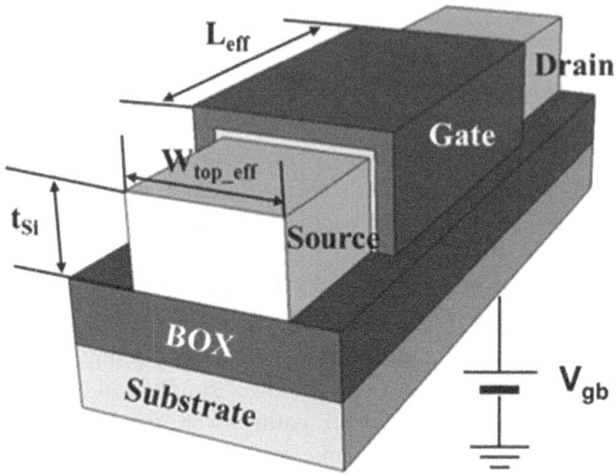

FIGURE 8.1 Pictorial representation of tri-gate structure.[42]

medication administration, safety and explosive identification.[3,12] Metal oxide semiconductor field-effect transistors (MOSFETs) can be used to create FET biosensors by changing the gate through different biomolecules.[8] These devices detect changes in charge or potential, and a key aspect of their fabrication is converting the biochemical reaction into a legible output through the transducer component.

Although there are various BioFETs with different structures and fabrication methods, the fundamental principle involves making an area of excess charge with an electric field in the semiconducting channel, which affects local conductivity.[10]

8.3 PRINCIPLE OF BIOSENSOR OPERATIONS

In FET-type biosensors, the targets are detected by placing the sensing nanomaterial on a substrate linked to the drain and source terminals. The channel conductivity of the sensing element was varied by applying bias potential at the gate electrode via accumulation or depletion of charge carriers. From the source and the drain, the voltage terminals control the amount of current flowing at the gate terminal.[13] Application of gate voltage establishes the electric field in a vertical direction, which regulates the channel's conductivity and charge density. The minimum gate voltage required to make a channel conducting via accumulating sufficient charge is known as threshold voltage (V_{th}). The sub-threshold zone, when the voltage at the gate is less than V_{th} of the typical FET transfer characteristics, and when the gate voltage is greater V_{th}, are the two main regions shown in a typical FET transfer curve. The sensing range of the FET sensor, that is, the change in current that occurs when the target substance interacts with the sensor, is greatly influenced by the threshold voltage (V_{th}).

FETs are categorized into two types based on the main charge carriers: n-type channels, which utilize electrons, and p-type channels, which utilize holes. The type of channel (n or p) determines the nature of the current carrier. These devices' electric fields, which regulate current, are created when a voltage that is positive is delivered to the gate terminal.[14] When a p-type channel is involved, there is a positive gates bias that repels the holes in the channel, generating a depletion region. The depletion region becomes filled with acceptor atoms' exposed negative charges. Additionally, the negatively charged electrons on a substrate are drawn to the positive gate voltage, creating a substrate surface inversion to an n-type channel. The voltages applied across the source and drain terminals then control the n-type channel's current, which is referred to as an n-FET. With p-type semiconductors, gate voltage has the exact opposite effect that can be observed: a positive gate voltage depletes carrier density and hence reduces conductance, whereas a negative gate voltage accumulates carriers and subsequently results in increased conductivity.[15] It has to be noted that the threshold voltage (V_{th}) of n-channel FETs is positive, whereas it is negative for p-channel FETs. Therefore, FETs have been explored as biosensing applications by adding charged biomolecules or an ion-selective membrane to the gate terminal. At the gate terminal, this alteration causes charge accumulation or depletion. The interaction between these charged molecules and the gate creates an electric field akin to an applied gate voltage, influencing the channel conductance. This change in conductance can be examined in order to determine the target analyte.[8]

Since the concentration of analytes directly affects the change in conductance, the framework can be used as a sensor. According to reports, when a BioFET operates in a sub-threshold region, its sensitivity reaches its maximum.[6] Shoorideh and Chui worked on enhancing the sensitivity of FET-based biosensors.[14] In order to investigate the underlying physical processes that affect sensitivity, they simulated the FET biosensor with MATLAB and examined the n-channel accumulation model's electronics. The bias point, or ideal voltage for a certain substrate contact and solutions electrode, was found through simulations.[10] Their model suggests that there are three stages to the FET-based biosensor's operation: a) Analyte concentration variations impact the charge along the biosensor interface, denoted as $d\sigma$; b) consequently, this charge shift results in a quantifiable effective gate voltage change ($dV_{g.eff}$); and c) the change in effective gate voltage results in variations in the drain current (dI).[10] Equation (2) provides a mathematical expression for the sensor's sensitivity.

$$\frac{dI}{I} = dc.\frac{d\sigma}{dc}.\frac{dV_{g.eff}}{d\sigma}.\frac{dI\,/\,dV_{g.eff}}{I_O}, \tag{8.2}$$

where dc represents a: the little variation in the sample's overall analyte concentration, and Io stands for the drain current as measured using the reference sample. Permission shift effect and charged interaction are the two documented processes for detection utilizing electrical FETs.[10]

8.4 THE ROLE OF NANOSTRUCTURES IN BIOSENSORS

Nanomaterials have attracted great attention in regard to developing highly selective and sensitive biosensors owing to their significant optical, thermal, catalytic and electrical properties.[16,17] At the nanoscale dimension, material properties are different in terms of stability, surface area and electrical and optical performance from their bulk counterparts. The high surface-to-volume ratio and adjustable surface properties of nanomaterials make them biosensors that can achieve low limits of detection because they are highly sensitive to alterations on the sensor surface.[18] Developments in nanotechnology have made it easier to combine different biosensors with nanoparticles for sensing uses. Innovative designs allow different biomolecules, such as nucleotides, antioxidants, antibodies and peptides, to be immobilized on nanotransducers.[18] Many nanomaterials have been investigated to employ them as semiconductor channels such as nanowires, nanotubes, organic semiconductors and so forth.

8.5 METAL OXIDE-BASED FET BIOSENSORS

Metal oxides (MOs) have recently garnered attention for biosensor development due to their remarkable properties.[10,19] Commonly used metal oxides for biosensors include zinc, iron, cerium, tin, zirconium, titanium and magnesium. These materials exhibit significant nano-morphological, functional, biocompatible, non-toxic and catalytic qualities. Additionally, their excellent adsorption capacity and enhanced electron-transfer kinetics make them ideal for biomolecule immobilization, thus improving electron transfer and biosensing capabilities.[19,20,21] Various methods can be employed to create different MO morphologies, such as RF sputtering and soft templating for the nanorods and nanofibers for rough nanostructures, hydrothermal deposition for shaped nanoparticles, and sol-gel techniques for creating 3D ordered rough nanos tructures.[5,21,22,23,24] Each of these nanostructured MOs has potential applications in biosensors. Recent studies have shown that incorporating nanoparticles of conducting or semiconducting materials such as carbon nanotubes, graphene, gold and silver into nanostructured MOs enhances their optical, electrical and magnetic properties.[10,22] This integration benefits the biosensor's performance. Using nanostructured MOs can lead to the development of new biosensing devices with improved signal amplification, coding techniques for bio-affinity tests, and effective electrical interaction with redox proteins and enzymes, potentially meeting future diagnostic requirements.

Nanostructured MOs offer significant potential for merging biological recognition events with electrical signal propagation, paving the way for innovative bioelectronics devices. The careful development of nanostructured MOs is essential for advancing biosensor technology. Understanding the potential and the challenges of using nanostructured MOs in biosensing devices is critical for developing new approaches to effective electrical communication and bio-affinity tests.[12,23] Various metal-oxide semiconductor compounds are used as active components in FETs for biosensing applications. Common examples include zinc oxide, indium oxide and tin oxide, chosen for their specific electrical properties and stability. When selecting metal-oxide semiconductors for chemical or biosensor FETs, several factors must be considered, including the intended application, environmental conditions (e.g. emergency departments or remote areas), sample types (e.g. saliva, blood, urine, sweat), target concentration levels, and diagnostically relevant cut-offs.[15, 17] Additionally, the commercial availability of these materials and their ability to maintain electrical and mechanical stability in various environments are crucial. For instance, the materials must withstand high temperatures for gas or organic compound detection in breath analysers, while being stable in specific solutions for wet-chemical and biological sensing.[17,20] Light sensitivity and temperature tolerance are also important; components sensitive to visible light require strict packaging to isolate them or use in dark environments. High bandgap semiconductors are generally less sensitive to temperature and light. Finally, compatibility with advanced manufacturing processes is essential for real-world application.[17, 22,25] Due to their superior electrical properties, chemical stability and ease of production, indium oxide, zinc oxide and tin oxide are among the most commonly used metal-oxide semiconductor materials for FET-based biosensors.

8.6 POLYMER-BASED BIOSENSORS

Conjugated polymers with strong electrical conductivity and ion-transferring ability are known as electrically conducting polymers.[34–40] Some characteristics include simple biomolecule immobilization, high porosity, conductivity of electricity and non-hostile organic milieu. These materials are very appealing for the creation of new biosensors since they contain biomolecules and can be used as transducers and sensing elements. One-dimensional polymer nanostructures have received a lot of attention lately. Examples of these include electrically conducting polymer nanowires, which can be used for ultrasensitive biological species detection.[35–37] A significant conductivity shift related to molecules binding is caused by the one-dimensional nanostructure's high sensitivity and high surface-to-volume ratio. Numerous techniques, including hydrodynamic focusing, direct electrochemical synthesis, photolithography, e-beam lithography, dip-pen nanolithography and template-assisted synthesis, are used to create ECPNWs.[26, 38,39] Compared to existing industry standards, Chen et al.'s polymer-based biosensor holds great potential for advancement.[35] If an antibody or particular oligonucleotide could be connected to the quencher, it might be feasible to quickly and precisely identify antigens such as proteins, carbohydrates, nucleic acids or others in a range of settings. The method's key benefit is the fact that the polymer-based sensor functions as a homogeneous test.[27,34,35] The polymer is kept in an 'off' state by the quencher ligand, and fluorescence is only detected when the target molecule attaches to and displaces the quencher ligand.[36,37] Thus, the assay does not need to distinguish between bound and unbound antibody; only the bound antigen causes a signal (fluorescence) and eliminates the quencher within the polymer, negating the necessity for cleaning steps (i.e. in ELISAs). It appears that the produced fluorescence signal can be easily quantified with a typical fluorimeter.[35,37,38] Lastly, and maybe most significantly, the response happens almost immediately. This characteristic would be very helpful for 'real time' detection of potentially fatal poisonous compounds in air or water, in addition to saving time (and hence money).[39,40] Notwithstanding the intriguing nature of these ideas, many issues remain concerning the practical uses of such polymer-based biosensors. What impact do the size of the ligand and charge have on the ability of the quencher ligand to suppress polymer fluorescence?[28–33] If the ligand connected to the quencher is significantly larger than the biotin (i.e. antigens or antigen subunits with a molecular weight of 26,000–150,000 Da), the quencher's ability to bond with the chain of polymers may be hindered. Antibodies can also range significantly in terms of the amount of amino acids they contain, which affects their overall charge.[35,38] Quenching may be ineffective if the ligand's charge or size prevents it from interacting with the polymer in an efficient manner. Finding the restrictions on antibody charge and size will also be crucial.[38,39] It is probable that avidin's enormous size prevents the interaction between the polymer and avidin-bound quencher biotin.[34-38] The presented experiments have further shown that there is just one relationship between avidin and biotin, under ideal experimental conditions.[40] As previously shown, the affinity of normal antigen–antibody interactions is significantly lower [Kd of 1025–10210 (17)], while the avidin–biotin interaction has a strong affinity, with a 10215 binding constant (Kd). Consequently, whether a quencher antigen can be effectively removed from the polymer in order to provide a detectable signal is still unknown.[35,38–40] Similar doubts are expressed about this method's ability to identify interactions between DNA molecules. Furthermore, the test needs to be evaluated under less than optimal conditions for a biosensor to be genuinely beneficial, i.e. significant molecules need to be identified from unpurified ones.[40]

8.7 INFLUENCE OF BACK-GATE BIASING

To detect targets, the sensing element of a conventional FET is positioned on a semiconductor route that is coupled to the source and drain terminals, as shown in Figure 8.2. Applying and adjusting bias voltage to the gate electrode, the third electrode, is the process.[41–44] The channel's conductivity is affected by the gate potential when mobile charge carriers build up or disappear.[45–48] Applying voltage between the drain and source terminals controls the amount of current flowing at the gate terminal. The vertical electric field created by the provided gate voltage regulates the channel's conductivity and charge density. Threshold voltage (V_{th}) is the amount of voltage needed to build up enough charge for a channel to conduct.[48,49] As a result, two separate zones arise inside the typical FET transfer curve: a linear region when the gate voltage is larger than V_{th}, and a sub-threshold region when it is less than V_{th}. In terms of the FET sensor's sensitivity, which is determined by a change in current that occurs when the target analyte attaches to the sensor, V_{th} is crucial.[42] FETs come in two

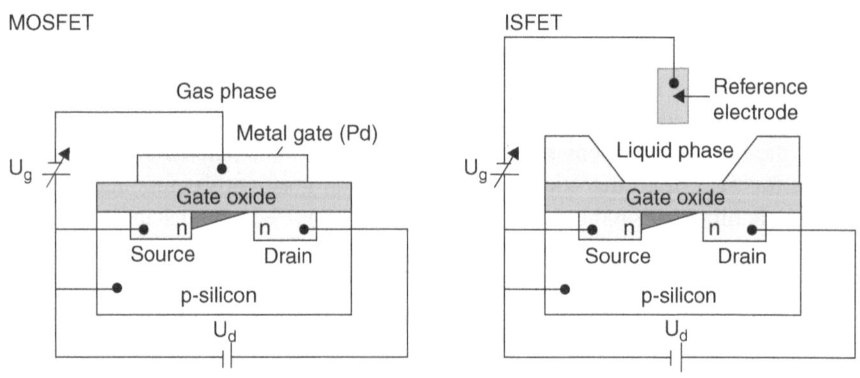

FIGURE 8.2 Schematic structure of MOSFET and ISFET.[50]

varieties: n-type channels that have electrons are the primary charge carriers in p-type channels, where holes are the main charge carriers.[42–45] The kind of channel, n or p, used in FETs determines the characteristics of the current carriers. These devices' current is controlled by an electric field created when a positive voltage is applied to the gate terminal.[46,47] A carrier-depleted zone is created when the gate receives a positive voltage, which opposes the free holes in that area.[47,48] The resulting depletion region is filled with acceptor atoms' exposed negative charges as holes are driven into the substrate by the positive voltage that is provided. In addition, electrons that are negatively charged on the substrate are drawn to the positive gate voltage, which changes the substrate's surface into a n-type channel. A voltage applied across the drain and source terminals can now cause current to flow through an n-type channel, which is known as an n-FET.[46–49] Only in p-type semiconductors is the reverse impact of the gate voltage visible: a positive gate voltage reduces carrier density and hence lowers conductance, whereas a negative gate voltage accumulates carriers and results in increased conductivity.[44–46] It has been noted that the V_{th} of n-channel FETs is positive, whereas the V_{th} of p-channel FETs is reversed.[45,46] These qualities allow FETs to be investigated for biosensing applications, such as altering the gate terminal with an ion-specific membrane or charged biomolecules, charge accumulation takes place at the gate terminal. This is as a result of the molecular interactions that are charged and the gate's generating an electric field that is similar to the applied gate voltage, affecting the channel's conductance, which can be examined to identify the analytical of interest.[45–47] Since the number of analytes directly affects the change in conductance, the framework can be used as a sensor. According to reports, BioFETs' highest sensitivity occurs when they are biased in the area below the threshold.[47,48] In 2012, Shoorideh and Chui conducted some groundbreaking research on enhancing the sensitivity of FET-based biosensors.[48] In order to investigate the underlying physical processes that affect sensitivity, the FET biosensor has been emulated using MATLAB, and the electronics of the n-channel accumulation model was examined.[44–46] The bias point, which provides the ideal voltage for a specific substrate contact or solution electrode, has been studied using simulations.[46,47] According to Shoorideh and Chui's derivation, there are three steps to the FET-based sensing operation model: a) The charge around the biosensor's interface changes as a result of variations in analyte concentration, as shown by dσ; b) Charge modification results in a measurably different change in the effective gate voltage (d$V_{g'eff'}$); c) Drain current variations (dI) result from the effective gate voltage change.[44–47] Mathematically, sensor sensitivity can be represented as $(I_g\text{-}I_a)/I_a$*100. Here, Io represents the drain current recorded under the reference sample, and dc is the smallest change in the sample's overall analyte concentration.[45–48] Two mechanisms for detection utilizing electrical FETs have been reported: the permittivity shift impact and the charge interaction.

8.8 CHALLENGES IN DEVICE DESIGN FOR HIGH-PERFORMANCE FET-BASED BIOSENSORS

For quick and precise biomolecule detection, nanomaterial design and selection are essential.[50,51] Thus, new nano-engineered segments with enhanced functionality will continue to be produced as long as techniques for material synthesis, enzyme/

protein engineering and immobilization/conjugation procedures are continuously improved.[52,53] Additionally, it is anticipated that a combination of recently developed sophisticated manufacturing technologies, such as three- and four-dimensional prints of multicomponent, multifunction nanostructures with a focus on nanoscale, will open up new possibilities for the current sensor design.[17,51,54] Research into advanced biological sensing to bio-functionalized multipurpose nanomaterials and the development of cost-effective biochip patterns using nanoscale-sensing materials can further pave the way for nano-biosensing locations and the creation of an accessible lab-on-a-chip substitute for actual or near real-time biomolecule sensing.[17,54,55]

It is anticipated that a broad variety of novel NMOs would further broaden the field of biosensors. Novel approaches to the production of a single-dimensional NMOs are probably going to yield new uses for bio-electronic sensing.[51–53] Novel approaches to the construction of biosensors may be made possible by the ability to modify NMOs with the appropriate groups that bind target biomolecules, and dope these with electrically active materials to produce increased charge transfer.[50–53] At the nano domain level, the production of MO nanoparticles of diverse morphologies may create an environment that is appropriate for achieving the targeted immobilization of desirable biomolecules, leading to the amplification of signals.[51–53] To create useful integrated devices, these NMOs can be produced and tested in specific configurations, including sensor arrays.[51,53] Multiple analytes might be simultaneously monitored in real time using the interface of NMO-based devices.[17,51,54,55] It will be intriguing to concentrate on novel techniques for producing cutting-edge biosensor arrays with the required characteristics.[17,51,54]

8.9 CONCLUSION AND FUTURE SCOPE

Nanostructured metal oxide (NMO) biosensors open new possibilities for innovative applications in both medical and non-clinical diagnostics. Future interdisciplinary research may produce a novel category of electrochemical biosensors due to the distinctive characteristics of NMOs. Scholars are concentrating on comprehending the relationships between diverse biomolecules and transducers using these fascinating nanomaterials. Promising research areas include particle-based carriers, surface plasmon reflection, nanoparticle-based chemiluminescence identification, and biocatalytic metallization for amplified analysis studies. Highly sensitive bio-detection techniques based on NMOs facilitate the early identification of threats and diseases, enabling the development of biosensors for biological and pathogenic agents as well as disease markers. Multi-analyte detection on nanoparticulate MO devices is expected to benefit from DNA-, enzyme-, and antibody-based bioassays. Exploring the variations in size and quality of MO nanoparticles across novel applications will be particularly interesting.

REFERENCES

1. Liu, S., & Guo, X. (2012). Carbon nanomaterials field-effect-transistor-based biosensors. *NPG Asia Materials,* 4(e23), 1–10. https://doi.org/10.1038/am.2012.42
2. Adzhri, R., MdArshad, M. K., Subash, C. B., Ruslinda, A. R., Fathil, M. F. M., Ayub, R. M., Mohd Nor, M. N., & Voon, C. H. (2016). High-performance integrated field-effect

transistor-based sensors. *AnalyticaChimicaActa*, 917, 1–18. https://doi.org/10.1016/j.aca.2016.02.042.
3. PoghossianIngebrandt, A. S., Abouzar, M., & Schoning, M. J. (2007). Label-free detection of charged macromolecules by using a field-effect-based sensor platform: Experiments and possible mechanisms of signal generation. *Applied. Physics,* A87(3), 517–524. DOI: 10.1007/s00339-007-3919-7
4. Kaisti, M. (2017). Detection principles of biological and chemical FET sensors. *Biosensors and Bioelectronics,* 98, 437–448.
5. Kwak, Y. H., Choi, D. S., Kim, Y. N., Kim H., Yoon, D. H., Ahn, S.-S., Yang, J.-W., Yang, W. S., & Seo, S. (2012). Flexible glucose sensor using CVD-grown graphene-based field effect transistor. *Biosensors and Bioelectronics*, 37(1), 82–87. https://doi.org/10.1016/j.bios.2012.04.042
6. Tran, T. T., & Mulchandani, A. (2016). Carbon nanotubes and graphenenano field-effect transistor-based biosensors. *TrAC Trends in Analytical Chemistry*, 79, 222–232.
7. Dai, X., Vo, R., Hsu, H. H., Deng, P., Zhang, Y., & Jiang, X. (2019). Modularized field-effect transistor biosensors. *Nano Letters*, 19(9), 6658–6664.
8. Syu, Y. C., Hsu, W. E., & Lin, C. T. (2018). Field-effect transistor biosensing: Devices and clinical applications. *ECS Journal of Solid State Science and Technology*, 7(7), Q3196.
9. Chen, Y., Ren, R., Pu, H., Guo, X., Chang, J., Zhou, G., ...& Chen, J. (2017). Field-effect transistor biosensor for rapid detection of Ebola antigen. *Scientific Reports*, 7(1), 10974.
10. Samota, S., Rani, R., Chakraverty, S., & Kaushik, A. (2022). Biosensors for simplistic detection of pathogenic bacteria: A review with special focus on field-effect transistors. *Materials Science in Semiconductor Processing*, 141, 106404.
11. Salami, M., Abadi, M. H. S., Sawan, M., & Abadi, N. S. K. (2019). BioFET-based integrated platform for accurate and rapid detection of E. coli bacteria: A review. *Biosensors and Bioelectronics,* 10, 266.
12. Chen, X., Liu, C., & Mao, S. (2020). Environmental analysis with 2D transition-metal dichalcogenide-based field-effect transistors. *Nano-Micro Letters*, 12, 1–24.
13. Schöning, M. J., & Poghossian, A. (2002). Recent advances in biologically sensitive field-effect transistors (BioFETs). *Analyst*, 127(9), 1137–1151.
14. Shoorideh, K., & Chui, C. O. (2012). Optimization of the sensitivity of FET-based biosensors via biasing and surface charge engineering. *IEEE Transactions on Electron Devices*, 59(11), 3104–3110.
15. Park, J., Nguyen, H. H., Woubit, A., & Kim, M. (2014). Applications of field-effect transistor (FET)-type biosensors. *Applied Science and Convergence Technology*, 23(2), 61–71.
16. Schwirn, K., Tietjen, L., & Beer, I. (2014). Why are nanomaterials different and how can they be appropriately regulated under REACH? *Environmental Sciences Europe*, 26(1), 1–9.
17. Nehra, A., & Singh, K. P. (2015). Current trends in nanomaterial embedded field effect transistor-based biosensor. *Biosensors and Bioelectronics*, 74, 731–743.
18. Masatoshi, H. S., & Daisuke, I. Y. (2013). *Applications of Nanomaterials in Sensors and Diagnostics*. Springer Series on Chemical Sensors and Biosensors, Springer, Berlin Heidelberg.
19. Solanki, P. R., Kaushik, A., Agrawal, V. V., & Malhotra, B. D. (2011). Nanostructured metal oxide-based biosensors. *NPG Asia Materials*, 3(1), 17–24.
20. Rahman, M. M., Ahammad, A. J., Jin, J. H., Ahn, S. J., & Lee, J. J. (2010). A comprehensive review of glucose biosensors based on nanostructured metal-oxides. *Sensors*, 10(5), 4855–4886.

21. Amen, M. T., Pham, T. T. T., Cheah, E., Tran, D. P., & Thierry, B. (2022). Metal-oxide FET biosensor for point-of-care testing: Overview and perspective. *Molecules*, 27(22), 7952.
22. Hahn, Y. B., Ahmad, R., & Tripathy, N. (2012). Chemical and biological sensors based on metal oxide nanostructures. *Chemical Communications*, 48(84), 10369–10385.
23. Nunes, D., Pimentel, A., Gonçalves, A., Pereira, S., Branquinho, R., Barquinha, P., & Martins, R. (2019). Metal oxide nanostructures for sensor applications. *Semiconductor Science and Technology*, 34(4), 043001.
24. Ansari, A. A., Solanki, P. R., & Malhotra, B. D. (2008). Sol-gel derived nanostructured cerium oxide film for glucose sensor. *Applied Physics Letters*, 92(26). DOI: 10.1002/elan.200804499
25. Tenne, R. (2006). Inorganic nanotubes and fullerene-like nanoparticles. *Journal of Materials Research*, 21(11), 2726–2743. DOI: 10.1038/nnano.2006.62
26. Liu, J., Li, Y., Huang, X., & Zhu, Z. (2010). Tin oxide nanorod array-based electrochemical hydrogen peroxide biosensor. *Nanoscale Research Letters*, 5, 1177–1181. DOI: 10.1007/s11671-010-9622-1
27. Choi, H. N., Kim, M. A., & Lee, W. Y. (2005). Amperometric glucose biosensor based on sol–gel-derived metal oxide/Nafion composite films. *AnalyticaChimicaActa*, 537(1–2), 179–187.https://doi.org/10.1016/S1452-3981(23)18002-3
28. Singh, S. P., Arya, S. K., Pandey, P., Malhotra, B. D., Saha, S., Sreenivas, K., & Gupta, V. (2007). Cholesterol biosensor based on rf sputtered zinc oxide nanoporous thin film. *Applied Physics Letters*, *91*(6).https://doi.org/10.1063/1.2768302
29. Zhao, Z., Lei, W., Zhang, X., Wang, B., & Jiang, H. (2010). ZnO-based amperometric enzyme biosensors. *Sensors*, 10(2), 1216–1231. https://doi.org/10.3390/s100201216
30. Wang, J. (2005). Nanomaterial-based amplified transduction of biomolecular interactions. *Small*, 1(11), 1036–1043. https://doi.org/10.1002/smll.200500214
31. Wang, J. (2005). Nanomaterial-based electrochemical biosensors. *Analyst*, 130(4), 421–426. http://dx.doi.org/10.1039/b414248a
32. McNeil, S. E. (2005). Nanotechnology for the biologist. *Journal of Leukocyte Biology*, 78(3), 585–594. https://doi.org/10.1189/jlb.0205074
33. Pan, C., & Zhu, J. (2009). The syntheses, properties and applications of Si, ZnO, metal, and heterojunction nanowires. *Journal of Materials Chemistry*, 19(7), 869–884. https://doi.org/10.1039/B816463K
34. Zamani, F. G., Moulahoum, H., Ak, M., Demirkol, D. O., & Timur, S. (2019). Current trends in the development of conducting polymers-based biosensors. *TrAC Trends in Analytical Chemistry*, 118, 264–276. https://doi.org/10.1016/j.trac.2019.05.031
35. Heeger, P. S., & Heeger, A. J. (1999). Making sense of polymer-based biosensors. *Proceedings of the National Academy of Sciences*, 96(22), 12219–12221. https://doi.org/10.1073/pnas.96.22.12219
36. Valcárcel, M., Cárdenas, S., & Gallego, M. (1999). Sample screening systems in analytical chemistry. *TrAC Trends in Analytical Chemistry*, 18(11), 685–694. https://doi.org/10.1016/S0165-9936(99)00167-3
37. Lee, K., Povlich, L. K., & Kim, J. (2010). Recent advances in fluorescent and colorimetric conjugated polymer-based biosensors. *Analyst*, 135(9), 2179–2189. https://doi.org/10.1039/C0AN00239A
38. Teles, F. R. R., & Fonseca, L. P. (2008). Applications of polymers for biomolecule immobilization in electrochemical biosensors. *Materials Science and Engineering: C*, 28(8), 1530–1543. https://doi.org/10.1016/j.msec.2008.04.010
39. Naseri, M., Fotouhi, L., & Ehsani, A. (2018). Recent Progress in the development of conducting polymer-based nanocomposites for electrochemical biosensors

applications: a mini-review. *The Chemical Record*, 18(6), 599–618. https://doi.org/10.1002/tcr.201700101

40. Kanungo, S., Majumdar, B., Mukhopadhyay, S., Som, D., Chattopadhyay, S., Rahaman, H. (2020). Investigation on the effects of substrate, back-gate bias and front-gate engineering on the performance of DMTFET-based biosensors. *IEEE Sensors Journal*, 20(18), 10405–10414. DOI: 10.1109/JSEN.2020.2994295
41. Zhou, Q., Chen, B., Jin, Y., Huang, S., Wei, K., Liu, X., ... & Zhang, B. (2015). High-Performance Enhancement-Mode Al 2 O 3/AlGaN/GaN-on-Si MISFETs with 626 Figure of Merit. *IEEE Transactions on Electron Devices*, 62(3), 776–781. DOI: 10.1109/TED.2014.2385062
42. Park, S. J., Jeon, D. Y., Montès, L., Barraud, S., Kim, G. T., & Ghibaudo, G. (2014). Impact of channel width on back biasing effect in tri-gate MOSFET. *Microelectronic Engineering*, 114, 91–97. https://doi.org/10.1016/j.mee.2013.09.016
43. Wei, J., Liu, S., Li, B., Tang, X., Lu, Y., Liu, C., & Chen, K. J. (2015). Low on-resistance normally-off GaN double-channel metal–oxide–semiconductor high-electron-mobility transistor. *IEEE Electron Device Letters*, 36(12), 1287–1290. DOI: 10.1109/LED.2015.2489228
44. Masahara, M., Liu, Y., Sakamoto, K., Endo, K., Matsukawa, T., Ishii, K., Suzuki, E. (2005). Demonstration, analysis, and device design considerations for independent DG MOSFETs. *IEEE Transactions on Electron Devices*, 52(9), 2046–2053. DOI: 10.1109/TED.2005.855063
45. Biswas, A., & Bhattacherjee, S. (2013). Accurate modeling of the influence of back gate bias and interface roughness on the threshold voltage of nanoscale DG MOSFETs. *Microelectronics Reliability*, 53(3), 363–370. https://doi.org/10.1016/j.microrel.2012.09.005
46. Mehrad, M., & Ghadi, E. S. (2017, April). C-shape silicon window nano MOSFET for reducing the short channel effects. In *2017 Joint International EUROSOI Workshop and International Conference on Ultimate Integration on Silicon (EUROSOI-ULIS)* (pp. 164–167). IEEE. DOI: 10.1109/ULIS.2017.7962572
47. Kuo, J. B., Lee, W. C., & Sim, J. H. (1992). Back-gate bias effect on the subthreshold behavior and the switching performance in an ultrathin SOI CMOS inverter operating at 77 and 300 K. *IEEE Transactions on Electron Devices*, 39(12), 2781–2790. DOI: 10.1109/16.168724
48. Yang, Y., Yang, X., Zou, X., Wu, S., Wan, D., Cao, A., ... & Duan, X. (2017). Ultrafine graphene nanomesh with large on/off ratio for high-performance flexible biosensors. *Advanced Functional Materials*, 27(19), 1604096. https://doi.org/10.1002/adfm.201604096
49. Dao, B. A., Hoang, T. T., Le, A. T., Tsukamoto, A., Suzaki, K., & Pham, C. K. (2021). Exploiting the back-gate biasing technique as a countermeasure against power analysis attacks. *IEEE Access*, 9, 24768–24786. DOI: 10.1109/ACCESS.2021.3057369
50. Sadighbayan, D., Hasanzadeh, M., & Ghafar-Zadeh, E. (2020). Biosensing based on field-effect transistors (FET): Recent progress and challenges. *TrAC Trends in Analytical Chemistry*, 133, 116067. https://doi.org/10.1016/j.trac.2020.116067
51. Matsumoto, A., & Miyahara, Y. (2013). Current and emerging challenges of field effect transistor based bio-sensing. *Nanoscale*, 5(22), 10702–10718. https://doi.org/10.1039/C3NR02703A
52. Yoon, H., Lee, S. H., Kwon, O. S., Song, H. S., Oh, E. H., Park, T. H., & Jang, J. (2009). Polypyrrole nanotubes conjugated with human olfactory receptors: high-performance transducers for FET-type bioelectronic noses. *Angewandte Chemie*, 121(15), 2793–2796.https://doi.org/10.1002/ange.200805171

53. Wadhwa, G., & Raj, B. (2020). An analytical modeling of charge plasma based Tunnel Field Effect Transistor with impacts of gate underlap region. *Superlattices and Microstructures*, 142, 106512. https://doi.org/10.1016/j.spmi.2020.106512
54. Vu, C. A., & Chen, W. Y. (2019). Field-effect transistor biosensors for biomedical applications: Recent advances and future prospects. *Sensors*, 19(19), 4214.https://doi.org/10.3390/s19194214
55. Lei, Y. M., Xiao, M. M., Li, Y. T., Xu, L., Zhang, H., Zhang, Z. Y., & Zhang, G. J. (2017). Detection of heart failure-related biomarker in whole blood with graphene field effect transistor biosensor. *Biosensors and Bioelectronics*, 91, 1–7. https://doi.org/10.1016/j.bios.2016.12.018

9 MEMS-Based SAW Sensors Using Nanorods to Discover Volatile Gases

Aman Raj, Baruna Kumar Turuk, and Basudeba Behera

9.1 INTRODUCTION

A surface acoustic wave (SAW) travels along just the surface of materials, while along the depth of the device or material, the amplitude of this SAW wave decays exponentially.[1] A SAW device generally comprises a piezoelectric substrate upon which interdigitated transducers (IDTs) are present. The principles on which SAW devices are based can be either the delay-line model, or the single-port resonator or two-port resonator model.[2] As shown in Figure 9.1, a SAW delay-line device consists of a piezoelectric substrate, IDTs and an absorber.[3] Generally, the piezoelectric substrate may be lithium niobate, lithium tantalate or quartz. There is a pair of IDTs, i.e. input and output, which are parted by a distance and consist of absorbers at the device's ends. These IDTs that are present on the piezoelectric substrate are a comb-like structure. Voltage is applied at the input IDT alternately, meaning one IDT is given a voltage supply, and the other is grounded, due to which a wave travels along the surface of the substrate, as a result of the inverse piezoelectric effect.[4] The inverse piezoelectric effect occurs when a mechanical wave or stress is formed on the surface due to an electric charge being applied. This wave is then received by the output IDT after travelling through the delay path between the input and the output IDT.[5] The delay line or path is the surface between the input and the output IDT. The output IDT works on the piezoelectric effect, resulting in an electrical response from mechanical waves received through the piezoelectric substrate. This wave travelling through the sensing area is detected at the output IDT as a phase change is observed due to the perturbation in velocity it undergoes.[6] The telecommunication industry uses SAW delay-line mode devices for high-frequency filter designing.[7] SAW delay-line models are also used in gas and biological sensors where health-related issues are addressed and in the development of chemicals.[8]

Plessis et al. (2013) presented a study where ZnO nanopillars were added to the sensing area of a two-port delay-line structure to increase sensitivity and were simulated using COMSOL Multiphysics. The best height of the IDTs to obtain better results for the electrodes was also determined.[9] Hasanuddin et al. (2016) focused

DOI: 10.1201/9781003453635-9

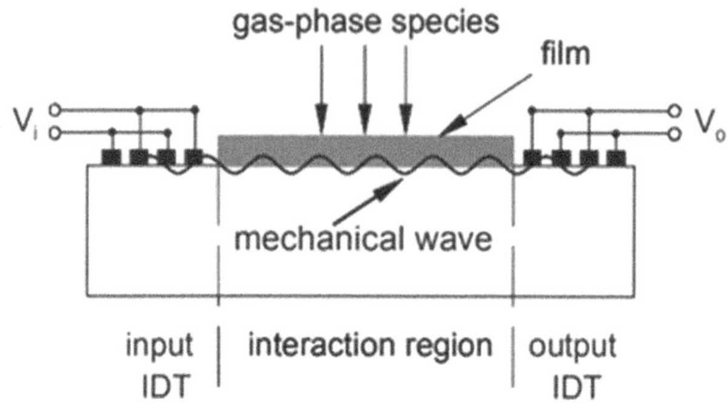

FIGURE 9.1 Sensing mechanism of the SAW gas sensor.

on the efficiency of various metal oxides that act as sensing layers for the SAW-based gas sensor. The study was conducted for sensing C_2H_4 gas as present for fruit ripening detection. The study offers a sensing layer of metal oxides with 0.5μm thickness laid upon lithium niobate ($LiNbO_3$) as the substrate for ethylene gas sensing to differentiate the fruit's maturity level. The optimal operational frequency was 867 MHz, taking the sensing layer as zinc oxide.[10] Kabir et al. (2016) studied SAW structures with $LiNbO_3$ as the substrates and two- and four-electrode finger pairs in both IDTs.[11] The results showed that when one of the electrodes, considered input, is set to zero volts, a better similarity in the actual and theoretical frequency response is seen. While calculating frequency response, all the simulated SAW device structures' current density components were considered. Hasan et al. (2017) designed, simulated and fabricated nanostructure-based and layered SAW-based gas sensors. A substrate of $LiNbO_3$ and ZnO nano-multilayer structures was projected to boost the gas sensor's performance. For optimisation purposes, diverse materials are considered an intermediate layer. The sensor's sensitivity may be enhanced because of the increased active surface area, which can be gained by changing the nanorods' aspect ratio, the IDT's thinness, and the distance between the electrodes. Thus, the nanorod-based SAW gas sensor presented a better sensing response than the layered SAW gas sensor regarding sensitivity performance.[12] Gupta et al. (2017) compared a single- and multilayered SAW-based sensor for detecting di-chloromethane (DCM) gas. Single-layered sensors with PIB/$LiTaO_3$ and multilayered sensors using PIB/Al_2O_3/$LiTaO_3$ are simulated in COMSOL Multiphysics. The study showed higher sensitivity exhibited by the multilayered SAW-based sensor than that of the resonant frequency towards DCM. This study similarly studied the effect of variations of dimensions, such as the thickness of the intermediate layer and atmospheric conditions on frequency shift.[13]

N. Harathi et al. (2019) performed a simulation in which a 2D SAW sensor was designed to sense the presence of hydrogen gas. The sensor is designed with a sensing layer of zinc oxide to improve sensitivity. The performance was assessed concerning the displacement and the operating frequency in the presence and absence of the testing gas. The analysis of the sensor was carried out with the help of finite element modelling.[14] S. Omar et al. (2020) aimed to examine the sensitivity of polymer-coated

SAW-based gas sensors to volatile organic compounds (VOCs). Here, the substrate is piezoelectric, while the sensing layers were polymeric materials, which were considered because of their properties such as low energy consumption, high sensitivity, performance at room temperature and short response time. The materials which were chosen were polydimethylsiloxane (PDMS), polyisobutylene (PIB), phenylmethyl diphenyl silicone (OV25), polyimide (PI), and polyisoprene (PIP). The VOCs selected for investigation were dichloromethane (DCM), trichloroethylene (TCE), 1,2-dichloroethylene (DCE), and carbon tetrachloride (CCl_4). The performance of these various types of substrates, which have different properties, was tested for the multiple gases considered.[14] Moustafa et al. (2021) simulated and discussed the SAW-based gas sensors to identify DCM. The gas sensor response was studied using five different piezoelectric materials considered substrates. After the simulation was performed, $LiNbO_3$ reported the uppermost resonance frequency. This study showed different effects of varying the thickness of the sensing layer and also indicated the frequency shift takes place as there is a change in the concentration of gas in the surrounding area that comes in contact with the sensing layer of the sensor.[14] The LN and diamond layers that make up the SAW resonator's two-layer construction result from removing the silicon dioxide layer from a three-layered structure, and we discovered a high-quality factor in the outcome. Two-layer structures were shown to have a higher quality factor than three-layer structures. This indicates that the layered complexity is decreased and acquired without a performance impact.[15]

S. Trivedi et al. (2015) developed a 3D love wave-based SAW-based gas sensor on a 36° YX lithium tantalate substrate that serves as the device's waveguide and is made of SiO_2. Insertion loss and mass sensitivity were noted. The instrument produced a mass sensitivity of 83.22 m^2/kg. This was in line with earlier findings.[16] There are two ways to build the resonator: a single IDT with a finite electrode over a piezoelectric substrate, or a short IDT with a reflecting grating at the end. SAW capabilities include filters, resonators, signal processors and delay lines between sensors and actuators. When a voltage is given to the electrode, a mechanical wave known as a SAW is excited on a piezoelectric substrate. The IDT determines the frequency of the propagating wave based on the distance between the two electrodes.[17] G. Matthews et al. (2016), investigated a novel method for analysing SAW devices using a series of 2D FEM simulations – two SAW architectures with varying numbers of electrodes in their IDT and $LiNbO_3$ as the substrate. Then, SAW devices were created based on those calculations, and it was discovered that the measured frequency responses were entirely consistent with the results of the simulations.[11] Zhang et al. (2011) simulated a 3D SAW-based gas sensor for DCM. The substrate is made of $LiNbO_3$, and the sensing layer is made of polyisobutylene (PIB). The selective adsorption of DCM by PIB from air increases the bulk of the PIB film. Resonance frequency shifts as a result and is somewhat lowered. So, this sensor can be used to detect DCM.[18]

This chapter uses COMSOL Multiphysics to simulate a one-port SAW resonator. It shows how the interdigital transducer (IDT) and the piezoelectric substrate's geometrical structure affect resonance frequency, and simulation is then used to optimise the electrode's height and width. After a thorough investigation that involved adjusting the depth's length and watching its impact on the nature of the resonance frequency, the depth of the substrate was chosen. Al was selected as the material for the electrode, while $LiNbO_3$ was employed as the substrate. Last, a one-port SAW resonator with optimal geometry was simulated, and mechanical deformation and electric potential distribution were recorded. This is highly helpful for modelling any SAW devices.[19]

9.2 WORKING PRINCIPLE

The working theory behind the operation of the sensors is based on gas adsorption on a surface, which causes a change in the mass, modulus and conductivity of the sensing layer, which can be accurately detected using SAW-based gas sensors. To sense hydrogen (H_2) gas, various geometries of SAW-based gas sensors are investigated, constructed and simulated in the current thesis study. Using frequency shift and displacement plots, it is observed that the nanorod-based structure showed better sensitivity than the other structure without nanorods. Various materials are considered interdigitated transducer (IDT) materials for optimised purposes. Similarly, the height of nanorods is varied to decide the optimum length of the nanorods. Also, the material in the intermediate layer was varied to see the effects and find a better-resulting model.[20]

The SAW gas sensor's operating frequency depends upon the wave's velocity (v_0) travelling through the substrate.[21] As the substrate changes, the velocity with which the wave travels across the device also changes. In a delay line model of the SAW gas sensor, the wavelength of a wave is determined using the dimensions of the IDT[22]:

$$\lambda=2(W_e + W_{sp}), \tag{9.1}$$

where λ 'is the wavelength, 'W_e' is the width of each electrode and 'W_{sp}' is the distance between each electrode. Using λ, operating frequency of the device can be found out. The relation between λ and the operating frequency is as follows:

$$f_0 = \frac{v_0}{\lambda}, \tag{9.2}$$

where 'v_0' is the propagation velocity of the wave in the piezoelectric substrate, 'λ' is the wavelength of the wave, and 'f_0' is the operating frequency. The input IDTs are provided with a voltage supply at alternate IDTs. When the sensing layer comes in contact with the gas that is to be detected, the properties of the material change. Adsorption of the gas molecules on the layer causes changes in the device's frequency. The primary mechanism for the SAW sensors is based on thermodynamics equilibrium.[23] The frequency downshift can be calculated by:

$$\Delta f = f - f_0. \tag{9.3}$$

where, f_0 *and* f is the resonant frequency for the device before and after exposure to gas. Δf is considered a simple mass load attached to the SAW surface as the thin layer formed by the adsorbed gas is thin:[24]

$$\Delta f = (k_1 - k_2) f_0^2 \rho h \tag{9.4}$$

and

$$\Delta f = (k_1 - k_2) f_0^2 \rho h\text{- k2f02}\rho \frac{4\mu}{v_0^2}\left\{\frac{\lambda+\mu}{(\lambda+2\mu)}\right\}, \tag{9.5}$$

where k_1 *and* k_2 are the coupling constant; 'ρ' is the density of the sensing layer, which changes after the adsorption of the gas; ' h' is the height of the sensing layer; 'λ' is the bulk modulus; and 'μ' is the shear modulus.

9.3 DEVICE DESCRIPTION

The multilayered nanorod-based SAW gas sensor consists of a piezoelectric substrate, input and output IDTs, an intermediate layer and nanorods. The most commonly used materials for IDTs are metals such as aluminium, copper, gold and tungsten. Generally, metal oxides such as ZnO, SnO_2 and Al_2O_3 are used as sensing layers. In this case, nanorods perform this role. In the proposed model, we have observed the results for a device with $LiNbO_3$ as a piezoelectric substrate and ZnO material used for nanorods. The height of the nanorods varied from 0.05 μm to 0.3 μm by taking a difference of 0.05μm to obtain the optimised height to get the best results. Also, ZnO and palladium (Pd)[25] are used as intermediate layers to find which material's presence gives optimised results. The wave velocity, i.e. the velocity with which the wave travels across the lithium niobate substrate, is 3488 m/s^2.[20] So here, we have taken 'λ' as 4 μm for the device simulation. For 2D simulations, we have considered 2 IDTs of 1μm width separated by 1μm of distance. The thickness of the IDTs is 0.02 μm. The intermediate layer is taken as 0.2 μm of thickness. The height of the substrate is taken as '5λ', which means 20 μm for this model, shown in Figure 9.2. The results are obtained by varying the intermediate layer materials, changing the nanorods' height, and altering ID materials. Simulation for a 2D model was done for a device with a length of 4 μm and height of 8μm. The material used for the substrate is lithium niobate; ZnO is used for the nanorod structure; and IDT is made of aluminium. Simulation using the above materials is done by changing the intermediate layer's material.

9.4 STEPS TO MODEL IN COMSOL MULTIPHYSICS

The basic steps to create the primary-based gas sensor are as follows:

i. Input the parameters of the materials to be used for the sensor.
ii. Create the physical components for the sensor with the required dimensions.
iii. Define properties of the materials like piezoelectric materials etc.
iv. Assign the materials for each component and enter the accurate values for the used materials.

v. Define the exact conditions for the produced structure of the sensor.
vi. Construct a mesh by defining the accurate values.
vii. Set up a parametric sweep to solve the model with and without the adsorbed species on the sensor.
viii. Plot the graph, and by properly adjusting the deformed wave plot, see the effect.
ix. The eigenfrequency values can be seen in the results, which show values before and after exposure to the gas.

Simulation Methodology

For lithium niobate, the velocity of the wave through it is 3488m/s.[2] The values of the piezoelectric substrate required for simulation are as follows:[11]

$$C = \begin{pmatrix} c_{11} & c_{12} & c_{13} & c_{14} & 0 & 0 \\ c_{12} & c_{11} & c_{13} & -c_{14} & 0 & 0 \\ c_{13} & c_{11} & c_{33} & 0 & 0 & 0 \\ c_{14} & -c_{14} & 0 & c_{44} & 0 & 0 \\ 0 & 0 & 0 & 0 & c_{44} & c_{14} \\ 0 & 0 & 0 & 0 & c_{14} & (c_{11} - c_{12})/2 \end{pmatrix}, \tag{9.6}$$

and

$$e = \begin{pmatrix} 0 & 0 & 0 & 0 & e_{15} & -e_{15} \\ -e_{22} & e_{22} & 0 & e_{15} & 0 & 0 \\ e_{31} & e_{31} & e_{33} & 0 & 0 & 0 \end{pmatrix}, \tag{9.7}$$

$$\varepsilon = \begin{pmatrix} \varepsilon_{11} & 0 & 0 \\ 0 & \varepsilon_{11} & 0 \\ 0 & 0 & \varepsilon_{33} \end{pmatrix}, \tag{9.8}$$

where C: Elasticity Matrix, e: piezoelectric Constants, and ε: Dielectric Constants.

For the simulation process, we considered lithium niobate as a piezoelectric substrate, so the values for the same are as shown in Table 9.1.

After assigning the materials its values, we assign one IDT with an input voltage of 1V and another IDT as ground. The summary of boundary conditions for the 2D model used in the simulation is given in Table 9.2.

We perform eigenfrequency analysis on the model to find out the resonant and anti-resonant frequencies of the devices. Resonant frequency generally refers to the frequency when the displacement obtained is maximum for that material. We also observed the displacement of the device using frequency and time-dependent analysis. We observed the difference in the frequency in the absence and presence of the gas to be detected. The model used to determine these frequencies consists of a piezoelectric substrate upon which IDTs are placed.

TABLE 9.1
Properties of lithium niobate

Dielectric Constant	ε_{11}	85.2
	ε_{33}	28.7
Elastic Constants	c_{11}	2.03
(x 10^{11} N/m^2)	c_{12}	0.573
	c_{13}	0.752
	c_{14}	0.085
	c_{33}	2.424
	c_{44}	0.595
Piezoelectric Strain Constants (C/m^2)	e_{15}	3.7
	e_{22}	2.5
	e_{31}	0.23
	e_{33}	1.33
Crystal Density (kg/m^3)	ρ	4.647 x 10^3

TABLE 9.2
Solid mechanics condition for the model

Name	Mechanical Boundary Condition
Side 1, Side 2, Plane 2	Continuity /Periodic Condition
Plane 1	Antiperiodic Condition
Side 3, Plane 3	Fixed Constraint

9.5 RESULTS AND DISCUSSION

9.5.1 2D-Based Device Structure

Various graphs are obtained after the simulation of the proposed model. Time analysis and frequency domain study are used to obtain the graph output eigenfrequency study. The results obtained are based on the simulation of the proposed model using eigenfrequency, time domain and frequency domain analysis. The meshing done in this 2D environment is physics-controlled. Fine meshing is used to obtain more accurate results.

Figure 9.2 shows the displacement profile of the proposed device at the resonant frequency. The resonance frequency of the device is found to be 841.57 MHz. The above figure shows that the displacement observed at IDT is minimum, while that around the IDTs can be maximum. The output above is obtained using the eigenfrequency study. We can observe that the displacement at IDT is maximum at the resonant frequency and is reduced around the IDTs.

Figure 9.3 shows the displacement graph at the input IDT points for nanorod-based SAW gas sensor models with nanorods of varying heights. We can observe from the above graph the displacement versus time plot for models with nanorods of height 0.05 μm, 0.1 μm, 0.15 μm, 0.2 μm, 0.25 μm and 0.3 μm. The above graph shows that the displacement for a model with a nanorod height of 0.25 μm is maximum.

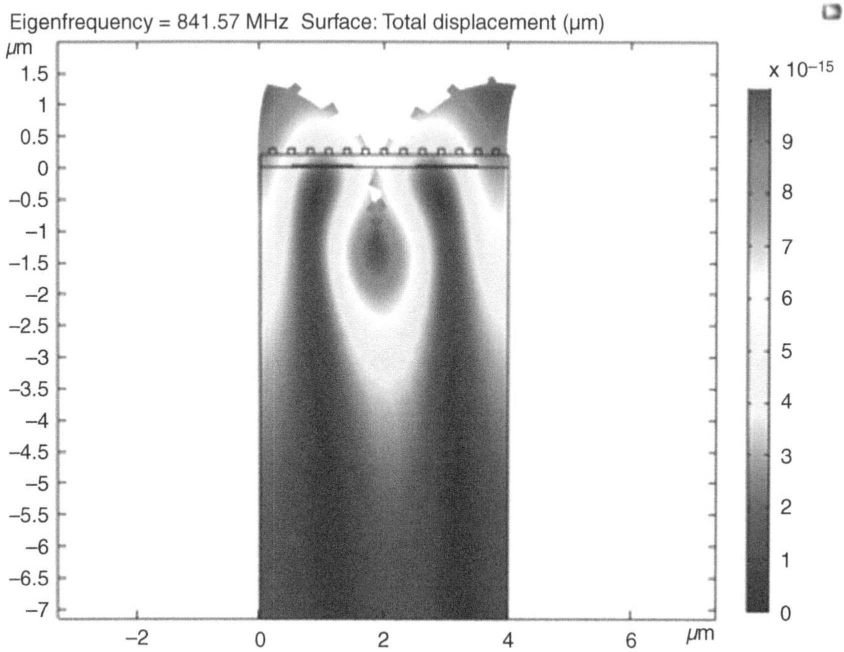

FIGURE 9.2 Displacement profile at resonance resonant.

Using previous graph information, we observed that the 0.25 μm height nanorod model had maximum displacement, so we performed time analysis on the same model by varying the IDT materials. We observed the graph outputs for aluminium, copper and tungsten. Figure 9.4 shows the time analysis displacement versus the time graph for the 0.25 μm height nanorod model with varying IDT materials. From the above graph, we can conclude that aluminium material, when used as IDT, shows maximum displacement.

9.5.2 Comparison between SAW Gas Sensors with and without Nanorods

The plots in Figure 9.5 show the displacement plot comparison between SAW-based gas sensors with and without nanorods in time domain analysis for the surface of the SAW-based gas sensor. The graph indicates that the displacement observed in the nanorod-based structure is better than the SAW gas structure without nanorods. Thus, the addition of nanorods to the structure has improved the sensitivity of the SAW gas sensor.

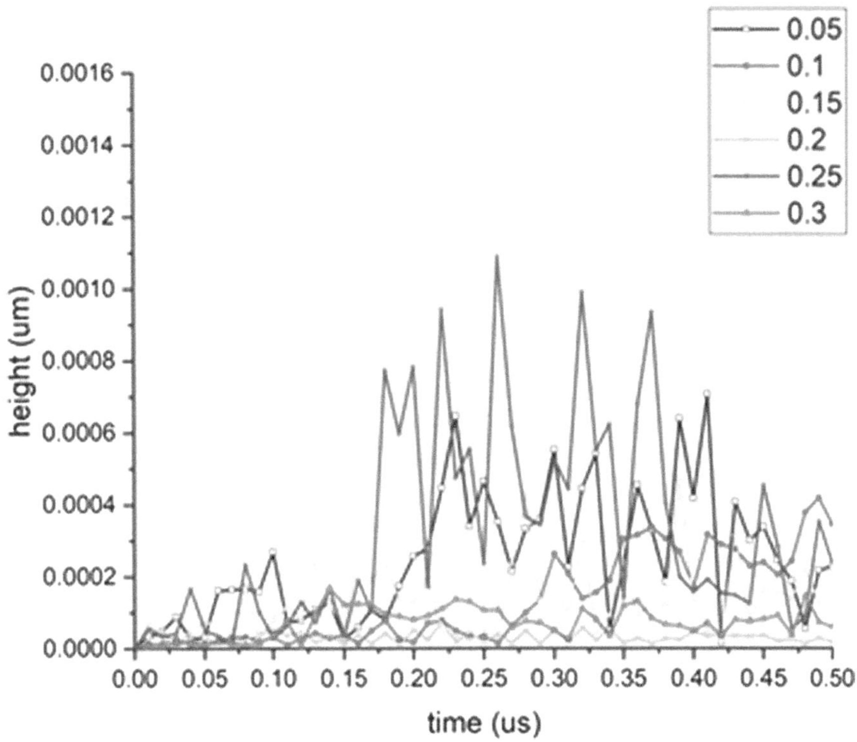

FIGURE 9.3 Time analysis graph for various nanorod heights.

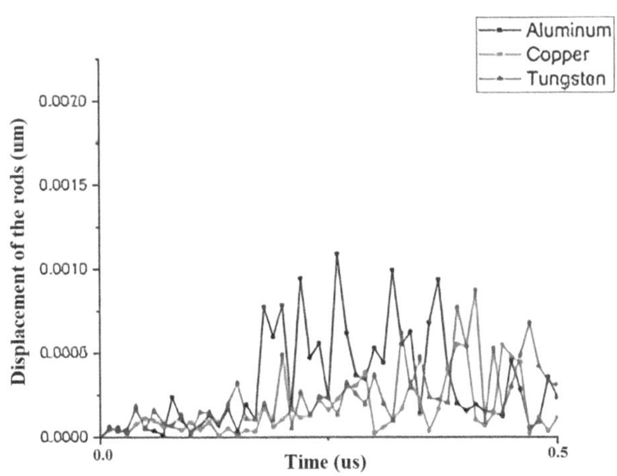

FIGURE 9.4 Time analysis graph after varying IDT materials.

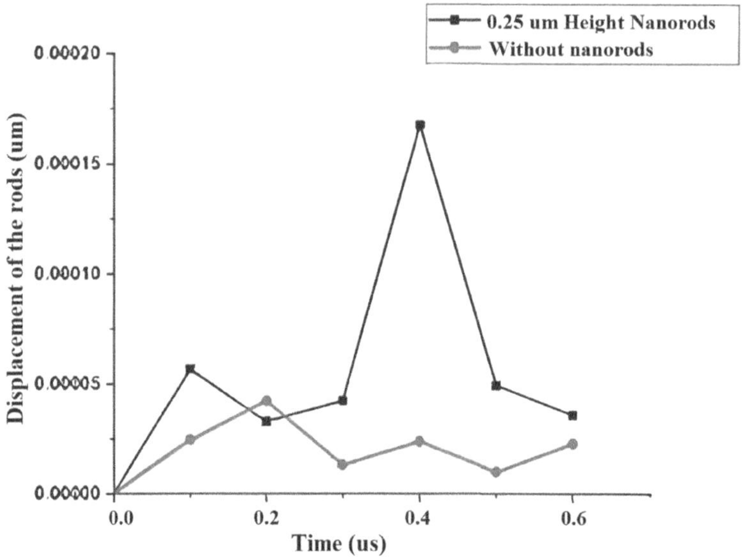

FIGURE 9.5 Displacement versus time plot for SAW gas sensor with and without nanorods.

9.6 CONCLUSION

Initially, various literature surveys were presented on the gas sensor. Then, the proposed sensor's construction, working principle and simulation methodology were presented. 2D-based nanorod-based SAW gas sensor simulation were presented. It is observed that when there is a change in the IDT material and intermediate material, the displacement graph changes accordingly. The results were obtained for frequency domain analysis and time domain analysis, with the height of the nanorod varying. The maximum displacement of the SAW gas sensor was observed for a model with 0.25 μm nanorod height on the surface. It was observed that nanorods of height 0.25 μm showed better results. Basically, from the simulation and analysis of the SAW gas sensor, it is concluded that it can detect volatile gases.

REFERENCES

1. Auld B. A. (1973), *Acoustic Fields and Waves in Solids* (Vol. I), First Edition. Stanford, California: Wiley Publication, p. 431, ISBN 0-47I-03700-I (V.1).
2. Turuk, B. K., Behera, B. (2024), "A review of one port SAW resonator for application in the communication systems," *IOP Eng. Res. Express*, vol. 6, pp. 1–17, DOI:10.1088/2631-8695/ad4234.
3. Behera, B., Nemade H. B., and Trivedi, S. (2016), "Modelling and finite element simulation of a Dual friction-drive SAW motor using flat slider," IEEE International Ultrasonics Symposium, pp. 1–4, DOI:10.1109/ULTSYM.2016.7728378.
4. Behera, B., and Nemade H. B. (2016) "Modelling and finite element simulation of a surface Acoustic wave driven linear motor," *Procedia Eng.*, vol. 144, pp. 1411–1418, DOI:10.1016/j.proeng.2016.05.172.

5. Behera, B., and Nemade H. B. (2018) "Dual friction-drive SAW motor utilising cylindrical shaft as external load driver," *Ferroelectr. Lett. Sect.*, vol. 45, no. 1–3, pp. 8–13, DOI:10.1080/07315171.2018.1499359.
6. Tiwary, A., Kumar, J., and Behera, B. (2023), "Analysis of CNT-based SAW sensor for the detection of volatile organic compounds," *Phys. B Condens. Matter*, vol. 669, pp. 415279, DOI:10.1016/j.physb.2023.415279.
7. Turuk, B. K., Behera, B. (2023), "Design of heterostructure one-port SAW resonator with improved bandwidth using Si and LiNbO3 layer," *Ferroelectrics*, vol. 615, no. 1, pp. 132–142, DOI:10.1080/00150193.2023.2243563.
8. Tiwary A., Pradhan, S., Kumar, J., and Behera, B. (2023), "Polyisobutylene and carbon nanotube-based heterostructure SAW sensor for detection of TCM gas," 2023 5th Int. Conf. Electr. Comput. Commun. Technol. ICECCT 2023, pp. 1–7, DOI:10.1109/ICECCT56650.2023.10179629.
9. Plessis, H. G., and Perold, W. J. (2013), "Simulation of ZnO Enhanced SAW Gas Sensor," Proc. 2013 COMSOL, pp. 1–4.
10. Hasanuddin N. H., and Wahid, M. H. A. (2016), "Metal oxide based surface acoustic wave sensors for fruits maturity detection," 2016 3rd Int. Conf. Electron. Des. ICED 2016, no. 1, pp. 52–55, DOI:10.1109/ICED.2016.7804605.
11. Kabir, K. M. M., Matthews, G. I., Sabri, Y. M., Russo, S. P., Ippolito, S. J., and Bhargava, S. K. (2016), "Development and experimental verification of a finite element method for accurate analysis of a surface acoustic wave device," *Smart Mater. Struct.*, vol. 25, no. 3, p. 035040, DOI:10.1088/0964-1726/25/3/035040.
12. Hasan, M. N., Maity, S., Sarkar, A., Bhunia, C. T., Acharjee, D., and Joseph, A. M. (2017), "Simulation and fabrication of SAW-based gas sensor with modified surface state of active layer and electrode orientation for enhanced H2 gas sensing," *J. Electron. Mater.*, vol. 46, no. 2, pp. 679–686, https://doi.org/10.1007/s11664-016-5128-7.
13. Gupta, A., Kumar, P., and Pandey, S. (2017), "Analysis of multilayered SAW based gas sensor," Proc. – Int. Conf. Trends Electron. Informatics, ICEI 2017, vol. 2018, pp. 239–242, DOI:10.1109/ICOEI.2017.8300924.
14. Harathi N., Kavitha, S., and Sarkar, A. (2020), "ZnO nanostructured 2D layered SAW based hydrogen gas sensor with enhanced sensitivity," *Mater. Today Proc.*, vol. 33, pp. 2621–2625, DOI:10.1016/j.matpr.2020.01.150.
15. Turuk, B. K., Behera, B. (2022), "Investigation to various parameters of a multilayered one-port surface acoustic wave resonator," *Mater. Today Proc.*, vol. 56, no. 2, pp. 883–888, https://doi.org/10.1016/j.matpr.2022.02.527.
16. Trivedi, S. and Nemade, H. B. (2015), "Finite element simulation of love wave based SAW delay line using COMSOL multiphysics," Excerpt from Proc. 2015 COMSOL Conf. Pune, vol. 1, pp. 2–5.
17. Turuk, B. K., Sahoo, G., Nagmani, A. K., and Behera, B. (2021), "Different parametric analysis of one port SAW resonator using COMSOL," In Dr. Prashant Kumar, Dr. Jayendra Kumar and Dr. Mrutyunjay Rout (Eds.), *A Collection of Contemporary Research Articles in Electronics, Communication and Computation*, pp. 337–340, https://doi.org/10.47531/MANTECH/ECC.2021.51.
18. Rao, Y. L., and Zhang, G. (2007), "3-D modeling of surface-acoustic-wave based sensor," Excerpt from Proc. 2007 COMSOL Conf. Newton, MA, pp. 1–5.
19. Nagmani, A. K., Turuk, B. K., and Behera, B. (2021), "Simulation and optimization of the geometrical structure of one-port SAW resonator using FEM," *AIP Conf. Proc.*, vol. 2341, pp. 020043-1–7, https://doi.org/10.1063/5.0049919.
20. Tiwary, A., Rout, S. S., and Behera, B. (2022), "Design and analysis of various characteristics of a MEMS-based PIB/ CNT/LiNbO3 multilayered SAW sensor for

CO_2 gas detection," *Trans. Electr. Electron. Mater.*, vol. 669, 415279, pp. 415279-1–9, https://doi.org/10.1016/j.physb.2023.415279.

21. Soloman, S. (2010), *Sensors Handbook*, Second Edition. New York, McGraw-Hill Companies, ISBN: 978-0-07-160571-7.
22. Morgan D. (2007), *Surface Acoustic Wave Filters with Applications to Electronic Communications and Signal Processing*, Second Edition. Oxford: Elsevier Academics Press eBook, ISBN: 9780080550138.
23. Ho, C. K., Lindgren, E. R., Rawlinson, K. S., McGrath, L. K., and Wright, J. L. (2003), "Development of a surface acoustic wave sensor for in-situ monitoring of volatile organic compounds," *Sensors*, vol. 3, no. 7, pp. 236–247, https://doi.org/10.3390/s30700236.
24. Moustafa, M., Laouini, G., ElNaggar, M., and AlZoubi, T. (2021), "Investigation into surface acoustic wave sensor for DCM gas detection using COMSOL multiphysics," *Ferroelectrics*, vol. 572, no. 1, pp. 94–105, https://doi.org/10.1080/00150193.2020.1868875.
25. Akshaya, S., and Juliet, A. V. (2021), "A computational study of a chemical gas sensor utilising Pd–rGO composite on SnO2 thin film for the detection of NOx," *Sci. Rep.*, vol. 11, no. 1, pp. 1–18, https://doi.org/10.1038/s41598-020-78586-7.

10 Wearable Sensors
Revolutionizing Healthcare

Prashant Kumar, Pankaj Kumar, Polychetty Veenasheela Rao, and Mrutyunjay Rout

10.1 INTRODUCTION

Healthcare is crucial in life. The increasing elderly population and the corresponding growth in chronic diseases are burdening recent healthcare systems, leading to demand high for resources like hospital beds, doctors and nurses. Clearly, a solution is needed to improve the strain on healthcare systems and maintain top-notch care for vulnerable patients. A wearable device is a piece of technology that can be worn on the body, typically integrated into items such as watches, clothing, glasses or even jewellery. These devices often have various sensors and connectivity capabilities to monitor and track data related to health, fitness and other activities. Examples of wearable devices include smartwatches, fitness trackers, smart glasses, and other health monitors like continuous glucose monitors for diabetics or heart rate monitors for patients with cardiac conditions. Wearable devices erase the partition between consumer electronics and medical devices.[1]

Wearable devices have brought about a paradigm shift in the landscape of health monitoring, offering unmatched opportunities for individuals to track and manage their well-being in real time. In recent years, wearable devices have become increasingly popular among urban consumers. Figure 10.1 shows the range of wearable devices, typically worn on the body or incorporated into clothing. These electronic gadgets are equipped with sensors capable of collecting various physiological data such as heart rate, blood pressure, activity levels, and sleep patterns. The integration of advanced data analytics, deep learning, and machine learning algorithms further boosts the capabilities of these devices, enabling them to provide actionable insights into an individual's health status.[2]

The market for wearable technology is predicted to rise at a CAGR of 12.9%. According to a report, the market size of global wearable technology is expected to rise to around USD $142.4 billion by 2030. As shown in Figure 10.2, the wearable device market shows an exponential growth pattern. This means a lot; there is a definite chance of bio-electronics booming.

Sensors made from flexible materials are attached to a person, with integration with an embedded system, to monitor constraints and transfer this set of data to the monitoring system unit for further analysis. Wearable sensors have played a

DOI: 10.1201/9781003453635-10

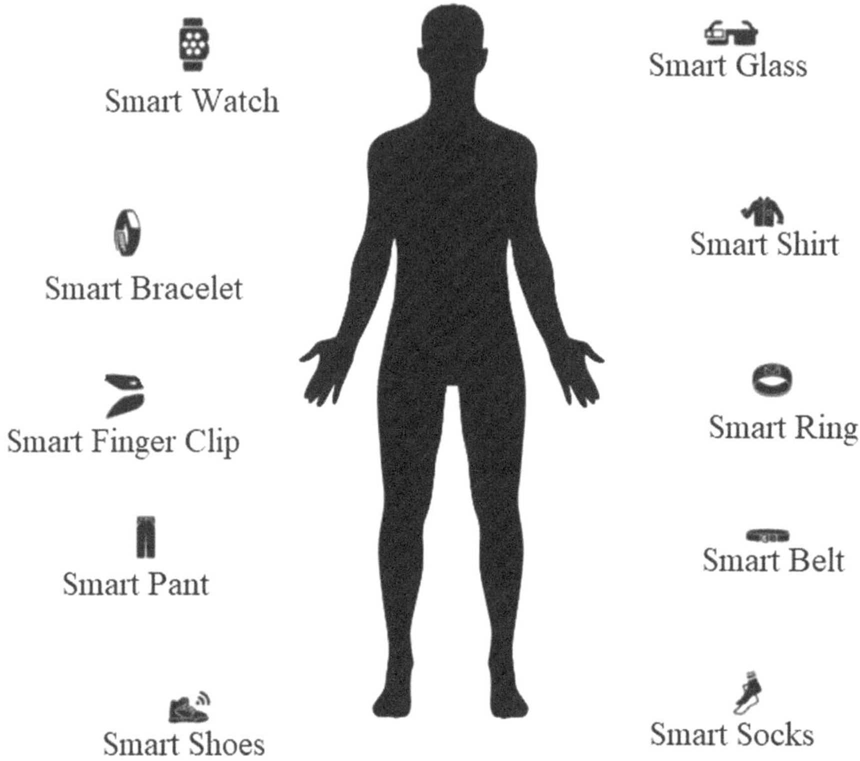

FIGURE 10.1 Gamut of wearable devices surrounding the whole body.

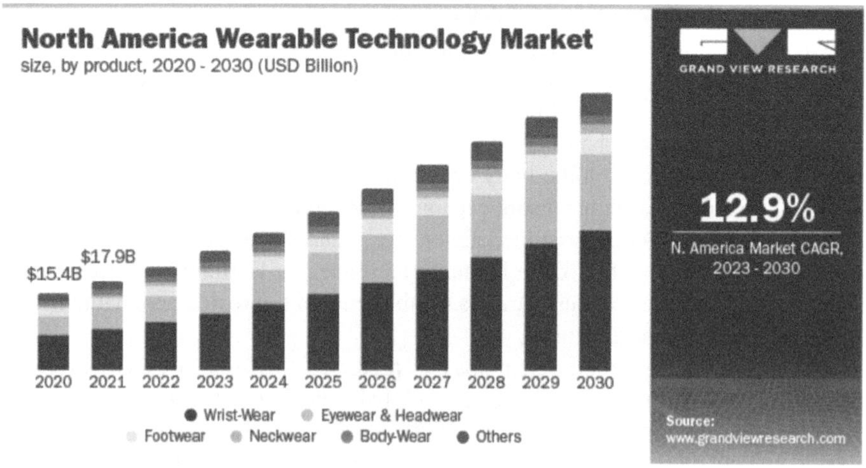

FIGURE 10.2 Wearable technology market growth. (www.grandviewresearch.com) The market of wearable technology is predicted to rise at a CAGR of 12.9%. According to a report, the market size of global wearable technology is expected to rise to around USD$142.4 billion by 2030. As observed from Figure 10.2, the wearable device market shows an exponential growth pattern. This means there is a definite chance of bio-electronics booming.

significant role in monitoring the physiological parameters of individuals to minimize any bodily malfunctions for the user.

The rest of this chapter is organized as follows: Section 10.2 reviews the different type of materials used for manufacturing sensors for wearable devices, and Section 10.3 discusses how wearable devices influence biomedical engineering and how it has revolutionized medical sciences. Section 10.4 discusses the different types of wearable devices and their applications. The next section 10.5 brief examines how wearable devices are conceived based on requirements and implemented in real time, with a preview of their latest versions available in the market. Section 10.6 discusses the radiation hazards of wearable devices and the precautions that need to be taken to minimize and ill-effects, and finally the last section 10.7 concludes the chapter.

10.2 MATERIALS USED FOR WEARABLE FLEXIBLE SENSORS

The material used for the fabrication of sensors is determined by various aspects such as the sensor's applications, availability, total manufacturing cost and so forth. Organic electronics are a key sector in the materials field that has been significantly developed for the production of flexible wearable devices. One of the main advantages of using organic materials in sensor fabrication is their flexibility, which allows for the creation of lightweight and conformable devices that can be easily integrated into clothing or accessories.[3] Additionally, organic electronics offer the potential for low-cost manufacturing processes, making them an attractive option for the mass production of wearable sensors. With ongoing research and advancements in this field, the future looks promising for the development of innovative and efficient sensor technologies. Organic and large area electronics (OLAE) is a process for developing electronic devices printed in very thin layers using purposeful inks. The substrates used for these operations are mainly popular due to transparency and lower cost compared to other organic polymers. Polyethylene terephthalate (PET) and Polyethylene naphthalate (PEN) are widely used.[4,5] The carbon compounds consist of graphene, carbon nanotubes (CNTs), carbon fibres and so forth along with metallic nanoparticles. Silver, gold and nickel are the most commonly used materials in flexible wearable sensors. Some most commonly used insulating substrates for the development of flexible sensors are Polydimethylsiloxane (PDMS), PET, PEN and Polyimide (PI). Some conductive polymers for the development of flexible sensors are poly(3,4- ethylenedioxythiophene) polystyrene sulfonate (PEDOT: PSS), Polyacetylene, and polyaniline, which conduct electricity due to their lower bandgap as compared to their insulating complements. These materials are also widely used for developing solar cells, batteries, LCD (liquid crystal display), and so forth.

Wearable flexible sensors have been implemented in types of sensing in human daily life. These differ in structural design and properties based on the type of sensing required. Electrochemical sensing is one of the most common types of sensing performed with flexibles sensors. Flex sensors, with their distinctive chemical and electronic properties, have been outstanding in performing different types of biochemical sensing. Electrochemical sensing involves monitoring of glucose, pH, cholesterol and more. Glucose and pH sensors are made from carbon nanotubes (CNT) because of their curvature sidewalls and hydrophobic nature that provide a strong interaction by π-bonding.[6] To develop some of these sensors, the use of a layer- by-layer (LBL)

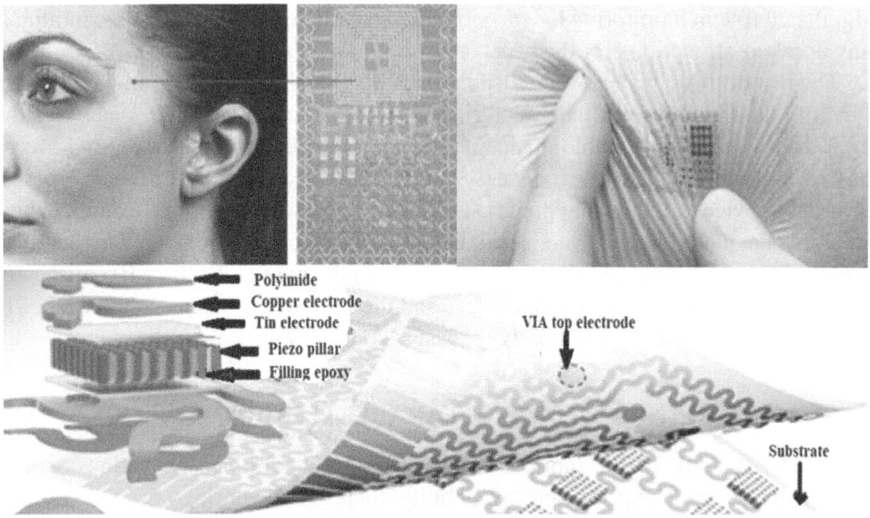

FIGURE 10.3 On-body screen-printed wearable device using organic compounds.

structure was employed to give it a more robust structure. Two kinds of polymers, that is, PDDA and PET, were used to develop the substrate. Self-powered electronics can be accomplished by harvesting energy from thermal, acoustic, photic (light), or near-field sources. Among these, photo energy is particularly promising as an energy source for wearable electronic devices due its high efficiency and compatibility with photovoltaic material and the ubiquity of light emission from solar artefacts. Organic photovoltaics (OPVs) have been extensively researched in the last two decades due to their thin structure, light weight, high flexibility and inexpensive solution methods for synthesis, and the environment around us, making them desirable. Shape architectures are promising for extensive integration. Figure 10.3 shows how flexible and ultra-thin wearables can be printed on the human body or used as stickers for sensing. OPVs, known for their light weight, high flexibility and scalability, have achieved electricity conversion efficiencies of over 17% today and are the best candidate for the next generation of these wearable devices.[7]

10.3 WEARABLE DEVICES FOR BIOMEDICAL ENGINEERING AND THE INTERNET OF MEDICAL THINGS (IOMT)

Prior to the concept of wearable devices, the medical field relied on conventional diagnosis techniques followed by treatment by the doctor. Biomedical engineering involves the measurement of medical parameters by modern techniques such as invasive, non-invasive, non-contact and imaging (X-ray, ultrasound). Invasive involves puncturing the skin for diagnosis; non-invasive is done by only contact to the skin; and in imaging we use X-ray, electromagnetic and ultrasound to visualize the defects.[8] With the advent of electronic wearable devices, there has been a paradigm shift. These wearables use miniature sensors and embedded circuits and establish wireless communication to medical servers though the internet via infrastructure-based networks. Wearable devices used for medical support can be called wearable medical devices (WMDs).[9] Once diagnosis is done, recommendations for treatment can be undertaken by expert systems which utilize artificial intelligence, machine learning and neural networks.

Wearable devices form a crucial part of body area networks (BANs), which are systems of interconnected devices on or around the human body. Advances in telecommunication, particularly fifth-generation (5G) communication networks, enhance the performance and connectivity of these networks by offering faster data transmission and reduced latency. This synergy enables seamless communication between wearables and external systems, improves real-time health monitoring and data analysis, and supports the growth of sophisticated applications within the Internet of Things (IoT) ecosystem. In some cases, BAN is utilized to monitor and alert for early warning signs to the patient. The BAN can be part of a bigger network with powerful servers dedicated for medical condition monitoring and the issuing of prescriptions if it uses a cellular network. Such a system involving patients, highly connected WMDs, and backed by an ever updating huge database constitutes the Internet of Medical Things (IoMT). The IoMT requires high-quality networking. The recent advancements in 4G, 5G and likely to come 6G communication technology have contributed to the exponential growth of wearable devices and the IoMT.[10]

Wearable devices demand network connectivity capabilities so that they can collectively monitor human health. The process involves setting up a network of master and slave nodes within wearable devices, either of which collects health data from its wearer. The master node consolidates this data and transmits it to a medical server, allowing continuous monitoring of health metrics and facilitating early detection of any serious health conditions. In cases of critical health events, such as emergencies, the master node plays a pivotal role. It alerts the wearer, providing basic first-aid instructions when necessary. Moreover, it notifies nearby hospitals for potential emergency bookings and sends the user's location to ambulances for prompt assistance. While smartphones traditionally assist in connecting such devices to networks, their usability is limited, especially for children and elderly individuals who may find them cumbersome to carry and operate.[11] To address this issue, smart watches equipped with e-SIM or Nano-SIM capabilities emerge as a practical alternative. These smart watches can seamlessly connect to cellular networks, ensuring continuous transmission of sensor data to medical servers. In essence, by leveraging smart watches with advanced connectivity features, healthcare providers can establish a robust system for real-time health monitoring.[12] This system not only enhances user convenience but also significantly improves response times during critical health incidents, thereby potentially saving lives through timely intervention.

10.4 WHAT CAN A WEARABLE DEVICE DO?

Continuous Monitoring:
Wearable devices provide continuous monitoring of various health metrics such as sleep patterns, heart rate, activity levels and even stress levels. This non-stop data collection allows for a more comprehensive understanding of one's health status

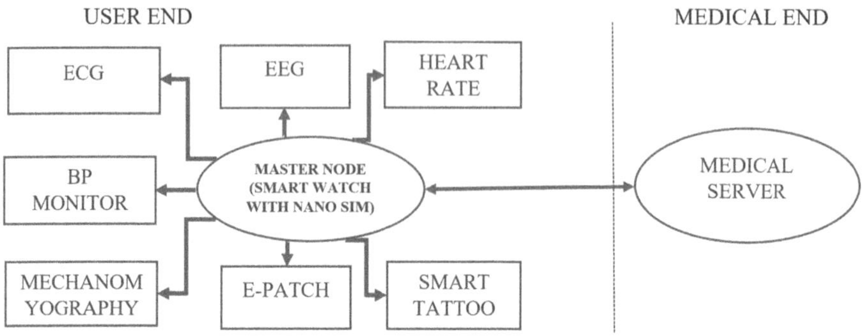

FIGURE 10.4 Non-stop transmission of wearable device data to medical servers.

compared to sporadic measurements done in traditional healthcare settings.[13] Moreover, wearable devices empower individuals to take a proactive approach to their health by providing personalized insights and recommendations based on their unique data. By integrating machine learning algorithms and artificial intelligence (AI)-driven analytics, wearable devices can offer actionable insights tailored to each user's specific health goals and needs.[14] Additionally, the convenience and accessibility of wearable devices make them valuable tools for remote monitoring and telemedicine. Healthcare providers can remotely monitor patients' health metrics in real time, allowing for timely intervention and personalized care management, especially for individuals with chronic conditions or those who require continuous monitoring. The whole process is shown in Figure 10.4.

Early Detection and Prevention:
By tracking vital signs and behaviours over time, wearable devices can detect subtle changes that may indicate the onset of health problems. This early detection enables proactive intervention and preventive measures, potentially averting more serious health issues.[15] For example, individuals at risk for cardiovascular disease can be alerted to changes in their heart rate or blood pressure, prompting them to seek medical attention or make lifestyle modifications to reduce their risk factors. Moreover, wearable devices can facilitate personalized health interventions by providing actionable insights and recommendations based on individual health data.[16] This could include reminders to engage in physical activity, maintain proper hydration, adhere to medication regimens or seek medical attention if certain thresholds are exceeded. By empowering individuals to take a proactive approach to their health, wearable devices have the potential not only to improve health outcomes but also reduce healthcare costs associated with preventable diseases and complications.

Personalized Insights: Wearable devices generate personalized insights based on individual data, offering tailored recommendations for lifestyle modifications, exercise routines and even dietary adjustments. This personalized approach to health management can lead to more effective outcomes compared to generic advice.[17]

Remote Monitoring and Telemedicine: Wearable devices facilitate remote monitoring, allowing healthcare providers to remotely track patients' health metrics without the need for frequent in-person visits. This is especially valuable for patients with chronic conditions or those recovering from surgeries, as it enables timely intervention and reduces the burden on healthcare resources.[18]

Behavioural Changes and Accountability: The feedback loop provided by wearable devices encourages users to adopt healthier behaviours. Features such as goal setting, progress tracking and social sharing promote accountability and motivation, fostering long-term adherence to healthy habits.

Research and Population Health: Aggregated data from wearable devices offer valuable insights for population health research, enabling epidemiologists and public health officials to identify trends, risk factors and patterns of disease at a large scale. This data can inform public health interventions and policy decisions aimed at improving community well-being.

Challenges and Considerations: Despite their benefits, wearable devices also pose challenges such as data privacy concerns, accuracy and reliability of measurements, and user engagement over the long term. Addressing these challenges requires collaboration among stakeholders including technology developers, healthcare providers, policymakers and users themselves.

10.5 TYPES OF WEARABLE DEVICES FOR HEALTH MONITORING

1. Fitness Trackers: These devices include sensors for tracking physical activity, such as step count, distance travelled and calories burned. They also often feature heart rate monitors and GPS functionality for more accurate tracking of exercise sessions and overall fitness levels.
2. Smartwatches: Smartwatches include features such as fitness tracking, heart rate monitoring and notifications from connected smartphones. Many smartwatches also include additional health-oriented features, such as electrocardiogram (ECG) monitoring, blood oxygen saturation (SpO2) measurement, and even fall detection for elderly users.[19]
3. Medical Wearables: Medical wearables are specifically designed for monitoring and managing chronic health conditions such as diabetes, hypertension and respiratory disorders. These devices may include continuous glucose monitors (CGMs) for diabetes management, blood pressure monitors and wearable ECG devices for monitoring heart health.[20]
4. Smart Clothing: Smart clothing incorporates sensors directly into garments such as shirts, socks, jeans and bras, allowing for continuous monitoring of various physiological parameters without the need for separate devices. This seamless integration of technology into everyday clothing holds great potential for improving user comfort and compliance with health monitoring routines. In this technology, we use garments made from carbon compounds and organic electronics. These are incorporated with general fabrics to make smart fabrics, and dresses made of these fabrics are smart clothes.[21]
5. Screen printing: This is also known as a body-flexible device. While it is like a tattoo on the body, it consists of many bio-electronic circuits that are capable of sensing as well as monitoring health issues. These are implemented on the body in such a manner that they will look like just a design mark. but they

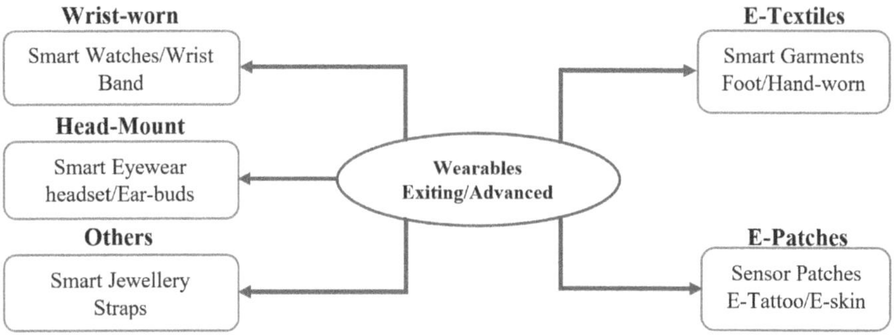

FIGURE 10.5 Most commonly used wearable devices.

are very capable of sensing and monitoring the health relevant to those areas where they are marked. Somehow they are implemented for energy harvesting because all sensors need energy for their functioning.[22]

Figure 10.5 shows the classification of most commonly used wearable devices based on the nature of their application.

10.6 APPLICATIONS OF WEARABLE DEVICES IN HEALTH MONITORING

Table 10.1 presents a comparison of the existing wearable system and its characteristic features.

Wearable devices have been included seamlessly in healthcare and medical monitoring, and can be found in all hospitals, clinics and nursing homes. Being cheap and easily available consumer electronics goods, these are readily available on most online shopping portals. Some of the popular applications of wearable devices are Disease Management, Remote Patient Monitoring, Fitness and Wellness, and Early Detection and Intervention.[23]

10.7 CHALLENGES AND LIMITATIONS OF WEARABLE DEVICES

Despite their immense potential, wearable devices for health monitoring face several challenges that need to be addressed:

1. Accuracy and Reliability: Factors such as sensor accuracy, data processing algorithms and user variability can affect the reliability of the obtained measurements.
2. Data Privacy and Security: Wearable devices collect sensitive health data, and misusing those data raises concerns about data privacy and security. Ensuring compliance with data protection regulations and implementing robust security measures is essential to protect users' personal health information.

TABLE 10.1
Comparison of the existing wearable system and its features

Product Name	Comfort Index	Usability	Accuracy	Sustainability	Physiological Index	Real Time
Smart Bracelet	More	Very Easy	Less	Yes	Simple	Yes
Smart Watch	More	Very Easy	Less	Yes	Simple	Yes
ECG Monitoring Instrument	Low	Hard	More	Yes	Simple	No
Heart Rate Monitor	Moderate	Easy	More	No	Simple	No
EEG Monitoring Instrument	Less	Hard	More	Yes	Simple	No
Screen Print Wearables	More	Easy	More	Yes	Complex	Yes
Fall Detection Device	Moderate	Easy	Less	No	No	Yes

3. User Adoption and Engagement: Encouraging widespread adoption of and long-term engagement with wearable health monitoring devices can be challenging. Factors such as device comfort, usability and the perceived value of the collected data influence users' willingness to incorporate these devices into their daily routines.
4. Integration with Healthcare Systems: Integrating wearable device data into existing healthcare systems and electronic health records (EHRs) presents technical and interoperability challenges. Seamless data exchange between wearable devices and healthcare providers' systems is essential for leveraging the full potential of these devices in clinical settings.[24,25]
5. Power Requirement: As these are electronic devices, they need an active power supply for their operation. But due to its small size, the battery size also gets reduced. That means there will be power limitations, so energy harvesting for these devices also deserves serious consideration. There are many ways to harvest energy such as solar, thermal, body movement and many more.[26,27]
6. Regulatory and Ethical Considerations: Regulatory frameworks must keep pace with advancements in wearable technology to ensure patient safety and data integrity. Clear guidelines for device approval, data privacy and ethical use of health data are essential to foster innovation while safeguarding patient interests. Collaboration among regulators, industry stakeholders and healthcare professionals is necessary to address regulatory gaps and ensure compliance with ethical standards.

Wearable Devices from Concept to Reality

Modelling an electronic wearable device involves multiple steps, from conceptualization to creating detailed designs. Following is a step-by-step guide:

The production of efficient and reliable wearable devices involves the following stages:

a. Conceptualization
   - Identify the primary function and purpose.
   - Identify the target user and their needs.
   - Analyse existing products and identify gaps or opportunities.

b. Requirements Gathering
   - Functional Requirements: List the features the device must have (e.g. heart rate monitoring, GPS, notifications).
   - Non-functional Requirements: Consider battery life, durability, comfort and aesthetic appeal.

c. Design Specifications
   - Hardware Components: Identify sensors, processors, displays, batteries and connectivity modules.
   - Software Components: Define the operating system, user interface and application software.

d. Prototyping
   - Electronic Design: Use CAD software like Eagle, Altium Designer or KiCad.
   - Mechanical Design: Design the housing and physical components using 3D modelling software like SolidWorks, Fusion 360 or Rhino.
   - Integration: Ensure the electronic components fit within the designed housing.

e. Simulation and Testing
   - Simulate Circuit Design: Use tools like LTspice or Multisim to simulate the electronic circuit.
   - Prototype Testing: Build a prototype and test all functionalities. Tools like Arduino or Raspberry Pi can be useful for initial prototyping.
   - User Testing: Conduct usability testing with real users to gather feedback on comfort and functionality.

f. Iterative Design
   - Feedback Integration: Modify the design based on testing feedback.
   - Optimize Components: Ensure all components are optimized for power consumption, performance and size.

g. Manufacturing Preparation
   - Final Design: Finalize the design with all components and dimensions.
   - Bill of Materials (BOM): Prepare a comprehensive list of all components required.
   - Supplier Identification: Find suppliers for components and manufacturing partners.

h. Compliance and Certification
   - Regulatory Compliance: Ensure the device meets regulatory requirements (e.g. FCC, CE).
   - Health Standards: Health monitoring devices must comply with relevant health standards.

i. Production
   - Pilot Production: Start with a small batch to test the production process.
   - Quality Control: Implement stringent quality control measures to ensure consistency and reliability.
   - Mass Production: Scale up production while maintaining quality standards.

j. Launch and Support
   - Marketing: Develop a marketing strategy to promote the device.
   - Customer Support: Set up support channels to assist users with any issues.

Important Considerations

Despite the challenges, wearable devices hold immense potential for transforming healthcare delivery and improving patient outcomes. Several promising developments are shaping the future of wearable health monitoring:

a. Advanced Sensor Technologies: Ongoing advancements in sensor technologies, including miniaturization, improved accuracy and new sensing modalities, will enhance the capabilities of wearable devices.
b. Artificial Intelligence (AI) and Machine Learning (ML): The integration of AI and ML algorithms enables wearable devices to analyse complex health data and provide personalized insights and recommendations tailored to individual users' needs, and to suggest primary treatment as per AI and ML.[28]
c. Hybrid Wearable Systems: Future wearable devices may combine multiple sensors and modalities, such as optical, biochemical and mechanical sensors, as well as some dedicated communication modules to provide a more comprehensive view of an individual's health status. Also, these wearable devices will inform the nearest hospital about the emergency medical services urgently needed.[29]
d. Healthcare Ecosystem Integration: Greater collaboration between wearable device manufacturers, healthcare providers and researchers will facilitate seamless integration of wearable device data into clinical workflows, leading to more personalized and effective healthcare interventions. The integration of electronic wearable devices into gold, platinum and silver jewellery has been a growing trend, combining fashion with functionality. These high-end wearables not only serve aesthetic purposes but also offer various technological benefits such as health monitoring, communication and personal security. Some notable examples of such devices available in market are listed below:

1. Oura Ring
   - The Oura Ring is available in several luxurious finishes, including silver, black and gold. It uses advanced sensors and provides users with detailed insights into their overall health and well-being.

2. Bellabeat Leaf
   - The Bellabeat Leaf comes in leaf-like shape in various finishes, including silver and rose gold. It tracks activity, sleep, stress levels and menstrual cycles. It also provides breathing exercises to help manage stress.

3. Motiv Ring
   - The Motiv Ring is available in finishes such as rose gold and silver. This smart ring tracks fitness metrics like steps, calories burned, distance, heart rate and sleep. It also includes security features like two-factor authentication and account access.

4. Helena Smart Necklace by Wisewear
   - The Helena Smart Necklace is available in finishes like 18k gold and palladium. This necklace includes fitness tracking, notifications and a distress messaging feature that sends your location to preselected contacts in an emergency.

5. Gucci x Oura Ring Collaboration
   - Gucci has collaborated with Oura to create a luxury version of the Oura Ring. It offers the same health-tracking capabilities as the standard Oura Ring, including sleep analysis, heart rate monitoring and activity tracking

6. Louis Vuitton Tambour Horizon Light Up

This smartwatch from Louis Vuitton features a variety of luxury materials, including cases in stainless steel and finishes in gold and platinum. The Tambour Horizon Light Up offers typical smartwatch functionalities such as notifications, fitness tracking and custom watch faces. It also integrates with the Louis Vuitton travel app for a seamless luxury travel experience.

7. Misfit Shine

The Misfit Shine, while not exclusively high-end, offers versions that can be paired with accessories made of premium materials like sterling silver and gold. This fitness tracker monitors activities such as walking, running, swimming, cycling and sleep patterns.

10.8 FUSION OF USEFUL WEARABLE DEVICES WITH ORNAMENTS

The integration of wearable technology into luxurious jewellery exemplifies the convergence of fashion and functionality. These devices offer the benefits of health monitoring, connectivity and personal security while maintaining the aesthetic appeal of traditional jewellery. As technology advances, we can expect even more innovative designs that cater to both tech enthusiasts and fashion-conscious consumers. Some of the key benefits are aesthetic appeal, durability and longevity, and hypoallergenic and antimicrobial properties. Further, this type of fusion enhances value proposition, discreet and seamless integration, and personalization and arouses sentimental value. In short, it can be concluded that as technology continues to evolve, the fusion of precious metals and wearable devices will likely become even more sophisticated, offering new opportunities for innovation and personalization in the wearable tech industry.

Radiation hazards of wearable devices:
Despite their advantages, there is growing concern about the potential radiation hazards associated with these devices. This section explores the types of radiation emitted by wearable devices, the potential health risks, and the ongoing debate within the scientific community.[30]

Types of Radiation
Wearable devices emit primarily non-ionizing radiation, which includes radiofrequency (RF) radiation and, in some cases, extremely low frequency (ELF) radiation. RF radiation is emitted by devices that communicate wirelessly, such as Bluetooth-enabled gadgets and smartwatches. ELF radiation, in contrast, can be emitted by the electrical circuits within these devices.[31]

Potential Health Risks
The primary concern with non-ionizing radiation is its potential to cause thermal and non-thermal biological effects. While non-ionizing radiation does not have enough energy to ionize atoms or molecules, it can still cause biological changes through other mechanisms.[32]

Thermal Effects: These effects result from the heating of tissues due to the absorption of RF radiation. Prolonged exposure to high levels of RF radiation can lead to tissue damage, but wearable devices typically emit radiation at levels considered too low to cause significant heating. Regulatory bodies, such as the US Federal Communications Commission (FCC), have established exposure limits to protect against thermal effects.

Non-Thermal Effects: There is ongoing debate about the non-thermal biological effects of RF radiation. Some studies have suggested potential links between long-term exposure to low-level RF radiation and adverse health outcomes, including an increased risk of cancer, neurological disorders and reproductive issues. However, these findings are not universally accepted, and many studies have found no conclusive evidence of harm.

Electromagnetic Hypersensitivity (EHS): A small subset of the population reports experiencing symptoms such as headaches, fatigue and dizziness when exposed to electromagnetic fields (EMFs) from devices, including wearables. While EHS is not recognized as a medical diagnosis by all health authorities, the symptoms are real and can significantly impact the quality of life for those affected.

Regulatory Standards and Guidelines:
To mitigate the potential risks, various international and national bodies have established guidelines and standards for the safe use of wearable devices. For instance, the International Commission on Non-Ionizing Radiation Protection (ICNIRP) and the FCC have set limits on the specific absorption rate (SAR), which measures the rate at which the body absorbs RF energy. Manufacturers of wearable devices must ensure their products comply with these standards to minimize potential health risks.[33]

The Scientific Debate:
The scientific community remains divided over the potential health hazards of non-ionizing radiation from wearable devices. Some researchers argue that the current evidence is insufficient to establish a clear link between RF radiation and adverse health effects, suggesting that the exposure levels from wearables are too low to cause harm. Others call for more rigorous long-term studies, citing the precautionary principle, which advocates for preventive action in the face of uncertain risks.

Wearable devices offer significant benefits, but concerns about the potential radiation hazards warrant careful consideration. While current regulations aim to protect users from thermal effects, the debate over non-thermal effects and long-term exposure continues. It is essential for ongoing research to address these uncertainties and for regulatory bodies to adapt their guidelines as new evidence emerges. In the meantime, consumers can take precautionary measures, such as limiting prolonged use and maintaining a safe distance from their devices when not in use, to mitigate potential risks.[34]

While wearable electronic devices are generally considered safe when used as intended, there have been instances where concerns about their potential health effects have arisen. These cases often involve long-term exposure or specific circumstances that may exacerbate the effects of electromagnetic radiation or other factors associated with these devices. Below are some significant examples involving humans and animals.

Examples Involving Humans:
Smartwatches and Skin Reactions: There have been reports of skin irritation, rashes and burns associated with prolonged use of smartwatches. For instance, in 2014, some users of the Fitbit Force reported experiencing skin rashes and irritation, which led to the recall of the product. This was attributed to the nickel and other materials used in the device's construction.

Bluetooth Headsets and Earbuds: Some individuals have reported experiencing headaches, dizziness and discomfort after prolonged use of Bluetooth headsets and earbuds.

Fitness Trackers and Sleep Disruption: There have been anecdotal reports of sleep disturbances related to the use of fitness trackers that monitor sleep patterns. The psychological impact of constant health monitoring can lead to stress and anxiety, which may indirectly affect sleep and overall well-being.

Examples Involving Animals:
RF Radiation Studies on Rats and Mice: National Toxicology Program (NTP) Study: A significant study conducted by the NTP in the United States involved exposing rats

and mice to high levels of RF radiation similar to that emitted by cell phones. The study found a potential association between high exposure levels and the development of tumours in male rats.

Implications: While the levels of radiation in this study were much higher than those emitted by wearable devices, the findings have raised concerns about long-term exposure and the need for further research.

Behavioural Changes in Animals:

Observation: Some studies have observed behavioural changes in animals exposed to RF radiation. For example, a study involving rodents found that prolonged exposure to RF fields could affect their cognitive functions and behaviour.

Scientific Debate: These studies often involve exposure levels higher than those typically encountered by wearable device users, and their relevance to human health is still under debate.

10.9 CONCLUSION

Wearable devices have the potential to revolutionize healthcare by providing continuous monitoring, early detection of health issues, and personalized insights into individuals' health status. From fitness trackers to medical wearables, these devices offer a wide range of applications in disease management, remote patient monitoring, fitness tracking and early intervention. However, addressing challenges such as accuracy, data privacy, user engagement and healthcare system integration is crucial to realizing the full potential of wearable devices in healthcare. Addressing these challenges requires collaboration among technology developers, healthcare providers, policymakers, regulators and users themselves. By working together, stakeholders can overcome barriers to adoption and harness the full potential of wearable technology to improve healthcare delivery, enhance patient outcomes and promote population health. At the same time, it is to be kept in mind that technology comes with some costs. Radiation hazards cannot be neglected; potential biological effects warrant further investigation. Regulatory bodies, manufacturers and consumers must remain aware of potential issues and continue to prioritize safety in the design and use of wearable technology.

Open research challenges in wearable devices include improving battery life to extend usage without frequent charging, enhancing data accuracy and sensor reliability, and ensuring data privacy and security to protect sensitive user information. Additionally, challenges involve advancing materials for greater comfort and durability, developing seamless user interfaces, and integrating more sophisticated health monitoring capabilities. Facilitating interoperability with other devices and systems, managing data overload, and addressing ethical considerations related to data use and consent also remain significant areas of research. Overcoming these challenges is crucial for the widespread adoption and effectiveness of wearable technology. Advances in telecommunication, especially the deployment of 5G networks, significantly influence the growth of wearable devices by providing faster, more reliable connectivity and lower latency. This enhances the real-time data exchange capabilities of wearables, enabling more responsive health monitoring, better fitness tracking, and improved user experiences in applications like augmented reality. Enhanced connectivity also supports the development of more sophisticated features and integration with other smart devices. Additionally, advancements in telecommunication infrastructure facilitate the expansion of IoT ecosystems, promoting seamless interconnectivity and driving innovation in wearable technology.

With ongoing advancements in sensor technologies, artificial intelligence and healthcare ecosystem integration, wearable devices are poised to play an increasingly significant role in shaping the future of healthcare delivery and improving patient outcomes, which opens a huge amount of investment not only in products but also in research and development.

REFERENCES

1. Iqbal, S.M.A., Mahgoub, I., Du, E., Leavitt, M.A., Asghar, W. (2021) Advances in Healthcare Wearable Devices, *Npj Flexible Electronics*, 5, 1–14. DOI 10.1038/s41528-021-00107-x.
2. Godfrey, A., Hetherington, V., Shum, H., Bonato, P., Lovell, N.H., Stuart, S. (2018) From A to Z: Wearable Technology Explained, *Maturitas*, 113, 40–47. DOI 10.1016/j.maturitas.2018.04.012.
3. Kaushal, J.B., Raut, P., Kumar, S. (2023) Organic Electronics in Biosensing: A Promising Frontier for Medical and Environmental Applications, *Biosensors*, 13, 976. DOI 10.3390/bios13110976.
4. Nag, A., Mukhopadhyay, S.C., Kosel, J. (2017) Wearable Flexible Sensors: A Review, *IEEE Sensors Journal*, 17, 3949–3960. DOI 10.1109/JSEN.2017.2705700.
5. Cruz, S.M.F., Rocha, L.A., Vian, J.C. (2018) Printing Technologies on Flexible Substrates for Printed Electronics, Flexible Electronics, *InTech.* DOI 10.5772/intechopen.76161.
6. Yao, X., Zhang, Y., Jin, W., Hu, Y., Cui, Y. (2021) Carbon Nanotube Field-Effect Transistor-Based Chemical and Biological Sensors, *Sensors*, 21, 995. DOI 10.3390/s21030995.
7. Yu, K., Rich, S., Lee, S., Fukuda, K., Yokota, T., Someya, T. (2019) Organic Photovoltaics: Toward Self-Powered Wearable Electronics, *IEEE*, 107, 2137–2154. DOI 10.1109/JPROC.2019.2929797.
8. Deng, Z., Guo, L., Chen, X., Wu, W. (2023) Smart Wearable Systems for Health Monitoring, *Sensors*, 23, 2479. DOI 10.3390/s23052479.
9. Haghi, M., Thurow, K., Stoll, R. (2017) Wearable Devices in Medical Internet of Things: Scientific Research and Commercially Available Devices, *Healthc Inform*, 23, 4–15. DOI 10.4258/hir.2017.23.1.4.
10. Shah, S.H.A., Koundal, D., Sai, V., Rani, S. (2022) 5G Edge computing enabled Internet of Medical Things, *IEEE Transactions on Industrial Informatics*, 18, 1–3. DOI 10.1109/TII.2022.3193708.
11. Amjad, A., Kordel, P., Fernandes, G. (2023) A Review on Innovation in Healthcare Sector (Telehealth) through Artificial Intelligence, *Sustainability*, 15, 6655. DOI 10.3390/su15086655.
12. Masoumian Hosseini, M., Masoumian Hosseini, S.T., Qayumi, K., Hosseinzadeh, S., Sajadi Tabar, S.S. (2023) Smartwatches in Healthcare Medicine: Assistance and Monitoring; A Scoping Review, *BMC Medical Informatics and Decision Making*, 23, 248. DOI 10.1186/s12911-023- 02350-w.
13. Hrabovska, N., Kajati, E., Zolotova, I. (2023) A Validation Study to Confirm the Accuracy of Wearable Devices Based on Health Data Analysis, *Electronics*, 12, 2536. DOI 10.3390/electronics12112536.

14. Secara, I.A., Hordiiuk, D. (2024) Personalized Health Monitoring Systems, Integrating Wearable and AI. *Journal of Intelligent Learning Systems and Applications*, 16, 44–52. DOI 10.4236/jilsa.2024.162004.
15. Haghi, M., Danyali, S., Ayasseh, S., Wang, J., Aazami, R., Deserno, T.M. (2021) Wearable Devices in Health Monitoring from the Environmental towards Multiple Domains: A Survey, *Sensors*, 21, 2130. DOI 10.3390/s21062130.
16. Bayoumy, K., Gaber, M., Elshafeey, A. (2021) Smart Wearable Devices in Cardiovascular Care: Where We are and How to Move Forward, *Nature Reviews Cardiology*, 18, 581–599. DOI 10.1038/s41569-021-00522-7.
17. Mukhopadhyay, S.C. (2015) Wearable Sensors for Human Activity Monitoring: A Review, *IEEE Sensors Journal*, 15, 1321–1330. DOI 10.1109/JSEN.2014.2370945.
18. Baker, S.B., Xiang, W., Atkinson I. (2017) Internet of Things for Smart Healthcare: Technologies, Challenges, and Opportunities, *IEEE Access*, 5, 26521–26544. DOI 10.1109/ACCESS.2017.2775180.
19. Gravina, R., Fortino G. (2021) Wearable Body Sensor Networks: State-of-the-Art and Research Directions, *IEEE Sensors Journal*, 21, 12511–12522. DOI 10.1109/JSEN.2020.3044447.
20. Yang, G., Pang, G., Pang, Z., Gu, Y., Mäntysalo, M., Yang, H. (2019) Non-Invasive Flexible and Stretchable Wearable Sensors with Nano-Based Enhancement for Chronic Disease Care, *IEEE Reviews in Biomedical Engineering*, 12, 34–71. DOI 10.1109/RBME.2018.2887301.
21. Ma, L.-Y., Soin, N. (2022) Recent Progress in Printed Physical Sensing Electronics for Wearable Health-Monitoring Devices: A Review, *IEEE Sensors Journal*, 22, 3844–3859. DOI 10.1109/JSEN.2022.3142328.
22. Yao, K., Yang, Y., Wu, P., Zhao, G., Wang, L., Yu, X. (2023) Recent Advances in Materials, Designs and Applications of Skin Electronics, *IEEE Open Journal of Nanotechnology*, 4, 55–70. DOI 10.1109/OJNANO.2022.3218960.
23. Putra, K.T. (2024) A Review on the Application of Internet of Medical Things in Wearable Personal Health Monitoring: A Cloud-Edge Artificial Intelligence Approach, *IEEE Access*, 12, 21437–21452. DOI 10.1109/ACCESS.2024.3358827.
24. Zheng, M., Liu, P.X., Gravina R., Fortino, G. (2018) An Emerging Wearable World: New Gadgetry Produces a Rising Tide of Changes and Challenges, *IEEE Systems, Man, and Cybernetics Magazine*, 4, 6–14. DOI 10.1109/MSMC.2018.2806565.
25. Meng, W., Cai, Y., Yang, L.T., Chiu, W.-Y. (2021) Hybrid Emotionaware Monitoring System Based on Brainwaves for Internet of Medical Things, *IEEE Internet Things Journal*, 8, 16014–16022. DOI 10.1109/JIOT.2021.3079461.
26. Mayer, P., Magno, M., Benini, L. (2021) Energy-Positive Activity Recognition – From Kinetic Energy Harvesting to Smart Self-Sustainable Wearable Devices, *IEEE Transactions on Biomedical Circuits and Systems*, 15, 5, 926–937. DOI 10.1109/TBCAS.2021.3115178.
27. Chong, Y.-W., Ismail, W., Ko, K., Lee, C.-Y. (2019) Energy Harvesting for Wearable Devices: A Review, *IEEE Sensors Journal*, 19, 9047–9062. DOI 10.1109/JSEN.2019.2925638.
28. Zang, Y., Zhang, F., Di, C., Zhu, D. (2015) Advances of Flexible Pressure Sensors Toward Artificial Intelligence and Health Care Applications, *Materials Horizons*, 2, 140–156. DOI 10.1039/C4MH00147H.
29. Liang T., Yuan, Y.J. (2016) Wearable Medical Monitoring Systems Based on Wireless Networks: A Review, *IEEE Sensors Journal*, 16, 8186–8199. DOI 10.1109/JSEN.2016.2597312.

30. Vahidnia, R., Dian, F.J. (2020) Radiation Safety Hazards of Cellular IoT Devices, 2020 11th IEEE Annual Information Technology, Electronics and Mobile Communication Conference (IEMCON), Vancouver, BC, Canada, pp. 0116–0120, DOI 10.1109/IEMCON51383.2020.9284868.
31. Tsantarliotis, P., Tsipouras, M.G., Giannakeas, N. (2018) Personalized UV Radiation Risk Monitoring Using Wearable Devices and Fuzzy Modeling, *Inventions*, 3, 26. DOI 10.3390/inventions3020026.
32. Prayitno, Shyu, C.R., Putra, K.T., Chen, H.C., Tsai, Y.Y., Hossain, K.S.M.T., Jiang, W., Shae, Z-Y. (2021) A Systematic Review of Federated Learning in the Healthcare Area: From the Perspective of Data Properties and Applications, *Application Science*, 11, 11191. DOI 10.3390/app112311191.
33. Brönneke, J.B., Müller, J., Mouratis, K., Hagen, J., Stern, A.D. (2021) Regulatory, Legal, and Market Aspects of Smart Wearables for Cardiac Monitoring, *Sensors*, 21, 4937. DOI 10.3390/s21144937.
34. Naren, Elhence, A., Chamola, V., Guizani, M. (2020) Electromagnetic Radiation due to Cellular, Wi-Fi and Bluetooth Technologies: How Safe Care We? *IEEE Access*, 8, 82980– 83000. DOI 10.1109/ACCESS.2020.2976434.

11 Electrochemical and Fluorescent Sensors for Healthcare Applications

Vikash Verma and Ashish Pathak

11.1 INTRODUCTION

This chapter gives insight onto the following topics:

A. Fabrication Aspects of Electrochemical and Fluorescent Sensors:
Nanomaterial integration: Incorporating nanomaterials such as carbon nanotubes or graphene into sensor fabrication enhances surface area, sensitivity, and selectivity, improving the sensor's performance.

1. Bio-functionalisation techniques: Utilising biomolecules like enzymes or antibodies allows for specific recognition of target analytes, enabling highly sensitive as well as selective detection in complex biological specimens.

B. Uses of Electrochemical Sensors for Contact Lenses:

1. Glucose monitoring: By incorporating electrochemical sensors into contact lenses, glucose levels in tear fluid can be continuously monitored, providing a non-invasive method for managing diabetes.
2. Drug release monitoring: Contact lenses equipped with electrochemical sensors can track drug release into the eye, enabling personalised treatment for conditions such as glaucoma or ocular infections.

C. Uses of Electrochemical Sensors for Cardiovascular Diseases:

1. Biomarker detection: Electrochemical sensors are used to detect cardiac biomarkers like troponin and B-type natriuretic peptide (BNP), assisting in the early diagnosis of myocardial infarction or heart failure.
2. Wearable cardiac monitoring devices: Miniaturised electrochemical sensors integrated into wearable devices enable continuous monitoring of biomarkers like lactate or pH in sweat, providing insights into cardiovascular health and performance.

DOI: 10.1201/9781003453635-11

D. Uses of Fluorescent Sensors for Glutathione Detection:

1. Oxidative stress assessment: Fluorescent sensors are utilised for the detection of glutathione, a key antioxidant, providing insights into cellular oxidative stress levels associated with various diseases like neurodegenerative disorders or cancer.
2. Environmental monitoring: Fluorescent sensors for glutathione detection are employed in environmental studies to assess pollutant-induced oxidative stress in organisms, contributing to ecological risk assessment.

E. Uses of Fluorescent Sensors for Proactive Prognosis of Covid-19:

1. Viral load quantification: Fluorescent sensors designed to detect viral RNA or proteins enable quick and sensitive measurement of SARS-CoV-2 viral load in patient samples, aiding in the prognosis of the disease and the monitoring of treatment.
2. Antigen detection: Fluorescent sensors are used to detect viral antigens in point-of-care testing, enabling early diagnosis and proactive management of Covid-19 outbreaks in communities.

F. Challenges for Designing Electrochemical and Fluorescent Sensors:

1. Sensitivity and selectivity optimisation: Achieving high sensitivity and selectivity while minimising interference from complex biological matrices remains a challenge in sensor design.
2. Stability and reproducibility: Ensuring long-term stability and reproducibility of sensor responses, particularly in real-world applications, requires robust materials and fabrication techniques to mitigate degradation and variability.

11.2 FABRICATION ASPECTS OF ELECTROCHEMICAL AND FLUORESCENT SENSORS

Electrochemical and fluorescent sensors play a crucial role in various healthcare applications, including diagnostics, biomarker monitoring, and disease state monitoring. When it comes to fabrication, choosing the appropriate electrode material is vital for the sensor's performance. Common electrode materials for electrochemical sensors include gold, platinum, and carbon-based materials such as *carbon nanotubes (CNT)* and graphene. These materials are known for their high conductivity, stability, and biocompatibility.[1,2]

Fabrication Aspects of Electrochemical Sensors:

1. **Fabrication of Electrodes**

   Material selection is the process of selecting the right electrode material, such as carbon nanotubes, gold, platinum or glassy carbon, based on stability, conductivity, and compatibility with the intended analyte. *Methods of Preparation*: Electrodes with the required surface properties and morphology are made using fabrication techniques such as electro-deposition, physical vapour deposition and screen printing.

2. **Surface Alteration**
   Functionalisation is the process of immobilising recognition elements (such as enzymes, antibodies or DNA probes) on the electrode surface by means of covalent attachment, physical adsorption or self-assembly. Surface engineering is the process of adding nanomaterials (such as nanowires or nanoparticles) to the electrode surface in order to increase its surface area, electron transfer kinetics and analyte binding affinity.
3. **Packaging and Encapsulation**
   Protection: To guard against fouling, chemical interference, and mechanical damage, the sensor components in biocompatible polymers or membranes are enclosed. Miniaturisation involves developing small, lightweight sensor platforms for point-of-care applications, often incorporating lab-on-a-chip or microfluidic systems.[1]
4. **Definition and Enhancement**
   Performance testing involves determining how well sensors function under various experimental settings, including sensitivity, selectivity, dynamic range, and response time.

Optimisation: To improve sensitivity, stability and repeatability, fabrication procedures, and sensor design are iteratively refined based on experimental data.

Fabrication Aspects of Fluorescent Sensors:

1. **Fluorophore Selection**
   This involves selecting fluorophore with desired emission wavelengths, quantum yields, photo stability, and sensitivity to environmental changes, which are all considered spectral properties.
2. **Conjugation of Recognition Elements:**
   This consists of covalently attaching fluorophores to recognition elements (such as molecular probes, aptamers or antibodies) in order to identify particular targets.[3]
3. **Transmission of Signals**
   Fluorescence modulation is the process of designing sensors in response to changes in analyte binding that occur in fluorescence intensity, lifetime or wavelength. These changes can be caused by quenching/enhancement or fluorescence resonance energy transfer (FRET).

   Signal Amplification: Sensor sensitivity is increased by utilising signal amplification techniques (such as enzyme-catalysed reactions and nanoparticle-based enhancements).[1–13]
4. **Detection and Set-Up for Optics**
   Setting up optical systems with the proper light sources, filters, detectors, and optics for the excitation and detection of fluorescence signals is known as the 'excitation and emission setup'. Establishing calibration curves and standard operating

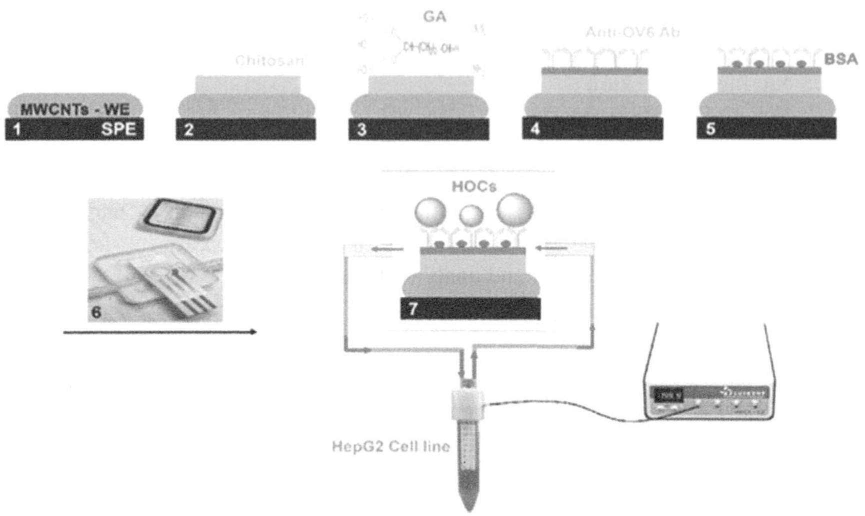

FIGURE 11.1 Electrochemical biosensor for detecting hepatic oval cells.[6]

procedures for the quantitative analysis of target analytes based on fluorescence measurements is known as calibration and standardisation.[4] Finally, a multidisciplinary approach involving materials science, surface chemistry, bioconjugation techniques, transducer integration, and signal processing is required for the fabrication of both electrochemical and fluorescent sensors in order to develop reliable and sensitive sensing platforms for a range of healthcare applications.

An embedded disposable functionalized electrochemical biosensor with a 3D-printed flow cell was developed for the detection of hepatic oval cells (HOCs), as illustrated in Figure 11.1.

11.3 USES OF ELECTROCHEMICAL SENSORS FOR CONTACT LENSES

Cutting-edge works and uses of electrochemical sensors for contact lenses are being discovered. From monitoring glucose levels to detecting pathogens, these sensors offer a new technicality in wearable technology. The advantages, applications, challenges, and latest advancements in this exciting field are being explored. Advantages of using electrochemical sensors for contact lenses are *Real-Time Monitoring*, which is the continuous monitoring of analyte levels that allows for timely intervention and better management of health conditions, and non-invasive and convenient which means the integration of sensors into contact lenses prevents the need for additional devices or invasive procedures, and electrochemical sensors improve accuracy, high sensitivity, and selectivity and provide accurate and reliable measurements. Its potential and applications are seen in many research studies, such as on health monitoring, continuous monitoring of glucose, lactate, and pH levels, and biomarkers in tears that enables early detection of diseases such as diabetes, ocular disorders, and drug delivery. Electrochemical sensors in contact lenses can facilitate controlled release of drugs, improving treatment effectiveness and patient compliance; environmental monitoring and detection of pollutants and toxic substances in tears can provide insights into outside exposure and assist in clarifying the health risks.[7]

Despite their advantages and uses, electrochemical sensors face many challenges such as integrity, complexity, stability, power management, and interpretation of data. Overcoming these limitations requires effort through collaboration and further research. The latest advancements and ongoing research in the field miniaturisation focuses on *developing smaller and more compact sensors, enhancing comfort and wearability*. Biosensing technology advancements in nanotechnology and bioengineering enable the creation of sensors with increased sensitivity and specificity. Smartphone integration means integrating sensors with smartphones allows for convenient data collection, data processing, and interpretation, enhancing usability and availability.[8] *Hydrogen-based contact* lenses are changing radically the world of vision correction, and these lenses have gained immense popularity due to their extraordinary comfort. These lenses are made from a water-absorbing polymer that creates a gel-like substance which imitates closely the natural moisture of the eye. They provide excellent vision correction, allowing persons to enjoy clear and crisp vision throughout the day without discomfort. The most-used sensor on contact lens is the *glucose sensor* employed for diabetes management.[9] Various techniques, including near-infrared glucose-sensitive photonic crystals, photonic microstructures, and fluorescent sensors, have been utilised in contact lenses to measure tear glucose. *Electrochemical biosensors* for glucose detection rely on highly selective enzymes such as glucose oxidase (GOx) and glucose dehydrogenase. *MET-based bioelectrodes* offer more reliable glucose sensors due to their increased tolerance to oxygen depletion and interference. Contact lenses are commonly used as ocular prosthetics or biomedical devices with multiple functionalities. Various electrochemical scanning probe techniques, including *scanning ion-conductance microscopy (SICM), scanning electrochemical microscopy (SECM), and SECM combined with atomic force microscopy (SECM-AFM)*, have attracted significant interest.[10] Glucose detection using a soft, smart contact lens is illustrated in Figure 11.2.

As the field of electrochemical sensors for contact lenses continues to evolve, emerging trends include personalised medicine, remote patient monitoring, and the integration of artificial intelligence. The future holds vast potential for improving healthcare outcomes through this remarkable technology. The idea of using hybrid materials in stretchable electronics has been explored previously. However, since heterogeneous materials possess different refractive indices (n), utilising hybrid substrates could significantly affect the optical properties of the resulting films, leading to low transparency and increased haze. For contact lenses, it is essential to maintain both high clarity and minimal haze to ensure they do not obstruct the wearer's vision. Electrochemical sensors integrated into contact lenses revolutionise healthcare by offering real-time, non-invasive monitoring of various analytics. The advantages, applications, and ongoing research demonstrate the tremendous potential of this technology in improving health outcomes. The future development of this field will undoubtedly transform the way we care for our eyes and overall well-being.

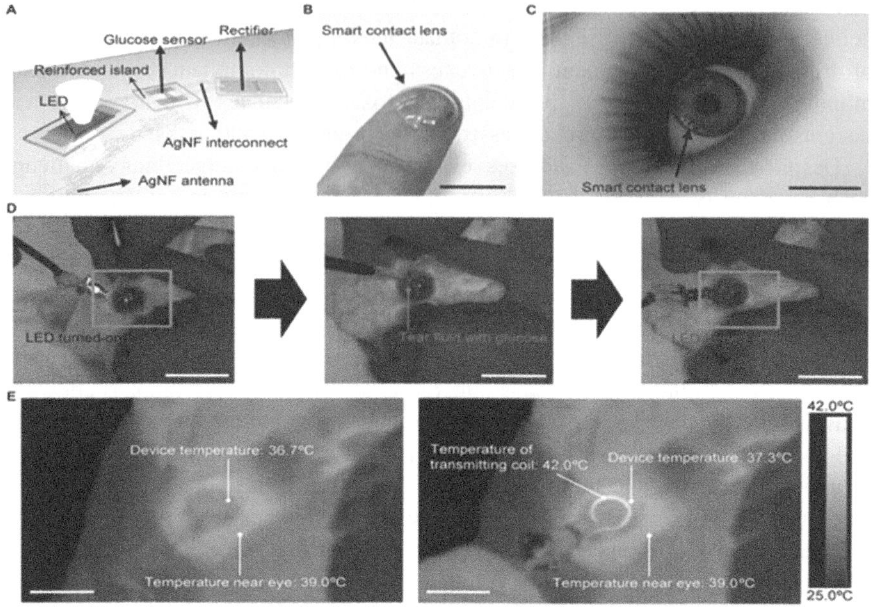

FIGURE 11.2 Glucose detection using soft, smart contact lens. (a) A schematic illustration of the soft, smart contact lens, (b) A photograph of the fabricated soft, smart contact lens, with a scale bar of 1 cm, (c) An image of the smart contact lens placed on a mannequin's eye, with a scale bar of 1 cm, (d) Photographs depicting an in vivo test on a live rabbit using the soft, smart contact lens. Left: LED is on when the lens is mounted on the rabbit's eye. Middle: Injection of tear fluids with a glucose concentration of 0.9 mM. Right: LED turns off after detecting the increased glucose concentration. Scale bars are 1 cm. € Heat tests conducted while a live rabbit wears the operating soft, smart contact lens, with scale bars of 1 cm.[30]

11.4 USES OF ELECTROCHEMICAL SENSORS FOR CARDIOVASCULAR DISEASES

In the process of detecting and monitoring cardiovascular illnesses (also known as CVDs), electrochemical sensors are a very useful tool. These sensors make use of the fundamentals of electrochemistry to identify certain proteins or markers that are related to one's cardiovascular health. The following is a list of some of the applications of electrochemical sensors concerning cardiovascular diseases.

A. **Identification of Biomarkers**
Electrochemical sensors are capable of detecting specific biomarkers associated with cardiovascular diseases, such as cardiac troponins, creatine kinase-MB (CK-MB), and brain natriuretic peptide (BNP). Elevated levels of these biomarkers suggest the presence of cardiac injury or dysfunction.[13]

B. **Testing Done at the Patient's Bedside**
Because of their mobility, quick reaction times, and user-friendliness, electrochemical sensors are an excellent choice for testing at the point of care. They

are versatile enough to be used in medical facilities such as clinics and emergency rooms, as well as in private homes, enabling the rapid and hassle-free monitoring of cardiovascular biomarkers.[13]

C. **Monitoring Glucose Levels for Diabetes Patients**
Diabetes is a major contributor to the increased risk of developing cardiovascular illnesses. Electrochemical sensors are quite common in the field of glucose monitoring. These sensors provide extremely helpful information that may be utilised to better manage diabetes and lower the risk of cardiovascular issues.[14]

D. **Cholesterol Levels Are Determined by**
Electrochemical sensors can measure cholesterol levels, including total cholesterol and specific fractions such as LDL and HDL. Elevated cholesterol levels are associated with a higher risk of atherosclerosis and coronary heart disease.[14]

E. **Ischemia May Be Detected by**
Ischemia is a condition that occurs when there is not enough blood flow going to the tissues in the body. It is a crucial component of cardiovascular illnesses. Electrochemical sensors have the potential to identify biomarkers that are linked with ischemia, hence assisting in the early identification of heart-related conditions.[15]

F. **Monitoring the Platelet Aggregation Process**
Platelet aggregation is an essential step in the production of blood clots, and aberrant platelet activity has been linked to some conditions that affect the cardiovascular system. The risk of thrombosis may be evaluated using electrochemical sensors, which can also be used to monitor the aggregation of platelets.[16]

G. **Monitoring the pH**
Alterations in pH levels may serve as early warning signs for a variety of cardiovascular diseases. Electrochemical sensors can monitor pH fluctuations in the blood as well as in other body fluids, which may provide information about the acid-base balance as well as probable metabolic disorders.[16]

H. **Keeping an Eye on the Levels of Nitric Oxide (NO)**
Nitric oxide is an extremely important component in the process of controlling how blood vessels work. Monitoring NO levels using electrochemical sensors may provide information on the health of the vascular system as well as the function of the endothelial cells.[17]

I. **Monitoring That Is Constantly Performed on Implantable Devices**
Electrochemical sensors are included in implanted devices such as pacemakers and defibrillators for continuous monitoring of physiological parameters. This assists in the management and immediate response to changes in the patient's cardiovascular health.[18]

Research on the cardiovascular system and the discovery of new drugs both make extensive use of electrochemical sensors. They make it possible for researchers to investigate the impact of new drugs on biomarkers or other physiological parameters, which contributes to the development of innovative therapies for cardiovascular illnesses. In summary, electrochemical sensors are widely used in the field of cardiovascular disease, from research and drug development to early detection and condition monitoring. Their high sensitivity and specificity, along with their suitability for point-of-care applications, make them valuable tools in combating cardiovascular diseases.

11.5 USES OF FLUORESCENT SENSORS FOR GLUTATHIONE DETECTION

Detection of *glutathione (GSH)* molecules in biological samples or other matrices is the process of identifying and quantifying them. Three amino acids make up the *tripeptide molecule glutathione: glutamate, cysteine, and glycine*. It is essential for several cellular functions, such as detoxification, antioxidant defence, and cellular redox balance management. Glutathione can be found via a variety of techniques, each having advantages and disadvantages.

Colorimetric Assays: These assays rely on the chemical interaction of glutathione with particular reagents, which produces a colour shift that can be detected using spectrophotometry. In comparison to other techniques, these assays may be less sensitive and specific, but they are comparatively easy to use and inexpensive.

Fluorescence Assays: These assays use fluorescent probes or sensors that bind to glutathione specifically and, as a result, produce a detectable fluorescence signal. Fluorescence detection is a widely used method for detecting glutathione in biological samples because of its great sensitivity and specificity. Quantifying glutathione in complex samples is often done using *high-performance liquid chromatography (HPLC)* combined with ultraviolet (UV) or fluorescence detection. This technique enables accurate quantification by separating glutathione from other sample components based on their chemical properties. *Mass spectrometry (MS)* provides sensitive and precise detection of glutathione by determining the mass-to-charge ratio of glutathione ions in the sample. Techniques such as *liquid chromatography-mass spectrometry (LC-MS)* and *gas chromatography-mass spectrometry (GC-MS)* are commonly used for glutathione analysis in research and clinical laboratories. *Enzyme-linked immunosorbent assay (ELISA)* kits employ specific antibodies that recognise glutathione, allowing for its quantitative measurement in biological samples. ELISA offers a convenient and high-throughput method for detecting glutathione across various sample types. Determining the presence of glutathione is crucial for comprehending its physiological functions, evaluating the degree of oxidative stress and exploring its role in many illnesses and metabolic conditions. Glutathione detection using fluorescent sensors has numerous significant uses in both clinical and research contexts.

Research on Oxidative Stress: An important antioxidant that helps shield cells from oxidative damage is glutathione (GSH). Changes in intracellular GSH levels can be tracked using fluorescent sensors, which can provide information on cellular redox conditions and oxidative stress reactions. With the use of these sensors, scientists can examine the dynamics of GSH metabolism and how it affects a number of physiological and pathological processes.

Drug Development: Changes in GSH levels have been linked to a number of illnesses, such as cancer, neurological conditions, and cardiovascular ailments. When looking for drugs, fluorescent sensors can be used to find substances that either target GSH-related pathways or modify GSH levels.

In the field of biomedical diagnostics, GSH has the potential to be used as a biomarker because abnormalities in its metabolism have been linked to the aetiology of certain disorders. One sensitive and specific way to measure GSH levels in biological samples – like blood, urine, and tissues – is with fluorescent sensors. These sensors can be used in diagnostic assays for GSH-related illnesses to help detect and track their progression early.

Environmental Monitoring: Glutathione aids in the detoxification of xenobiotics and environmental contaminants. GSH may be found in environmental samples using fluorescent sensors, which makes it easier to monitor GSH-dependent detoxification processes in organisms exposed to environmental pollutants and to evaluate environmental pollution.

Food and Nutritional Analysis: A vital component of cellular health, glutathione, can be found in a variety of foods. GSH levels in food products can be quantitatively analysed using fluorescent sensors, allowing for the evaluation of antioxidant and nutritional content. These sensors can also be used to track variations in GSH levels that occur during food preparation and storage, which can yield important data for quality assurance and food safety.

All things considered, fluorescence sensors for glutathione detection provide flexible instruments for researching GSH metabolism, evaluating oxidative stress and identifying illnesses linked to GSH in both applied and research contexts. Their great sensitivity, selectivity and usability make them useful tools in a variety of disciplines, from clinical practice to basic science.

11.6 USES OF FLUORESCENT SENSORS FOR PROACTIVE PROGNOSIS OF COVID-19

In the context of a medical condition or disease, *proactive prognosis* refers to the expectation or forecast of future events or developments. It entails seeing possible problems or hazards before they materialise completely in order to act quickly to stop or lessen unfavourable effects. Proactive prognosis in the context of Covid-19 refers to using different tactics and technology to predict how the disease will progress, identify those who are at risk of infection or serious illness, and put policies in place to reduce effect and transmission. Predictive modelling using epidemiological data, early case detection by testing and surveillance, and focused measures to manage outbreaks and safeguard vulnerable groups are important elements of the Covid-19 proactive prognosis comprise:

Early Detection: This involves finding Covid-19 cases as soon as feasible, including individuals who are asymptomatic or presymptomatic, by using surveillance systems and diagnostic tests. Risk assessment determines a person's likelihood of developing a serious illness or Covid-19 sequelae by looking at factors such age, underlying medical problems and socioeconomic determinants of health. Predictive modelling is the process of forecasting the virus's transmission, projecting future trends in case counts and healthcare demand, and guiding the selection of public health treatments. It involves using mathematical models and data analytic tools.

Measures to prevent the spread of Covid-19 and stop the spread of new infections include immunisation, mask use, physical separation and good hand cleanliness.

Targeted interventions: This involves adapting public health measures, including lockdowns and quarantines, to particular people or environments in response to epidemics, and taking into account risk factors, transmission dynamics, and epidemiological data. In general, proactive estimation is essential to controlling the Covid-19 pandemic because it facilitates early case discovery, directs public health actions and slows the virus's transmission to safeguard people and communities. Through numerous applications, fluorescent sensors can be vital to the proactive prognosis of Covid-19.

Quick Diagnostic Testing: Certain Covid-19 biomarkers, like viral RNA or proteins (like the spike protein), can be found using fluorescent sensors. These sensors can identify the virus quickly and accurately, allowing for a prompt diagnosis even in cases when the patient shows no symptoms, or in places where testing facilities are not easily accessible.

Point-of-Care Testing (POCT): Rapid and decentralised screening is made possible by the integration of fluorescent sensors into portable diagnostic instruments for on-site testing. This can be especially helpful in public places, airports or medical institutions where decisions about isolation and treatment must be made quickly and based on knowledge.

Monitoring and Monitoring: To keep an eye on the virus's prevalence in local populations, wastewater monitoring systems can be equipped with fluorescent sensors. These sensors work by identifying viral RNA fragments excreted in faeces, which enables public health officials to take targeted treatments and provide early warnings of possible outbreaks.

Environmental Monitoring: In high-traffic places like hospitals, schools, and public transportation, fluorescent sensors can be used to identify SARS-CoV-2 particles in the air or on surfaces. Assessing the risk of transmission and putting the right disinfection measures in place can be facilitated by ongoing monitoring of viral presence in the environment.

Testing for Vaccine Efficacy: In research settings, fluorescent sensors can be used to find neutralising antibodies in patient samples, which allows researchers to evaluate the efficacy of Covid-19 vaccines. This can help improve vaccination techniques and offer insightful information on the immunological response elicited by vaccination.

Proactive Screening in High-Risk Environments: Healthcare personnel, the elderly and people with underlying medical issues are examples of high-risk populations that can benefit from proactive screening using fluorescent sensors. Frequent screening with sensitive fluorescence assays can assist in locating carriers who do not exhibit symptoms and stop epidemics in susceptible environments.

Fluorescent sensors, in general, provide a flexible and sensitive platform for proactive Covid-19 prognosis, allowing for quick detection, observation and monitoring of the virus in a range of environments, incorporating these sensors into surveillance and diagnostic systems.

11.7 CHALLENGES FOR DESIGNING ELECTROCHEMICAL AND FLUORESCENT SENSORS

Electrochemical sensors are extensively utilised in industrial, environmental, and medical applications. These methods enable real-time and in situ investigation of various chemical compositions, and electrochemical detection is employed for identifying disease-related biomarkers in biological molecules. Recent advancements in aptasensors hold significant promise. Modern electrochemical aptasensors have the potential to revolutionise the digitisation of biosensors. Aptamer-based biomedical devices integrate various electronic technologies to detect pathogens, biomolecules, small molecules, ions, and the physical-chemical properties of nucleic acids. Fluorescent sensors, used for detecting ions and neutral analytes, find applications in fields such as biology, physiology, pharmacology, and environmental sciences. These fluorescent chemosensors are increasingly effective in identifying biologically and environmentally significant species.

Despite advancements in electrochemical and fluorescent sensors, several challenges remain. This presentation addresses these challenges, particularly those related to the design and development of surface-based biosensors. The multidisciplinary nature of electrochemical biosensors presents challenges in modifying surfaces with capture probes for detection, often using self-assembly methods. Complex surface structures can enhance probe-target interactions, though they typically affect only a small fraction of the overall surface, impacting sensor performance and reliability unpredictably.

Addressing these issues requires the use of imaging techniques across various length scales, such as optical microscopy and scanning probe microscopy, which offer insights into the diversity of surface structures and molecular environments at the sensing interface. In situ analytical methods, relevant to the sensing environment, are crucial. Representations of probe-decorated biosensing surfaces often use simplified cartoons, showing idealised probe-surface interactions and arrangements. However, these illustrations frequently omit numerous probe-surface interactions, leading to heterogeneous surface modifications. These undesired arrangements are usually not part of the sensor design and can affect performance.[14–21]

The advancement of medical diagnostic technology has significantly improved wearable biosensors. These sensors provide continuous, real-time physiological information through dynamic, non-invasive measurements of biochemicals in biofluids when in direct contact with the sampled fluids, without causing discomfort to the wearer. The rapid development of wearable chemical sensors facilitates the non-invasive detection of analytes in accessible biofluids, offering new insights into the overall dynamic biomolecular state of the human body.

Although there are many types of biofluids, they often present limitations for wearable sensing. Blood is a prime option for medical monitoring of analytes, but invasive sampling methods pose significant challenges, particularly for newborns, elderly individuals and those with a fear of needles. While urine is a valuable clinical sample, it is unsuitable for continuous monitoring. Human tears contain specific salts, enzymes, proteins, and lipids that can reveal eye diseases; however, collecting tear samples can cause irritation and reflex tears, affecting the sensor's accuracy.

Monitoring saliva also poses challenges due to impurities like food particles that can impact data reliability. The concentration of analytes in tissue fluids is similar to that in blood, making them important for monitoring, but fine needles or subcutaneous excitation currents used for sample collection can irritate the skin and cause discomfort.

Recent advancements in instrumentation and camera sensitivity have enhanced the molecular specificity of spectroscopy, especially when combined with imaging capabilities. Spatial resolution is limited by diffraction, approximately the wavelength of light used. IR microscopy applied to electrode surfaces using a synchrotron source achieves a spatial resolution of 10 μm. *Atomic force microscopy (AFM)* coupled with infrared (IR) (nano IR or near-field IR) overcomes the diffraction limit, providing a tool for molecular characterisation of biosensor surfaces. However, the primary limitation of nano IR is its poor compatibility with liquid environments, which prevents in situ measurements.

Fluorescent chemosensors are compounds that incorporate a binding site, a fluorophore and a mechanism for communication between the two sites. The structural representation of fluorescent chemosensors is illustrated in Figure 11.3.

Initially, most fluorescent chemosensors relied on coordination interactions between hosts and guests. For instance, two naphthalene-based chemosensors, 1 and 2, are used for detecting alkali metal ions. When these sensors form a complex with alkali metal chloride salts in 95% ethanol glass at 77 K, sensor 1 exhibits a decrease in fluorescence quantum yield and an increase in phosphorescence quantum yield, with a slight decrease in phosphorescence lifetime. Conversely, when sensor 2 forms a complex with potassium (K^+), rubidium (Rb^+), or cesium (Cs^+) chloride salts, it shows a significant increase in fluorescence quantum yield, a decrease in phosphorescence

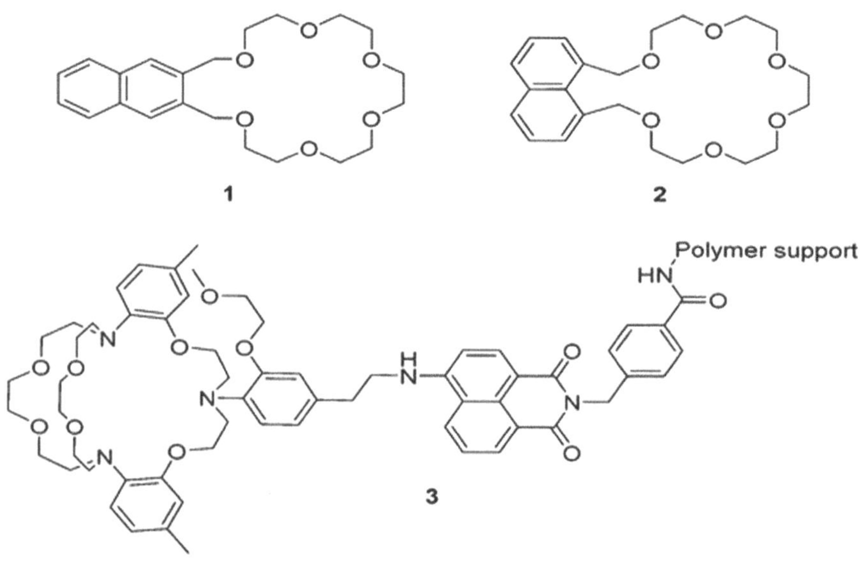

FIGURE 11.3 Structural representation of fluorescent chemosensors.[22]

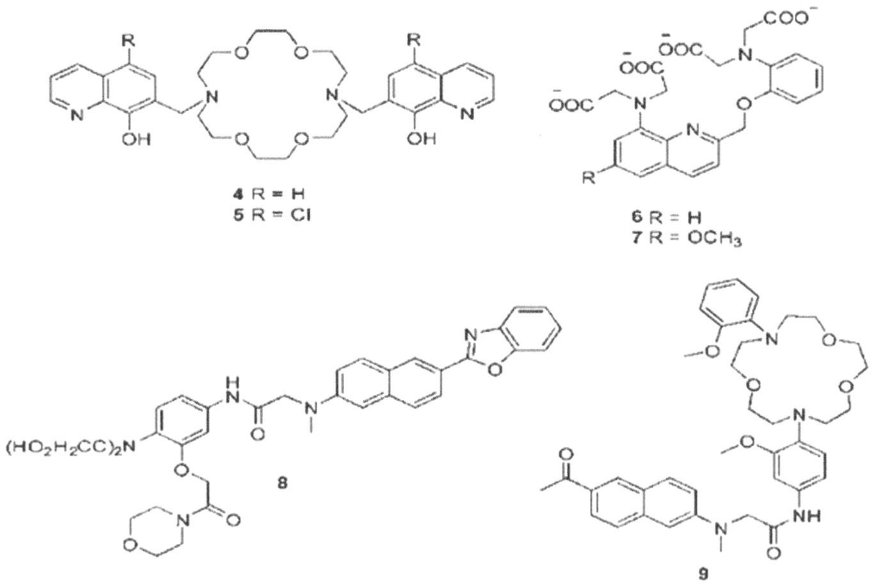

FIGURE 11.4 Structural representation of fluorescent chemosensors.[22]

quantum yield, and a substantial decrease in phosphorescence lifetime. These changes can be attributed to the heavy atom effect (for Rb^+ and Cs^+), a complexation-induced shift in triplet energy relative to the ground, and excited singlet state energies as well as rigidification and conformational effects. Figure 11.4 provides additional structural variations of fluorescent chemosensors.

Chemically and genetically encoded sensors were originally designed for rapid signalling processes such as *calcium spiking*. In contrast, dynamic changes for other ions, metabolites, and hormones occur at much slower rates. The steady-state level of a compound is influenced by multiple rate constants, which typically include the rates of uptake and efflux from a cell, rates of synthesis and degradation, buffering capacity of compartments, and rates of import into or release from subcellular compartments. While individual ion channels for calcium can conduct up to hundreds of millions of units per second, transporters for metabolites move only a few units per second. When measuring analyte levels, it is important to consider that the rate of uptake may be limiting and much slower than metabolism, where analyte levels do not increase. It is generally recommended to use multiple sensors with overlapping detection ranges to investigate the accumulation or depletion of analytes and to account for potential artifacts.[21–25]

11.8 FUTURE ASPECT

Three primary challenges must be addressed for electrochemical sensors to function effectively: *low limit of detection (LOD), sensitivity to real samples, and stability and repeatability*. These challenges are also the main hurdles in producing electrochemical biosensors. The LOD determines the smallest amount of an analyte that a sensor can detect, and optimal biosensors must have a very low LOD. In terms of both manufacturing and marketing, repeatability of sensor results is crucial; the results from one sensor must be reproducible across all similar sensors, as testing each sensor individually is impractical.

The application of sensors to actual samples is the most critical aspect. Sensors cannot be used for diagnosis if they fail during real sample testing. Typically, actual samples for electrochemical biosensors include saliva, blood, urine, sweat, tears, and other bodily fluids. Collecting genuine samples is challenging; several factors must be considered during collection to avoid the *matrix effect*, which can hamper sensor performance. Dilution is often necessary to mitigate this effect, but excessive dilution may lead to inaccurate results. An ideal electrochemical biosensor should detect real samples without requiring dilution or extensive processing. For instance, saliva samples often need to be diluted before detection, and pH variations in urine samples can alter peak positions. Tear samples, used to test for diabetes, present challenges due to pH fluctuations and varying analyte concentrations based on the cause of tear production. Additionally, species such as proteins and lipids in real samples can adsorb onto the sensor surface, affecting its sensitivity and repeatability. Researchers are exploring novel materials and methodologies to address these issues, including active approaches that use shear pressures to prevent other species from adhering to the sensor surface, and passive approaches that use hydrophilic polymers to prevent protein adsorption. Stability is also critical for electrochemical biosensors to withstand harsh environmental conditions.

Nanomaterials possess intriguing electromagnetic, optical, and piezoelectric properties due to their small size and large surface area, making them highly suitable for applications in bioimaging, medical diagnostics, and healthcare. Nanoparticles have a high affinity for biomolecules, facilitating the immobilisation of proteins, nucleic acids, enzymes, antibodies, and other clinically significant materials, thus paving the way for developing diverse sensing platforms like sandwich assays, aptasensors, immunosensors, and enzymatic sensors. Recent advancements in biosensing platforms have incorporated a variety of nanomaterials, from monomolecular nanomotors to larger nanocages. It is expected that nanomaterial-based biosensors, which offer rapid, cost-effective, and straightforward operating methods, will soon replace expensive traditional sensing systems.

Electrochemical sensor research is a promising field. Recognising that selection lies at the heart of most issues in this area is crucial. When direct detection in unaltered samples is feasible, the fast analytical speed and the ability of electrochemical sensors to detect minute amounts without significantly damaging the sample are highly desirable features. Future advancements in biosensor techniques may lead to the development of cell-friendly analytes for point-of-care testing and precision medical diagnostics. The introduction of electrochemical sensors has greatly benefited biological research. Future industries will gain from additional features such as sensitivity, selectivity, and processing speed. For disease detection and quality standards of stem-cell-based products, rapid, non-destructive, and tuneable electrochemical sensors could be utilised in complex large systems. Recent developments in this field include the creation of diverse biological recognition materials, breakthroughs in microelectronic industries, micro- and disposable sensors and the use of arrays to monitor a wide range of inorganic and organic contaminants. Online and flow-injection systems have also been developed to monitor various pollutants. Advancements in nanomaterials have enhanced sensor functionality, and sensor arrays that can detect multiple analytes are useful for pollution management, treatment monitoring and diagnosis.[25–29]

11.9 CONCLUSION

This study reviews recent advancements in electrochemical sensors for detecting small biomolecules (such as DNA, enzymes, and hormones) and monitoring a variety of inorganic and organic contaminants through electrochemical methods. Ongoing research focuses on developing new sensors using various chemical and biological sensing materials. Additionally, mass production technologies have enabled the creation of highly compact, reproducible, and cost-effective (disposable) sensor devices, ideal for the microelectronics industry. These devices are combined with lightweight, user-friendly microprocessor-based instruments. Further advancements include the development of selective and stable identification elements, such as 'smart' sensors, remote electrodes, molecular devices, multiparameter sensor arrays, micromachining, and nanotechnology, all of which are expected to significantly impact pollution management. Electrochemical sensors represent a major advancement in biological research, and future industries are likely to benefit from enhanced sensitivity, selectivity, and processing speed. Electrochemical techniques are fast, accurate, and non-destructive tools for analysing a wide range of targets. To enhance sensitivity, researchers have employed functional peptides, aptamers, and nanomaterials, including carbon nanotubes, graphene, graphene derivatives, metal nanoparticles, and gold nanoparticles. During electrochemical experiments, the interaction between the target and a specific probe or composite produces a detectable signal.

REFERENCES

1. Wei H., Pan D., & Han H. (2021). Electrochemical monitoring of marine nutrients: From principle to application. *TrAC Trends in Analytical Chemistry*, 138, 116242, https// doi:10.1016/j.trac.2021.116242
2. Caratelli V., Fillo S., D'Amore N., Rossetto O., Pirazzini M., Moccia M., Avitabile C., Moscone D., Lista F., & Arduini F. (2021). Paper-based electrochemical peptide sensor for on-site detection of botulinum neurotoxin serotype A and C. *Biosensors and Bioelectronics*, 183, https://doi.org/10.1016/j.bios.2021.113210
3. Ferreira P.C., Ataíde V., Chagas C., Angnes L., & Araujo W. (2019). Wearable electrochemical sensors for forensic and clinical applications. *TrAC Trends in Analytical Chemistry*, 119, 115622, doi:10.1016/j.trac.2019.115622
4. Bakker E., & Telting-Diaz M. (2002). Electrochemical sensors. *Analytical Chemistry*, 74(12), 2781–800, doi:10.1021/ac0202278
5. Kalambate P.K., Gadhari N.S., Li X., Rao Z., & Huang Y. (2019). Recent advances in MXene–based electrochemical sensors and biosensors. *TrAC Trends in Analytical Chemistry,* 120(5), 115643, doi:10.1016/j.trac.2019.115643

6. Damiati S., Peacock M., Leonhardt S., & Damiati L. (2018). Embedded disposable functionalized electrochemical biosensor with a 3D-Printed flow cell for detection of Hepatic Oval Cells (HOCs). *Genes,* 9(2), 89, doi:10.3390/genes9020089
7. Yokota T., Zalar P., Kaltenbrunner M., Jinno H., Matsuhisa N., Kitanosako H.,Tachibana Y., Yukita W., Koizumi M., & Someya T. (2016). Ultraflexible organic photonic skin. *Science Advances*, 2(1), https://doi/10.1126/sciadv.1501856
8. Park S.I., Brenner D.S., Shin G., Morgan C.D., Copits B.A., Chung H.U., Pullen M.Y., Noh K.N., Davidson S., Oh S.J., Yoon J., Jang K.I., Samineni V.K., Norman M., Grajales-Reyes J.G., Vogt S.K., Sundaram S.S., Wilson K.M., Ha J.S., Xu R., Pan T., Kim T.I., Huang Y., Montana M.C., Golden J.P., Bruchas M.R., Gereau IV R.W., & Rogers J.A. (2015). Soft, stretchable, fully implantable miniaturized optoelectronic systems for wireless optogenetics. *Science Advances*, 4(1), https://doi/full/10.1126/sciadv.aap9841
9. Yao H., Shum A.J., Cowan M., Lähdesmäki I., & Parviz B.A. (2011). A contact lens with embedded sensor for monitoring tear glucose level. *Biosensors and Bioelectronics*, 26(7), 3290–3296, doi:10.1016/j.bios.2010.12.042
10. Chen G.Z., Chan I.S., & Lam D.C.C. (2013). Capacitive contact lens sensor for continuous non-invasive intraocular pressure monitoring. *Sensors and Actuators A Physical,* 203, 112–118, https://doi.org/10.1016/j.sna.2013.08.029
11. Liu J., Agarwal M., & Varahramyan K. (2008). Glucose sensor based on organic thin film transistor using glucose oxidase and conducting polymer. *Sensors and Actuators B: Chemical*, 1(135), 195–199, https://doi.org/10.1016/j.snb.2008.08.009
12. Kim J., Cha E., & Park J.U. (2019). Recent advances in smart contact lenses. *Advanced Materials*, 5(1), https://doi.org/10.1002/admt.201900728
13. Bakirhan N.K., Ozcelikay G., & Ozkan S.A. (2018). Sensitive detection of cardiovascular disease markers by electrochemical-based biosensors. *Journal of Pharmaceutical and Biomedical Analysis,* 159, 406–424, doi:10.1016/j.jpba.2018.07.021
14. Dominguez -Renedo O., Marata Navarro-Cunado A., & Asunción Alonso-Lomillo M. (2023). Electrochemical devices for cholesterol detection. *Journal of Pharmaceutical and Biomedical Analysis*, 224. https://doi.org/10.1016/j.jpba.2022.115195
15. Fatemeh Y., Ahmad M., Zahra S.,Younes G., Shahram R., Ali Shobnani-Nasab., Soheila M., Behzad H., Samaneh R., & Amir S. (2020). Electrochemical-based biosensors: New diagnosis platforms for cardiovascular disease. *Current Medicinal Chemistry*, 27, 2550–2575, https://doi.org/10.2174/0929867326666191024114207
16. Altintas Z., Fakanya W.M., & Tothill I.E. (2014). Cardiovascular disease detection using bio-sensing techniques. *Talanta*, 128, 177–86, https://doi.org/10.1016/j.talanta.2014.04.060
17. Hu B., Li L., Zhao D., Jing Y., Li Z., & Sun S. (2023). Research progress of electrochemistry in detecting markers of cardiovascular disease, *Microchemical Journal*, 195, 0026–265X, https://doi.org/10.1016/j.microc.2023.109489
18. Szunerits S., Mishyn V., Grabowska I., & Boukherroub R. (2019). Electrochemical cardiovascular platforms: Current state of the art and beyond, *Biosensors and Bioelectronics* , 131, 287–298, https://doi.org/10.1016/j.bios.2019.02.010
19. Chan J.F.W., Lau S.K.P., To K.K.W., Cheng V.C.C., Woo P.C.Y., & Yuen K. (2015). Middle east respiratory syndrome coronavirus: Another Zoonotic Betacoronavirus causing SARS-like disease. *Clinical Microbiology Review*, 28, 465–522, https://doi.org/10.1128/cmr.00102-14
20. Mao Y., Du H., Wang X., Tian M., Wang Y., Liu L., Wei J., Xue F., Liu G., Zhang X., Yi T. (2019). A ratiometric fluorescent probe for rapidly detecting bio-thiols in vitro

and in living cells. *Dyes and Pigments*, 171, 0143–7208, https://doi.org/10.1016/j.dyepig.2019.107688
21. Bizzotto D., Burgess I.J., Doneux T., Sagara T., & Yu H.Z. (2017). Beyond simple cartoons: Challenges in characterizing electrochemical biosensor interfaces, *ACS*, 3, 5–12, https://doi.org/10.1021/acssensors.7b00840
22. Wu D., Sedgwick A. C., Gunnlaugsson T., Akkaya E.U., Yoon J., & James T.D. (2017). Fluorescent chemosensors: The past, present and future. *Chemical Society Reviews*, 46, 7105–7123, https://doi.org/10.1039/C7CS00240H
23. Rozenblum G.T., Pollitzer I.G., & Radrizzani M. (2019). Challenges in electrochemical aptasensors and current sensing architectures using flat gold surfaces. *Chemosensors*, 7(4), 57, https://doi.org/10.3390/chemosensors7040057
24. Monzó J., Insua I., Trillo F.F., & Rodriguez P. (2015). Fundamentals, achievements and challenges in the electrochemical sensing of pathogens. *Analyst*, 140, 7116–7128
25. Sadoine M., Ishikawa Y., Kleist T.J., Wudick M.M., Nakamura M., Grossmann G., Frommer W.B., & Ho C.H. (2021). Designs, applications, and limitations of genetically encoded fluorescent sensors to explore plant biology. *Plant Physiology*, 187(2), 485–503. doi:10.1093/plphys/kiab353
26. Gao F., Liu C., Zhang L., Liu T., Wang Z., Song Z., Cai H., Fang Z., Chen J., Wang J., Han M., Wang J., Lin K., Wang R., Li M., Mei Q., Ma X., Liang S., Gou G., & Xue N. (2023). Wearable and flexible electrochemical sensors for sweat analysis: A review. *Microsystems & Nanoengineering*, 9, 1. doi:10.1038/s41378-022-00443-6
27. Singh A., Sharma A., Ahmed A., Sundramoorthy A.K., Furukawa H., Arya S., & Khosla A. (2021). Recent advances in electrochemical biosensors: Applications, challenges, and future scope. *Biosensors (Basel)*, 11(9), 336, doi:10.3390/bios11090336
28. Pirzada M., & Altintas Z. (2019). Nanomaterials for healthcare biosensing applications. *Sensors*, 19(23), 5311, https://doi.org/10.3390/s19235311
29. Baranwal J., Barse B., Gatto G., Broncova G., & Kumar A. (2022). Electrochemical sensors and their applications: A review. *Chemosensors*, 10(9), 363. https://doi.org/10.3390/chemosensors10090363
30. Park J., Kim J., Kim S.Y., Cheong W.H., Jang J., Park Y.G., Na K., Kim Y.T., Heo J.H., Lee C.Y., Lee J.H., Bien F., & Park J.U. (2018). Soft, smart contact lenses with integrations of wireless circuits, glucose sensors, and displays. *Science Advances*, 4(1), doi:10.1126/sciadv.aap9841

12 Role of Nanotechnology for Diagnosis and Treatment of Covid-19

Basanta Bhowmik

12.1 INTRODUCTION

Covid-19, type of *Nidovirales*, family of *Coronaviridae*, and subfamily of *Orthocoronavirinae* started in Wuhan, China, in December 2019 and spread across the globe.[1] It was similar to the earlier SARS-2003 pandemic in mainland China, and the MERS-2012 pandemic in Saudi Arabia, which are also part of the class of *Coronaviridae*.[2] Due to the worldwide spread and the subsequent infection it imposes, it was called a pandemic. On 30 January 2020, the World Health Organization (WHO) declared the Severe Acute Respiratory Syndrome Coronavirus 2 (SARS-CoV-2) the sixth international human health emergency. Similar to Covid-19, prior pandemics declared by WHO were H1N1 in 2009, polio in 2014, Ebola in 2014 and Zika in 2016.[3] As per reported data up to 26 August 2020 by WHO, almost 222 countries were affected, with the total number of confirmed infected cases 208,305,870, and a death toll of 4,380,152. Among many other countries, the most severely affected were the USA, Brazil, India, Russia, and South Africa.[4] However, due to the rise of such a global major health issue, the inevitability of early detection, subsequent preventive measures and healing strategies were in high demand.

12.2 SYMPTOMS

Most infected cases are asymptomatic and a few are symptomatic. Among symptomatic patients, most cases are mild, with general symptoms like fever (83%–100%), cough (59%–82%), headache (7%–8%), fatigue, myalgia (11%–35%), diarrhoea (2%–10%), dyspnea, and so forth.[5] A few mild cases extend to severe acute respiratory syndrome. Similar to the Middle East Respiratory Syndrome Coronavirus (MERS-CoV), SARS-CoV-2 causes acute myocarditis and heart failure. In addition, hyperlipidaemia, lysophosphatidylcholine, lysophosphatidylethanolamine, glucose metabolism disorders, and phosphatidylglycerol are found to be prominent in Covid-19 patients.[6] The mortality rate is found to be higher in the case of Covid-19 patients having prior health abnormalities including cardiovascular disease, arrhythmia, hypertension, and diabetes.[1] Initially human-to-human transmission was ruled out, but at the later stage such transmission processes were confirmed. Reports further

DOI: 10.1201/9781003453635-12

revealed that the transmission capacity of the SARS-COV-2 is about ten times higher as compared to other SARS coronaviruses.[5]

The role of nanotechnology for point-of-care diagnosis with improved treatment for Covid-19 has immense importance. Kostas Kostarelos made a beautiful statement: 'viruses are the most beautiful, smart, and capable nanoparticles'.[7] One can easily gain facts about Covid-19 by taking advantage of technology, laboratory automation, and powerful design proposals. In today's world, nanotechnology has many roles, and it can be applied to all fields of study including multidisciplinary research such as nanoelectronics, biomaterials, nanomedicine, and so forth.[8,9] As far as Covid-19 is concerned, nanotechnology-enabled tools may offer early detection, prediction of the behaviour of RNA gene variation in host cells, and mechanisms of antibody and therapeutic agents against the viral genome.[7] In this short review, in addition to the Covid-19 discussion, a brief examination of the importance of nanotechnology in Covid-19 diagnosis and cure will be clarified. In the early 1990s, micro-cantilever-based sensor devices gained popularity for the detection of SARS-CoV.[10] At a later stage, electrochemical biosensors and mirofluidic devices were developed in order to improve sensitivity for more accurate detection of the influenza virus.[11] Similarly, another contribution from the scientific community was carbon dot nanoparticles having a diameter of ~2 nm in combination with glycoprotein gp120 as a preventive strategy for an epidemic like HIV-1.[12]

12.3 MORPHOLOGY OF COVID-19

In 1966, Tyrell and Bynoe first discovered the coronavirus in patients having symptoms of the common cold.[15] The coronavirus is a single-stranded RNA (+ssRNA) virus having an outer envelope that looks like a crown, which is derived from the Latin word *corona*, and thus name suggests 'corona virus'.[1–3] Transmission electron microscopy of Covid-19 confirms the RNA gene is encapsulated by four types of protein structures, viz. nucleoprotein (N), matrix protein (M), spike protein (S), and envelope protein (E).[13–16] However, this RNA gene has very strong mutation capability in the host cell within a short span of time. Frequent gene mutation ultimately results in prediction failure of the virus behaviour in the host cells and consequently fast spread.[17] A study by Minjin et al. on the ORF3a gene mutation and subsequent consequences recently received much attraction throughout the globe.[18] The single strain of RNA gene for Covid-19 was found to be ~ 29.9 KB, and the same for MERS; SARS-CoV is ~30.1 KB, ~27.9 KB, respectively.[1,14] However, from a structural point of view, ~80%, ~50% genome sequences of Covid-19 are similar to SARS and MERS-CoV, respectively.[19] Based on the genome structure, the coronavirus is categorised as alpha, beta, gamma, and delta.[1,14,18] The coronaviruses that originated from mammals are recognised as alpha and beta coronaviruses, whereas gamma and delta viruses evolved from pigs and birds, respectively.[15] Covid-19 is a type of beta coronavirus. HCoV-229E and HCoV-NL63 from the alpha-coronavirus and HCoV-HKU, HCoV-OC43, SARS-CoV and MERS-CoV from the beta-coronavirus are found to be the more vulnerable coronaviruses in human body.[1]

The receptor-binding domain (RBD) in the Covid-19 S-protein uses the receptor Angiotensin Converting Enzyme 2 (ACE2) for cell fusion and entry into the human cell.[16] It has a very high probability of attachment with the human receptor enzyme ACE2. ACE2 functions as receptor for SARS-CoV-2, which combines with the Spike(S) protein of the virus. The spike glycoprotein, which plays a major role in virus attachment and replication in the host cell, has two subunits, S1 and S2. The role of S1 is to find a suitable host species for attachment.[1,14–16] Upon attachment of the virus to the outer surface of the host cell, cell membrane fusion is initiated by subunit S2. Once the cell fusion is completed by the S2 subunit, protein encapsulation left over with only the RNA genome remains in the cytoplasm. Then the gene mutation process starts in the cytoplasm. Non-structural RNA-binding proteins NSP2 and NSP3 accelerate the mutation process and are possibly the influential factors for Covid-19 infection.[2,3,5,7] Details of the virus attachment, cell membrane fusion, and entry into the host cell are shown in Figure 12.2c. In a later section, we will show how this spike protein or protease 3CLpro can be used as a target protein by researchers for developing different antibodies, entry inhibitors or vaccines. Evidence suggests that the lungs contain maximum enzyme ACE2; however, it is more or less found in almost every organ of the human body including the arteries, kidneys, stomach, liver, intestines, brain, heart, and so forth.[6,17,23] Therefore, patients with abnormalities in the lungs are more likely to get infected due to excess secretion of ACE2.[6]

12.4 GENOME SEQUENCE OF COVID-19

The length of the Covid-19 genome sequence is about ~29.9 kb. The genome structure of Covid-19 is shown in Figure 12.1a. A single-strain genome comprises mainly

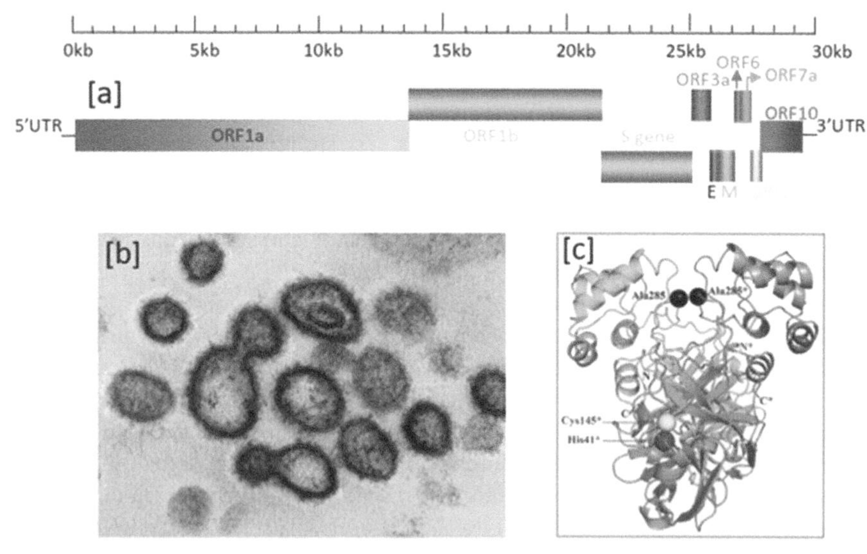

FIGURE 12.1 (a) Genome sequence of Covid-19; (b) TEM morphology of Covid-19, image is reproduced from National Institutes of Health from ref[13]; (c) three-dimensional crystal structure of Covid-19 main protease 3CLpro.[14]

three parts: open reading frames (ORFs), 5′ untranslated region (UTR) found at the starting section, and 3′ UTR at the end section.[16,20] Among the complete RNA, two-thirds of RNA in between ORF1a and ORF1b help to produce polyproteins (pp1a and pp1ab) and encode 16 non-structural proteins (NSP).[16] In contrast, the remaining one-third of RNA is useful for accessory proteins and structural proteins.[20] As a whole, Covid-19 has 27 structural and non-structural proteins. Based on the similarity of genome sequences, some research studies predicted the evolution of Covid-19 and its subsequent transmission to human beings. A similarity of ~96.2% between Covid-19 and bat coronavirus RaTG13 has been reported.[17] Manis et al. demonstrated ~73% genetic similarity between Covid-19 and pangolin.[18] However, the origin of the virus is still unclear and under debate among researchers.

12.5 MODE OF TRANSMISSION

Early research and development suggested the origin of Covid-19 had a direct link with the Huanan Seafood Wholesale Market.[5] However, later such a belief was ruled out because of the lack of evidence against a link between the first identified infected persons and the market. Further, at the early stages of the outbreak, it was believed that no direct human-to-human transmission was possible or occurred. However, based on the evidence, the WHO later declared transmission of the virus was possible through direct touching or inhalation of suspended airborne virus particles from coughing or sneezing of infected individuals.[21] Saliva (secreted from the salivary glands) contains maximum virus particles of the infected person. A model shows the role of saliva for transmission of Covid-19 from an infected person to healthy individuals (see Figure 12.2a). As reflected in Figure 12.2a, a person close to an infected person easily got infected either through small or large droplets of saliva. Frequent use of hand sanitiser or wearing a mask may reduce infection probability. However, a person

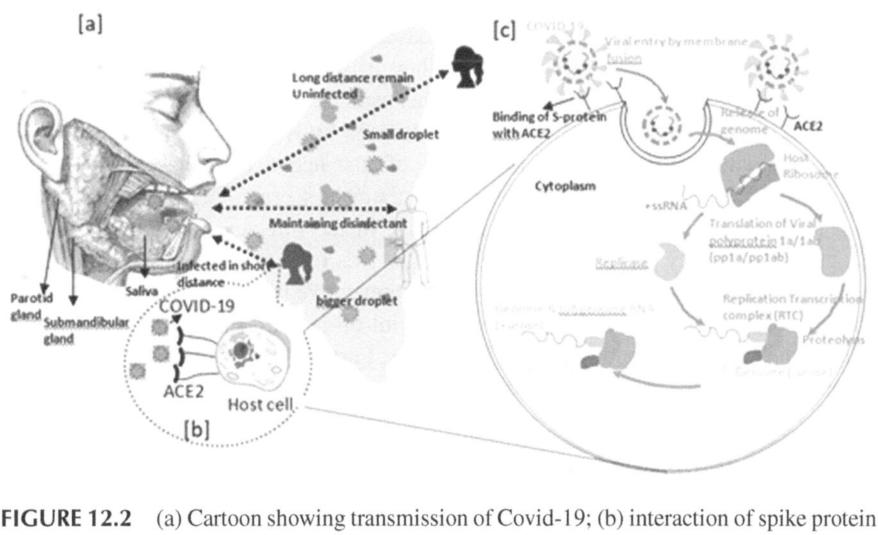

FIGURE 12.2 (a) Cartoon showing transmission of Covid-19; (b) interaction of spike protein and receptor ACE2; (c) enlarged view of (b) represents the replication process of RNA gene in host cell.

maintaining a distance of about > 1m may have less chance of infection. Apparently, the lungs and tongue are also hosts for transmission. In general, Covid-19 related to pulmonary disease is mostly airborne, depending on the size of the droplets. Smaller droplets becoming airborne can travel faster with the flow of air, whereas relatively larger droplets stick to surfaces like metal, glass, and so forth, and the virus remains alive for a longer period of time (~3–4 hours).[22] In such cases, one should follow an anti-disinfection process with 62%–71% ethanol, 0.1% sodium hypochlorite, or 0.5% hydrogen peroxide. Further, contact in the eyes, mouth, and nose can be a possible transmitter in the case of talking in close proximity with an infected person.[21] A recent study revealed water could be a potential transmitter of Covid-19.[1,2,21] Another study has also shown that the virus can be found in aerosols.[19] One can easily isolate symptomatic patients to stop spreading; however, asymptomatic patients are more dangerous and are silent transmitters of infection in a community.

12.6 ROLE OF NANOTECHNOLOGY

Nanotechnology or nanomaterial-enabled tools play a major role in fighting Covid-19. Weiss et al. proposed that nanotechnology may provide nanotheranostic approaches that can act as a bridge between diagnostics and therapy.[23] In fact, nanomaterial acts as a tool that can either stimulate or suppress the immune response.[23] As reported, some common symptoms including dry cough, sore throat, cold, fever, fatigue, dyspnea (breathing problem), and so on. were evident in the first stage of infection of a Covid-19 patient.[1,2] Therefore, one can use these as biomarkers for the newly developed techniques which possibly enable us to make quick decisions and develop subsequent preventive measures to stop community spread. The few existing invasive/non-invasive techniques such as reverse-transcription PCR (RT-PCR), CT imaging, metagenomic next-generation sequencing (mNGS), protein testing, nucleic acid testing, and so forth have been found to be laborious, with the lack of infrastructure in rural hospitals necessitating samples be sent to urban laboratories for examination.[19] Sample preparation for all the above-mentioned techniques increases the risk of getting infected due to the highly contagious nature of the virus. Further, it is worth to mentioning that during sample transportation, the protein capsid-encapsulated RNA gene may get damaged due to heat and transport conditions.[1] Such damage (in RNA gene) may reduce assay sensitivity, which contributes to a higher rate of false negative results.[8] Therefore, less accuracy, with high turnaround time makes them less preferable as far as the present pandemic is concerned. With an increase in the infected population, we must adopt high-accuracy, lower turnaround time devices for screening infectious individuals from healthy ones and apply subsequent preventive measures. In such scenarios, the nanotechnology community offers us point-of-care diagnostic techniques with nanoparticles as therapeutics or antiviral agents. Further, nanotechnology can be used to design and target protein-specific vaccines for Covid-19.

12.6.1 Detection or Screening Techniques

Nanotechnology plays a significant role in the detection of Covid-19.[21-40] The design of electrochemical biosensors for Covid-19 pathogen detection is one of the benefits of nanotechnology.[24–26] In an electrochemical approach, a bio-recognition element (sensing material) integrated with a transduction element has been used for biosensor performance.[24] In general, conducting or semiconducting materials were used to prepare transduction elements.[25] In contrast, nanoparticles (Au), nanoflowers, synthetic polymers, quantum dots, graphene oxide or different protein structures were employed as bio-recognition elements (sensing materials).[24] The chemical energy evolves from the binding of the target pathogen (containing complex matrix) and the bio-recognition element recognised by biosensor in the form of electrical energy.[24] This approach does not require any sample preparation, unlike PCR and ELISA methods.[24–27] Angiotensin-converting enzyme 2 (ACE2), a receptor function, helps spike protein entry and replication into the host cell. Therefore, reducing the level of ACE2 in the lungs may reduce lung injury for Covid-19 patients. However, such reduction of ACE2 in a specific organ might have other side effects as well.[8] Instead, one can employ ACE2-containing nanomaterials to prepare gloves, masks, nose filters or any kind of personal protective equipment (PPE) which might have a higher tendency to attract suspended airborne virus particles and possibly have less chance of entering the body.[25] Researchers have already developed quantum dots, nanoflowers or synthetic polymers coated with ACE2 material for the manufacturing of PPE.[25,26] These materials prevent Covid-19 pathogens from entering the body.

Plasmonic biosensors utilised by the effect of photothermal (PPT) and localised surface plasmon resonance (LSPR) have been employed for Covid-19 genome sequence detection.[28] For example, gold nanoislands (AuNIs) have been used as sensing material to functionalised cDNA (in S protein). The role of AuNIs is to enhance the specific target (oligonucleotides) genome sequence (of Covid-19) from a multistage mixture (i.e. decrease false negative detection rate) by thermoplasmonic heat treatment. Thermoplasmonic heat has been generated from laser-induced illumination in the sensing layer at plasmonic resonance frequency.[28] Graphene-coated field effect transistor (FET) sensors in association with a spike antibody as a receptor have shown successful Covid-19 detection in lower turnaround time.[29] The graphene sheet was first transferred on SiO_2/Si substrate, followed by fictionalisation of the device by spike antibody for 4 hours.[29] Figure 12.3b shows such a FET-based biosensor arrangement with computer interface for recognition of the spike antigen. Nasopharyngeal swabs of infected patients have shown spike antigen detection starting from concentrations of 100 fg/mL. This result (minimum concentrations of 100 fg/mL) suggests the method does not require any preparation/preprocessing of the clinical sample for Covid-19 detection. The previously mentioned sensors platform can easily be interfaced with a smartphone. Due to the availability of smartphones to most citizens and significant software and hardware compatibility with battery support from the phone itself to the interfacing circuitry or system, this arrangement (smart phone with sensor) is a popular point-of-care diagnostic method.[28,29] A smartphone-based sensor platform model for Covid-19 pathogen detection is shown in Figure 12.3a.

Gold nanoparticles in association with thiol-modified antisense oligonucleotides have been used for detecting SARS-CoV-2, specifically the N-gene. The turnaround time of the approach was found to be 10 min.[30] Chen et al. used lanthanide-doped polysterene nanoparticles (LNPs) for the detection of Covid-19 IgG in human serum.[33] They overcame photo bleaching and showed improved performance with

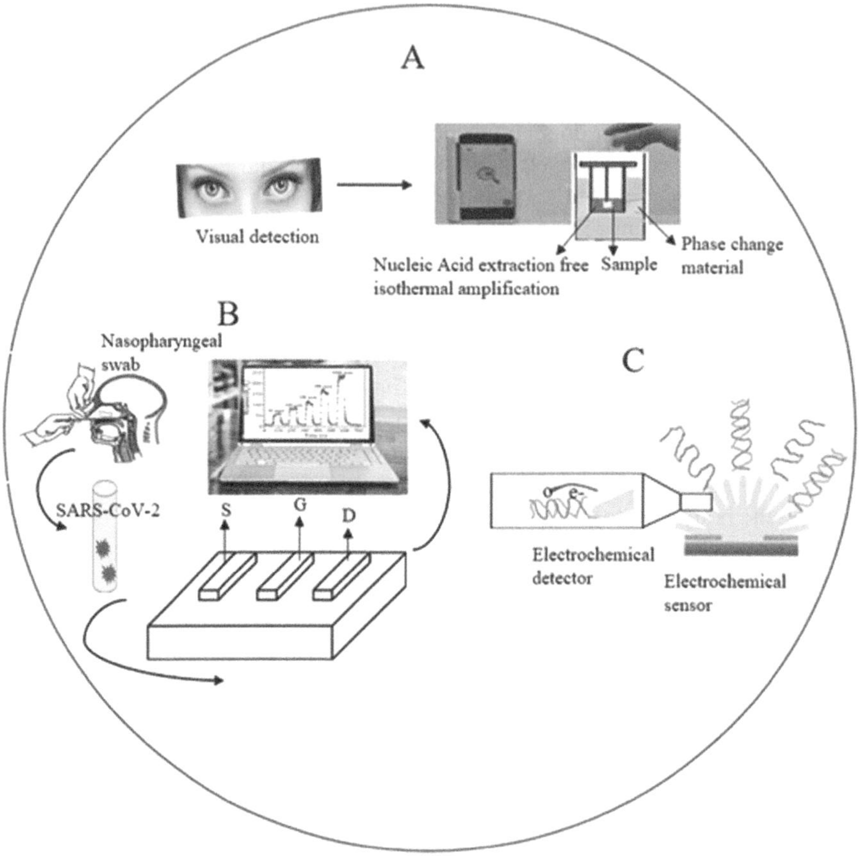

FIGURE 12.3 (a) home-made detection using visual or smartphone-based system; (b) nasopharyngeal swab as sample for FET-based sensor device; (c) electrochemical sensor detection.

rapid detection.[33] Mujawar et al. proposed the use of a biosensor as a diagnostic tool since it enables a fast sensing mechanism for a target biomarker.[34] It converts biomolecules into a measurable readout via optical, electrical, enzymatic, and other methods. A surface-plasmon resonance (SPR)-based biosensor measures the corona viral surface antigen.[35] According to Rabiee et al., gold (Au) nanoparticles can be used to inspect the virus because of its optical/electrical properties. They possess free electrons called plasmon that can cause SPR.[36] Pulse-controlled amplification relies on oligonucleotide-decorated gold nanoparticles where at optimum temperature bulk reaction takes place. Oligonucleotide-conjugated nanoparticles distributed homogeneously in the bulk solution by dissipating heat energy evolved from the laser pulse beam allowed to fall on it.[37] Magnetic bead-based immunoassay can be designed on a chip.[36] As demonstrated by Somvanshi et al., the fabrication of surface functionalised magnetic nanoparticles and viral RNA-extraction protocols can be used for detection of Covid-19.[40] The functionalised zinc ferrite nanoparticles were synthesised by the sol-gel auto-combustion method and it performed successful RNA extraction for the process of detection. The protocol is cost effective and helpful.[40] A recent investigation of prevotella found in the body due to chronic obstructive pulmonary disease demonstrated a correlation with Covid-19 patients.[31] The presence of prevotella releases many VOCs in exhaled breath. A nanomaterial-based array of sensors for detection of such a breath VOC pattern (biomarkers evident from abundance of Prevotella) can possibly provide a useful diagnosis of Covid-19. In addition, a patient with Covid-19 releases excessive levels of cytokines. Such a cytokine-induced interferon-γ inducible protein level can be another nanotechnology tool for spike protein diagnosis.[32]

Other diagnosis methods directly or indirectly related to nanotechnology are chest computed tomography (CT), the RT-PCR method, nucleic acid testing, protein testing, and so forth.[19] Artificial intelligence (AI) can also be used for diagnosis of Covid-19 where intelligent imaging platforms for infected region can be assessed. AI-empowered image acquisition workflows have made scanning procedures efficient, and they also provide protection for medical staff from Covid-19 infection.[38] The use of artificial intelligence techniques for analysis of cross-sectional chest CT images of infected patients at different angles reveals ground-glass opacity (GGO) as the most common abnormality.[3] Abdel-Basset et al. used a hybrid detection model based on an improved marine predators algorithm (IMPA) for X-ray image segmentation of Covid-19 patients.[39] The method follows a ranking-based diversity reduction (RDR) strategy to obtain better results in less iteration.[39] The WHO and the Centers for Disease Control and Prevention (CDC) have recognised molecular assay techniques for their successful implementation and test accuracy. A molecular assay technique such as Reverse Transcription Polymerase Chain Reaction (RT-PCR) involves a synthesis of complementary DNA (cDNA) carried out first, followed by nucleic acid amplification.[32] RNA extraction from collected nasal and throat samples was converted to complementary DNA (cDNA). Then the extracted genetic sequence from the sample was compared with reference to the genome sequence of Covid-19. Test results were assumed to be positive when the fluorescent signal amplified, whereas the absence of amplification signified results as negative. Few cases of pathogen detection based on a serological process where a blood sample collected by a finger prick of the infected patient are reported in literature.[1,2] Mainly in this type of detection process, antibodies were identified in the blood sample. A study revealed the presence of immunoglobulin M (IgM) as an indication of just Covid-19 infected (stage one) cases, whereas immunoglobulin G (IgG) confirmed infection since a long time back.[19] A rapid blood test employing gold nanoparticles coated with a virus lysate has been used for Zika virus pathogen detection.[27] Gold nanoparticles coated with a virus lysate form a pseudo pathogen on which plasma or serum samples dropped gradually.[27] Formation of nanoparticle agglomeration on the slide leads to positive infection cases. However, aggregations have been confirmed through highly sensitive and specific dynamic light scattering on the sample. Similar types of approaches may be adopted to detect Covid-19 provided we have knowledge of virus lysate.

12.6.2 Nanomedicine

Conventional vaccines are not enough for boosting the immune system against viral infection. These are prepared from a mixture of proteins and synthetic peptides.[41] Collaborative research by immunologists and nanotechnologists has demonstrated the effectiveness of nanoparticles as vaccine as well as drug delivery agents even to critical parts of the body.[41–48] Nanoparticle-based vaccines have been found to offer very high and stable immunity power, possibly due to the variant surface properties with controllable drug release kinetics.[42] Following Weiss et al., nanomaterials may be able to enhance the host immune response.[23] The biological and behavioural character of Covid-19 and subsequent gene mutation at different stages of infection limit the use of general vaccines.[41–45] Nanoparticle-based vaccines possibly inhibit such mutation and would be a potential approach, as suggested by Poon et al.[41] In general, two type of nanoparticles, mainly organic and inorganic compounds, have been employed for vaccine development based on their advantages and disadvantages.[41–48] Inorganic nanoparticles are small in size and highly stable as well as have good permeability, which offers high antigen loadings with a triggered release profile.[41–48] Mesoporous silica nanoparticles (MSNPs), gold nanoparticles (AuNPs), iron oxide nanoparticles (IONPs), and so forth are examples of inorganic nanoparticles.[41–45] A study by Gupta et al. reveals that iron-based magnetic nanoparticles (IMNPs) offer a good response against Covid-19.[49] They can interact with the virus because of same size regime and destroy or inhibit it. Carbon dots and graphene oxide-based nanomaterials can also stop viral spread by interacting with the negative charge present over it.[49] In contrast, organic nanoparticles are biocompatible and biodegradable with non-toxic characteristics like polymeric, liposomes, virus-like particles, and ferritin.[41] Biological nanocarriers like mesenchymal stem cell-derived exosomes may be used as efficient drugs for Covid-19. This type of nanocarrier expedites premature degradation of pathogens in the host cell and develops negative zeta potential in the body.[50]

Early detection in a robust system to trace infection, subsequent isolation and possible medical treatment might reduce the risk of community spread.[5] In recent scenarios, drug or vaccine development in the research stage and only the reuse of existing drugs (which were used for diseases like HIV, SARS-CoV, MERS-CoV etc.) based on clinical trials have been administered to Covid-19 patients. In addition, some earlier treatment methods, like for patients with refractory hypoxemia, might be suitable for oxygen therapy employing extracorporeal membrane oxygenation (ECMO).[1]

Apart from the role of nanotechnology, other approaches are illustrated in this section. Recently, medications such as chloroquine and its derivative hydroxychloroquine (HCQ) (familiar for its malaria inhibition) have been used on Covid-19 patient in countries like the USA and India.[42] However, there are many reports against the use of such medicines for curing Covid-19. However, chloroquine resists the virus replication process by splitting the spike protein in the host cell. Such splitting inhibits the growth of cathepsins and develops autophagosomes by changing the pH level of lysosomes.[43] On the other hand, hydroxychloroquine (HCQ) particles penetrate through the blood retinal barrier where they recombine with the retinal pigment epithelium which obstructs the viral entry.[33] The use of corticosteroids was found to be effective related to pulmonary inflammation, but one should pay attention to the dosage according to the level of infection.[44] Some of the frequently used medicines for Covid-19 treatment are listed in Table 12.1.[42–48] Among these significant drugs are remdesivir, arbidol, lopinavir (LPV)/ritonavir, ribavirin, and so on.[42] All these medications have antiviral functionality against RNA viruses. Clinical trials of remdesivir and IFN-β found them to be better options than LPV and ritonavir.[44] This combination (remdesivir and IFN-β) reduces the lung viral load.[46] Further, clinical trial revealed the effectiveness of the combination of remdesivir and chloroquine against Covid-19.

Recently, plasma therapy has been found to be a potential approach for curing Covid-19. In plasma therapy, plasma of a recovered patient is collected and injected into an infected patient.[47] The injection process continues until the patient develops immunity to combat Covid-19.[47] However, one can also increase the body's immunity power by changing food habits. Diet patterns may change the behaviour of microbiomes in the human body. Microbiomes like prevotella increase in excess amounts in the Covid-19 infected patient's body. Plant-based diets with fermented foods can possibly reduce the quantity of prevotella and cytokine in a Covid-19 patient's body.[31] Further, foods containing vitamin C are recommended to develop immunity against activation of effector cells of Covid-19 patients having lung injury.[48]

12.7 PREVENTION STRATEGIES

The nature of Covid-19 itself suggests how difficult it is for one to protect oneself from the virus. We cannot completely prevent the spreading of the virus, and that is why it is called a pandemic. However, we must obey some protocols to reduce the rate of spread. Infected people must be hospitalised, and care should be taken accordingly.[22,26] People who directly or indirectly come into contact with infected individuals must stay in quarantine at least for 14 days.[26] As suggested by many professionals, social distancing and lockdown are preventive measures to reduce community spread.[5] But how long can a lockdown continue? It hampers social life and kills economic growth. We can secure the health of the general public through the above-mentioned protocols, but what about frontline health professionals? Possibly improvements in PPE like face masks, gloves, boot covers, protective suits, goggles for eye protection and spill gowns may reduce the order of infection to some extent.[5,22] Further, it is advisable to use frequent sanitisation or wash the hands for reducing the risk of fomite transmission.[22,26]

12.8 CONCLUSION

In the present chapter, we have discussed the role of nanotechnology with a special emphasis on diagnosis and cure strategies. It has been observed that conventional methods of diagnosis like reverse-transcription PCR (RT-PCR), CT imaging, metagenomic next-generation sequencing (mNGS), protein testing, nucleic acid testing, and so forth require sample preparation with high infrastructure facilities. The turnaround time of these methods was found to vary from a few hours to an

TABLE 12.1
Reuse of medicine and its possible cure mechanism towards Covid-19

Sr No	Possible Drug	Structure	Application of earlier disease	Possible mechanism of use for COVID-19	Ref
1.	Chloroquine	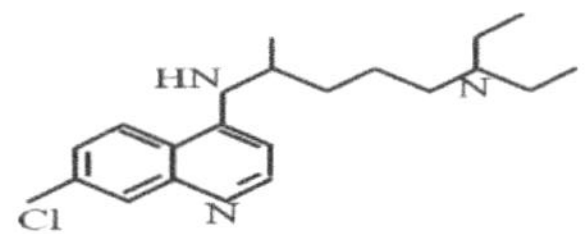	Malaria	Elevates pH level and interacts with receptor ACE2	[42]
2.	Hydroxychloroquine (HCQ)	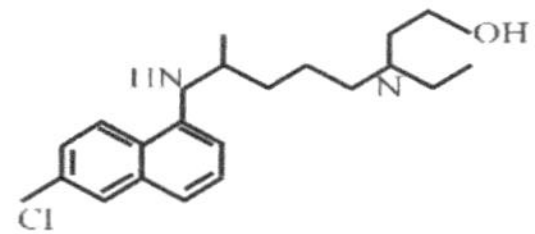	Malaria	Suppresses viral infection by changing pH level	[43]
3.	Lopinavir(LPV)	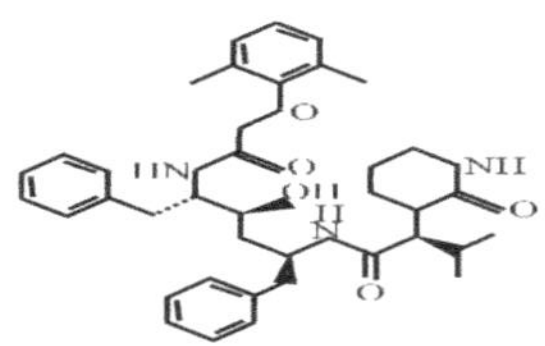	HIV-1, HIV-2	Protease inhibitor that produces viral proteases 3CLpro or PLpro	[46], [42]
4.	Ritonavir	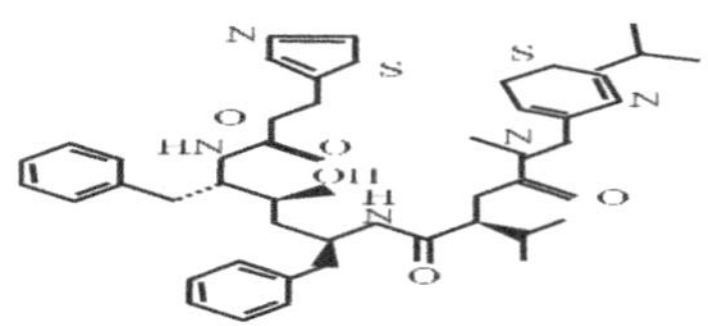	HIV-1, HIV-2	Inhibits action of 3CLpro and disrupts viral replication	[42,46]

5.	Oseltamivir	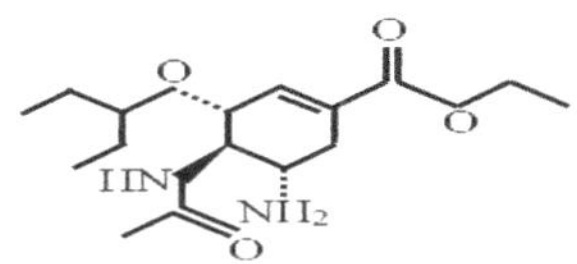	Influenza A, Influenza B	Neurominidase inhibitor	[5]
6.	Ribavirin	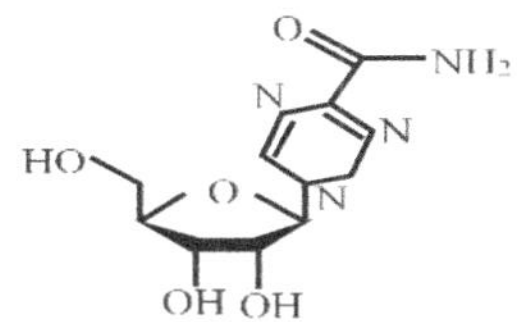	Hepatitis C,	Inhibits viral RNA replication and mRNA capping	[43]
7.	Remdesivir	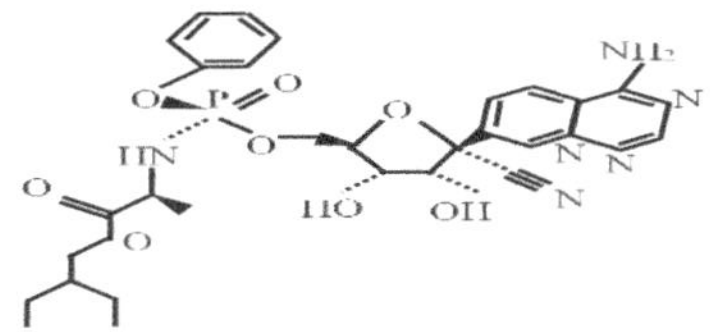	Ebola, Marburg	Nucleotide analogue that can produce RdRp, proteins essential for viral replication	[45-48]
8.	Arbidol (Umifenovir)	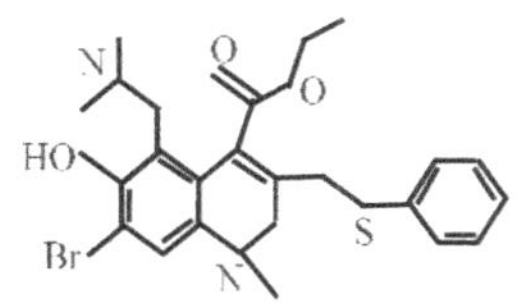	Influenza	An inhibitor that may disrupt the binding of a viral envelope protein to host cells and prevent viral entry to the target cell	[42]
9.	Favipiravir	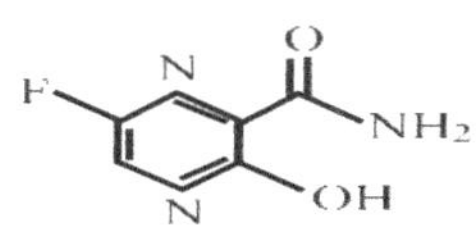	Influenza, Viral infections	RdRp inhibitor capable of blocking the replication of flavi-, alpha-, filo-, bunya-, arena-, noro-, and other RNA viruses	[43]

(*continued*)

TABLE 12.1 (Continued)
Reuse of medicine and its possible cure mechanism towards Covid-19

Sr No	Possible Drug	Structure	Application of earlier disease	Possible mechanism of use for COVID-19	Ref
10.	Darunavir	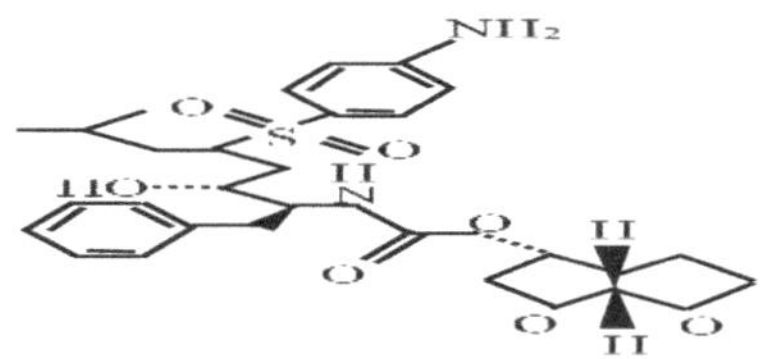	HIV-1	Protease inhibitor clears lower respiratory tract secretion	[45-48]
11.	Galidesivir	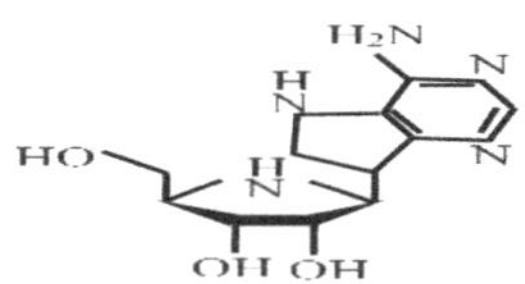	Hepatitis C, Ebola, Marburg	Disruption of Covid-19 viral RNA polymerase activity, i.e. premature termination of RNA strand	[42]
12.	Nitazoxanide	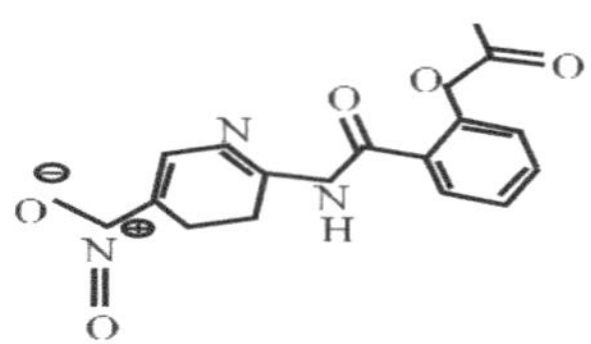	Various helminthic, protozoal, diarrhoea	Drug that may inhibit viral protein expression	[42]

13.	Imatinib(not an antiviral agent)	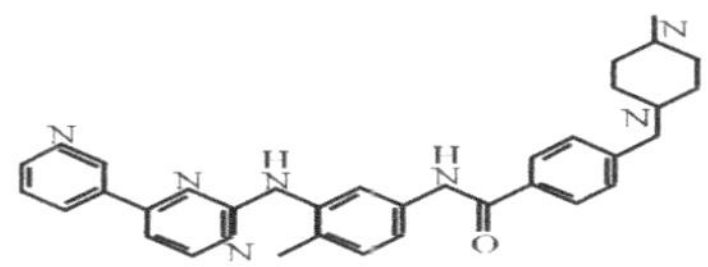	Cancer	Imatinib has anti-coronal activity primarily because it inhibits the fusion of virions with the endosomal membrane	[45,47]
14.	IFN-α	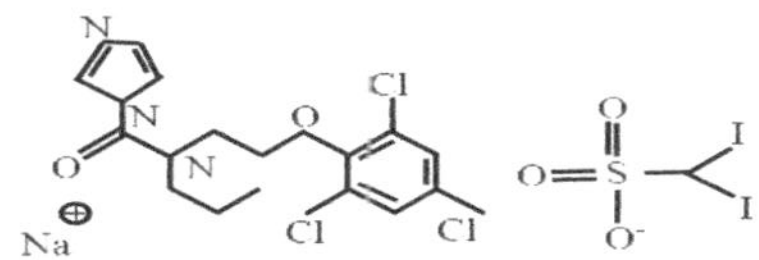	Hepatitis	Inhibits Covid-19 reproduction in host cell	[45]

entire day. However, nanomaterial-based detection was found to be faster, more accurate, and more cost effective. Further, cure strategies employing nanomaterials either for drug design or therapeutic agents or vaccine development, can possibly open a new pathway to the researcher. A brief discussion on transmission protocols and virus genome structure has also been presented in this chapter. This chapter will hopefully be a useful guideline for the scientific community working in the field of Covid-19.

REFERENCES

1. Guo YR, Cao QD, Hong ZS, Tan YY, et al. (2020). The origin, transmission and clinical therapies on coronavirus disease 2019 (COVID-19) outbreak – an update on the status. *Military Medical Research.* 7(11), 1–10. DOI: 10.1186/s40779-020-00240-0
2. Tang B, Bragazzi NL, Li Q, Tang S, Xiao Y, Wu J. (2020) An updated estimation of the risk of transmission of the novel coronavirus (2019-nCov). *Infectious Disease Modeling.* 5, 248–255. doi.org/10.1016/j.idm.2020.02.001
3. Lai CC, Shih TP, Ko WC, Tang HJ, Hsueh PR. (2020) Severe acute respiratory syndrome coronavirus 2 (SARS-CoV-2) and coronavirus disease-2019 (COVID-19): The epidemic and the challenges. *International Journal of Antimicrobial Agents.* 55(3). DOI: 10.1016/j.ijantimicag.2020.105924
4. WHO report, Coronavirus disease 2019 (COVID-19) Situation Report – 112. www.who.int/emergencies/diseases/novel-coronavirus-2019/situation-reports
5. Sohrabi C, Alsafi Z, O'Neill N, Khan M, Kerwan A, et al. (2020) World Health Organization declares Global Emergency: A review of the 2019 Novel Coronavirus (COVID-19). *International Journal of Surgery.* 76, 71–76. doi.org/10.1016/j.ijsu.2020.02.034
6. Zheng YY, Ma YT, Zhang JY, Xie X. (2020) COVID-19 and the cardiovascular system. *Nature Review Cardiology*, 17, 259–260.
7. Kostarelos K. (2020) Nanoscale nights of COVID-19. *Nature Nanotechnology.* 15, 343–344. doi:10.1038/s41565-020-0687-4
8. Chan WCW. (2020) Nano research for COVID-19. *ACS Nano.* 14, 3719–3720. https://doi.org/10.1021/acsnano.0c02540
9. Damhorst GL, Watkins NN, Bashir R. (2013) Micro- and nanotechnology for HIV/AIDS diagnostics in resource limited settings. *IEEE Transactions on Biomedical Engineering.* 60(3), 715–726. DOI: 10.1109/TBME.2013.2244894
10. Velanki S, Ji HF. (2006) Detection of feline coronavirus using microcantilever sensors. *Measures Science Technology.* 17(11), 2964–2968. DOI: 10.1088/0957-0233/17/11/015
11. Chen C, Zou Z, Chen L, Ji X, He Z. (2016) Functionalized magnetic microparticle based colorimetric platform for influenza A virus detection. *Nanotechnology.* 27(43), 435102–435108. DOI: 10.1088/0957-4484/27/43/435102
12. Fahmi MZ, Sukmayani W, Khairunisa SQ, et al. (2016) Design of boronic acid attributed carbon dots on inhibits HIV-1 entry. *RSC Advances.* 6(95), 92996–93002. https://doi.org/10.1039/C6RA21062G
13. COVID-19: Portraits of a virus. https://gulfnews.com/photos/news/ covid-19-portraits-of-a-virus-1.1583213015272?slide=2
14. Zhang L, Lin D, Sun X, Curth U, Drosten C, et al. (2020) Crystal structure of SARS-CoV-2 main protease provides a basis for design of improved α-ketoamide inhibitors. *Science.* 368(6489), 409–412. DOI: 10.1126/science.abb3405

15. Velavan TP, Meyer CG. (2020) The COVID-19 epidemic. *Tropical Medicine and International Health*. 25(3), 278–280. doi.org/10.1111/tmi.13383
16. Rokni M, Ghasemi V, Tavakoli Z. (2020) Immune responses and pathogenesis of SARS-CoV-2 during an outbreak in Iran: Comparison with SARS and MERS. *Reviews in Medical Virology*. 30(3), e2107. DOI: 10.1002/rmv.2107
17. Zhang C, Zheng W, Huang X, Bell EW, Zhou X, Zhang Y. (2020) Protein Structure and Sequence Reanalysis of 2019-nCoV Genome Refutes Snakes as Its Intermediate Host and the Unique Similarity between Its Spike Protein Insertions and HIV-1. *Journal of Proteome Research*. 19(4), 1351–1360. DOI: 10.1021/acs.jproteome.0c00129
18. Wang M, Li M, Ren R, et al. (2020) International expansion of a novel SARS-CoV-2 mutant. *Journal of Virology*. 94(12), e00567–202020. DOI: 10.1128/JVI.00567-20
19. Udugama B, Kadhiresan P, Kozlowski HN, et al. (2020) Diagnosing COVID-19: The disease and tools for detection. *ACS Nano*. 14(4), 3822–3835. DOI: 10.1021/acsnano.0c02624
20. Yan Y, Chang L, Wang L. (2020). Laboratory testing of SARS-CoV, MERS-CoV, and SARS-CoV-2 (2019-nCoV): Current status, challenges, and countermeasures. *Reviews in Medical Virology*. 30(3), e2106. DOI: 10.1002/rmv.2106
21. Xu R, Cui B, Duan X, Zhang P, Zhou X, Yuan Q. (2020) Saliva: Potential diagnostic value and transmission of 2019-nCoV. *International Journal of Oral Science*. 12(11), 1–6. https://doi.org/10.1038/s41368-020-0080-z
22. Huang H, Fan C, Li M, Nie HL, et al. (2020) COVID-19: A call for physical scientists and engineers. *ACS Nano*. 14(4), 3747–3754. DOI: 10.1021/acsnano.0c02618
23. Weiss C, Carriere M, Fusco L, Capua I, Regla-Nava JA, Pasquali M, Scott JA, Vitale F, et al. (2020) Toward nanotechnology-enabled approaches against the COVID-19 pandemic. *ACS Nano*. 14(6), 6383–6406 https://doi.org/10.1021/acsnano.0c03697
24. Cesewski E, Johnson BN. (2020) Electrochemical biosensors for pathogen detection. *Biosensors and Bioelectronics*. 14(6), 6383–6406. DOI: 10.1016/j.bios.2020.112214
25. Aydemir D, Ulusu NN. (2020) Angiotensin-converting enzyme 2 coated nanoparticles containing respiratory masks, chewing gums and nasal filters may be used for protection against COVID-19 infection. *Travel Medicine and Infectious Disease*. 1–4. DOI: 10.1016/j.tmaid.2020.101697
26. Cowling BJ, Aiello AE. (2020) Public health measures to slow community spread of Coronavirus disease 2019. *The Journal of Infectious Diseases*. 221(11), 1749–1751, https://doi.org/10.1093/infdis/jiaa123
27. Zheng T, Huo QA. (2020) Nanoparticle pseudo pathogen for rapid detection and diagnosis of virus infection. *Sensors International*. 2020, 1. DOI: 10.1016/j.sintl.2020.100010
28. Qiu G, Gai Z, Tao Y, Schmitt J, Kullak-Ublick GA, Wang J. (2020) Dual-Functional plasmonic photothermal biosensors for highly accurate severe acute respiratory syndrome coronavirus 2 detection. *ACS Nano*. 4(5), 5268–5277. https://doi.org/10.1021/acsnano.0c02439
29. Seo G, Lee G, Kim MJ, Baek SH, et al. (2020) Rapid detection of COVID-19 causative virus (SARS-CoV-2) in human nasopharyngeal Swab specimens using field-effect transistor based biosensor. *ACS Nano*. 14(4), 5135–5142. DOI: 10.1021/acsnano.0c02823
30. Moitra P, Alafeef M, Dighe K, Frieman MB, Pan D. (2020) Selective naked-eye detection of SARS-CoV-2 mediated by N gene targeted antisense Oligonucleotide capped plasmonic nanoparticles. *ACS Nano*. 14(6), 7617–7627. DOI: 10.1021/acsnano.0c03822
31. Kalantar-Zadeh K, Ward SA, Kalantar-Zadeh K, El-Omar EM. (2020) Considering the effects of microbiome and diet on SARS-CoV-2 infection: Nanotechnology roles. *ACS Nano*. 14(5), 5179–5182. DOI: 10.1021/acsnano.0c03402

32. Li X, Geng M, Peng Y, Meng L, Lu S. (2020) Molecular immune pathogenesis and diagnosis of COVID-19. *Journal of Pharmaceutical Analysis*. 10(2), 102–108. DOI: 10.1016/j.jpha.2020.03.001
33. Chen Z, Zhang Z, Zhai X, Li Y, Lin L, Zhao H, Bian L, Li P, et al. (2020) Rapid and sensitive detection of anti-SARS-CoV-2 IgG, using Lanthanide-Doped nanoparticles-based lateral flow immunoassay. *Analytical Chemistry*. 92(10), 7226–7231. doi/ 10.1021/acs.analchem.0c00784
34. Mujawar MA, Gohel H, Bhardwaj SK, Srinivasan S, Hickman N, Kaushik A. (2020) Nano-enabled biosensing systems for intelligent healthcare: Towards COVID-19 management. *Materials Today Chemistry*. 17, 100306–100315. https://doi.org/10.1016/ j.mtchem.2020.100306
35. Carter LJ, Garner LV, Smoot JW, Li Y, Zhou Q, Saveson CJ, Sasso JM, Gregg AC, Soares DJ, Beskid TR, et al. (2020) Assay techniques and test development for COVID-19 diagnosis. *ACS Central Science*. 6(5), 591–605. doi/10.1021/acscentsci.0c00501
36. Rabiee N, Bagherzadeh M, Ghasemi A, Zare H, Ahmadi S, Fatahi Y, et al. (2020) Point-of-use rapid detection of SARS-CoV-2: Nanotechnology-enabled solutions for the COVID-19 pandemic. *International Journal of Molecular Science*. 21(14), 5126–5150. DOI: 10.3390/ijms21145126
37. Zhang J, Gharizadeh B, Lu D, Yue J, Yu M, Liu Y, Zhou M. (2021) Navigating the pandemic response life cycle: Molecular diagnostics and immunoassays in the context of COVID-19 management. *IEEE Reviews in Biomedical Engineering*. 14, 30–47. DOI:10.1109/RBME.2020.2991444
38. Shi F, Wang J, Shi J, Wu Z, Wang Q, Tang Z, He K, Shi Y. (2020) Review of artificial intelligence techniques in imaging data acquisition, segmentation and diagnosis for COVID-19. *IEEE Reviews in Biomedical Engineering*. 14, 4–15. DOI: 10.1109/ RBME.2020.2987975.
39. Abdel-Basset M, Mohamed R, Elhoseny M, Chakrabortty RK, Ryan M. (2020) A hybrid COVID-19 detection model using an improved Marine predators algorithm and a ranking-based diversity reduction strategy. *IEEE Access*. 8, 79521–79540. DOI: 10.1109/ACCESS.2020.2990893
40. Somvanshi SB, Kharat PB, Saraf TS, Somwanshi SB, Shejul SB, Jadhav KM. (2020) Multifunctional nano-magnetic particles assisted viral RNA-extraction protocol for potential detection of COVID-19. *Materials Research Innovations*. 25, 169–174. DOI: 10.1080/14328917.2020.1769350
41. Poon C, Patel AA. (2020) Organic and inorganic nanoparticle vaccines for prevention of infectious diseases. *Nano Express*. 1(1). DOI: 10.1088/2632-959X/ab8075
42. Liu C, Zhou Q, Li Y, et al. (2020) Research and development on therapeutic agents and vaccines for COVID-19 and related human coronavirus diseases. *ACS Central Science*. 6(3), 315–331. doi.org/10.1021/acscentsci.0c00272
43. Singh AK, Singh A, Shaikh A, Singh R, Misra A. (2020) Chloroquine and hydroxychloroquine in the treatment of COVID-19 with or without diabetes: A systematic search and a narrative review with a special reference to India and other developing countries. *Diabetes & Metabolic Syndrome*. 14(3), 241–246. DOI: 10.1016/ j.dsx.2020.03.011
44. Zhai P, Ding Y, Wu X, Long J, Zhong Y, Li Y. (2020) The epidemiology, diagnosis and treatment of COVID-19. *International Journal of Antimicrobial Agents*. 55(5), 105955. DOI: 10.1016/j.ijantimicag.2020.105955
45. Dong L, Hu S, Gao J. (2020) Discovering drugs to treat coronavirus disease 2019 (COVID-19). *Drug Discoveries & Therapeutics*. 14(1), 58–60. DOI: 10.5582/ ddt.2020.01012

46. Scavone C, Brusco S, Bertini M, et al. (2020) Current pharmacological treatments for COVID-19: What's next? *British Journal of Pharmacology*. 177, 4813–4824. DOI: 10.1111/bph.15072
47. Bloch EM, Shoham S, Casadevall A, et al. (2020) Deployment of convalescent plasma for the prevention and treatment of COVID-19. *Journal of Clinical Investigation*. 130(6), 2757–2765. DOI: 10.1172/JCI138745
48. Mungroo MR, Khan NA, Siddiqui R. (2020) Novel coronavirus: Current understanding of clinical features, diagnosis, pathogenesis, and treatment options. *Pathogens*. 9(24), 297–308. DOI: 10.3390/pathogens9040297
49. Gupta A, Kumar S, Kumar R, Choudhary AK, et al. (2020) COVID-19: Emergence of infectious diseases, nanotechnology aspects, challenges, and future perspectives. *Materials Science inc. Nanomaterials & Polymers*. 5(25), 7521–7533. doi: 10.1002/slct.202001709
50. Pinky, Gupta S, Krishnakumar V, Sharma Y, Dinda AK, Mohanty S. (2020) Mesenchymal stem cell derived exosomes: A nano platform for therapeutics and drug delivery in combating COVID-19. *Stem Cell Reviews and Reports*. 17(1), 33–43. DOI: 10.1007/s12015-020-10002-z

Index

For Product Safety Concerns and Information please contact our EU representative GPSR@taylorandfrancis.com Taylor & Francis Verlag GmbH, Kaufingerstraße 24, 80331 München, Germany